The Essential Oils

BY

ERNEST GUENTHER, Ph.D.
Vice President and Technical Director
Fritzsche Brothers, Inc., New York, N. Y.

VOLUME ONE

HISTORY—ORIGIN IN PLANTS PRODUCTION—ANALYSIS

1948

D. VAN NOSTRAND COMPANY, Inc.

TORONTO NEW YORK LONDON

NEW YORK

D. Van Nostrand Company, Inc., 250 Fourth Avenue, New York 3

TORONTO

D. Van Nostrand Company, (Canada), Ltd., 228 Bloor Street, Toronto 8

LONDON

Macmillan & Company, Ltd., St. Martin's Street, London, W.C. 2

PRINTED IN THE UNITED STATES OF AMERICA

BY LANCASTER PRESS, INC., LANCASTER, PA.

Dedicated

to Mr. Frederick H. Leonhardt,

President of

Fritzsche Brothers, Inc.,

whose vision and generosity made this

work possible

AUTHORS OF CHAPTERS

ERNEST GUENTHER, Ph.D.
Vice President and Technical Director of Fritzsche Brothers, Inc., New York, N. Y.

A. J. HAAGEN-SMIT, Ph.D.
Professor, Bio-organic Chemistry, California Institute of Technology, Pasadena, California.

EDWARD E. LANGENAU, B.S.
Director of Analytical Laboratories, Fritzsche Brothers, Inc., New York, N. Y.

GEORGE URDANG, Ph.G., D.Sc. Nat., Sc.D.
Director, American Institute of the History of Pharmacy, Madison, Wisconsin.

PREFACE

The industry and science of essential oils have undergone within the last two decades more changes than could have been anticipated by those writers who, during the 1920's, contributed so valuably to our knowledge of this field. It is, therefore, no derogation of the works of Gildemeister and Hoffmann, Parry, Finnemore, and others, to state that the time is past due for bringing the whole subject up-to-date. This is the author's motive for the present treatise on the production, chemistry, analysis and application of these interesting and important products. It seems particularly fitting that this work be published in the United States, which, as the largest user of essential oils, has the most vital concern in the progress and expansion of the essential oil industry.

Within the last ten years—and largely as a result of World War II—there have taken place fundamental developments within the field, especially with respect to new sources of supply in the Western Hemisphere. It has been the author's rare privilege to witness most of these developments at first hand. His travels for more than twenty years have taken him the length and breadth of Europe, through Africa, Asia, Australia, into the new producing centers of North, Central and South America—in all of which places he surveyed the production of essential oils at their source.

The original purpose of this systematic survey was to supply Fritzsche Brothers, Inc., of New York, the essential oil industry in general, and Government agencies with exhaustive data on the production of essential oils. In the course of this work the author collected countless samples of authentic oils, produced under his supervision and of guaranteed purity. These were shipped to his headquarters in New York and submitted to careful analysis, which permitted the establishment of certain criteria of purity for many hitherto dubious oils. The results of such work have appeared in a series of articles published in leading trade journals during the last twenty years, for the information and guidance of those engaged in the production and in the use of these materials.

As the work progressed, the author was repeatedly urged to publish his data in their entirety, so that they might be readily accessible in comprehensive and up-to-date form. He has at last, therefore, attempted to combine in this work his own observations in the field, the seventy-five years' experience of Fritzsche Brothers, Inc., as producers, investigators and distributors of essential oils, and the data resulting from more than one hundred years of research by scientists all over the world, as described in the technical literature.

The author has been most fortunate in securing the collaboration of a number of outstanding organic chemists, specialists in the chemistry and analysis of essential oils. Their valuable contributions have enabled him to carry the first part of this work far beyond his original conception.

It is the plan of this first volume to describe, from a general point of view, the history, chemistry, biological origin and functions of the essential oils, method of production and analysis. The second volume will deal with the chemical constituents of essential oils. Succeeding volumes will be devoted to individual oils, their botanical and geographical origin, specific methods of production, physicochemical properties, assay and use.

The completion of this series represents the culmination of a life's work. Much painstaking toil has gone into it. It is an attempt to present basic facts in the science of essential oils, so far as these are now known. May it be received as a modest effort toward that end.

ERNEST GUENTHER

New York, N. Y.
September, 1947

ACKNOWLEDGMENT

A thorough treatise on essential oils must of necessity include some discussion of the ancillary sciences of botany, biochemistry, organic and analytical chemistry, and pharmacology—to mention only a few. It would, therefore, be impossible for a single individual to write a reliable and complete account of essential oils without the assistance of experts in associated fields.

The author has made use of such assistance, and takes this opportunity to thank those who have worked so hard and so long with him on a difficult project:

Dr. A. J. Haagen-Smit for his chapter, "The Chemistry, Origin and Function of Essential Oils in Plant Life."

Mr. Edward E. Langenau for his chapter,"The Examination and Analysis of Essential Oils, Synthetics, and Isolates," and for his advice on other phases of the work.

Dr. George Urdang for his chapter, "The Origin and Development of the Essential Oil Industry."

The late Dr. Philip W. Schutz, Professor of Chemistry, University of California, Berkeley, California, for his help with the chapter on "Distillation."

Dr. Darrell Althausen, Manager of Fritzsche Brothers, Inc., Clifton Factory, for his valuable suggestions on the entire manuscript.

Mr. Carl J. Koenig of Fritzsche Brothers, Inc., Clifton Factory, for his assistance with the drawings in the chapter on "Distillation."

Dr. Frances S. Sterrett of Fritzsche Brothers, Inc., New York, for her assistance with various parts of the manuscript.

Members of the staff of Fritzsche Brothers, Inc., New York, among them Mr. W. P. Leidy, Chief Librarian, and Mr. Anthony Hansen, Jr., Librarian, for their painstaking bibliographic work; and Mrs. Ann Blake Hencken, Miss Catherine McGuire, Mrs. Agnes Clancy Melody and Mrs. Elizabeth Campbell Adelmann for their patient, diligent and exact work in transcribing the author's and his collaborators' very difficult and oft-revised manuscripts.

In addition to the above, the author wishes to acknowledge most gratefully help from many friends and producers throughout the world, especially Mr. Pierre Chauvet, Seillans (Var), France, who supplied him with a great deal of information about latest developments in France.

ERNEST GUENTHER

NOTE

All temperatures given in this work are expressed in degrees Centigrade unless otherwise specified in the text.

CONTENTS

ILLUSTRATIONS AND DIAGRAMS

Chapter 1

THE ORIGIN AND DEVELOPMENT OF THE ESSENTIAL OIL INDUSTRY

BY

George Urdang

Note. All temperatures in this book are given in degrees centigrade unless otherwise noted.

CHAPTER 1

THE ORIGIN AND DEVELOPMENT OF THE ESSENTIAL OIL INDUSTRY

I. THE BEGINNING

Ex Oriente Lux—"The sun rises in the East."—Symbolically this old saying glorifies the East as the cradle of civilization. In the East also began the history of essential oils; for the process of distillation—the technical basis of the essential oil industry—was conceived and first employed in the Orient, especially in Egypt, Persia and India. As in many other fields of human endeavor, it was in the Occident, however, that these first attempts reached their full development. If oriental meditation kindled the light, occidental genius and industry kept it burning!

Data on the methods, objectives and results of distillation in ancient times are scarce and extremely vague. Indeed, it appears that the only essential oil of which the preparation (by a somewhat crude distillation) has been definitely established is oil of turpentine and, if we care to mention it in connection with essential oils, camphor. The great Greek historian, Herodotus (484–425 B.C.), as well as the Roman historian of natural history, Pliny (23–79), and his contemporary, Dioscorides—the author of the treatise "De Materia Medica" which dominated therapy for more than 1,500 years—mention oil of turpentine and give partial information about methods of producing it. They do not describe any other oil.

Until the early Middle Ages (and even later) the art of distillation was used primarily for the preparation of distilled waters. Where this process resulted in a precipitation of essential oils, as in the crystallization of rose oil on the surface of distilled rose water, it is likely that the oil was regarded as an undesired by-product rather than as a new and welcome one.

An extensive trade in odoriferous oils and ointments was carried on in the ancient countries of the Orient and in ancient Greece and Rome.[1] The oils used, however, were not essential oils, nor were they produced by mixing the latter with fatty oils; they were obtained by placing flowers, roots, etc., into a fatty oil of best quality, submitting the glass bottles containing these mixtures to the warming influence of the sun and, finally, separating the

[1] Urdang, "Pharmacy in Ancient Greece and Rome," *Am. J. Pharm. Educ.* 7 (1943), 169.

odoriferous oil from the solid constituents. Sometimes the flowers, etc., were macerated with wine before the fatty oil was added, and the product obtained by digestion filtered and then boiled down to honey consistency.

The same way of preparing odoriferous oils is described in the "Grabaddin" written by the somewhat mysterious Joannes Mesue, and published probably in the middle of the thirteenth century. This very widely used book did not list a single essential oil. However, two oils prepared by destructive distillation (oil of juniper wood or cade, and oil of asphaltum) are mentioned.

The first authentic description of the distillation of real essential oils has been generally ascribed to the Catalan physician, Arnald de Villanova (1235(?)–1311) who, by including products of distillation other than oil of turpentine, may be said to have introduced the art of distillation into recognized European therapy. However, it is by no means certain whether the "distilled" oils of rosemary and sage listed in the 1505 Venetian edition of his "Opera Omnia" were really mentioned in the original manuscript (written about two hundred years earlier) or were interpolated at some later time. Furthermore, it should be kept in mind that the term "distilled" in ancient and medieval writings did not have the exclusive and particular meaning it has today. It was, as E. Kremers pointed out in his translation of Fr. Hoffmann's historical introduction to E. Gildemeister and Fr. Hoffmann's "The Volatile Oils,"[2] "a collective term, implying the preparation of vegetable and animal extracts according to the rules of the art, or rectification and separation."

Nevertheless, whether Arnald de Villanova actually had prepared real distilled oils or not, his praise of the remedial qualities of distilled waters resulted in the process of distillation becoming a specialty of medieval and post-medieval European pharmacies—a specialty artfully executed and subjected to practical research, as well as to the theories of the time. Distillation being a means of separating the essential from the crude and nonessential with the help of fire, it met in an almost ideal way the definition of a "chymical" process valid until about the end of the seventeenth century and given a special meaning by the great Swiss medical reformer, Bombastus Paracelsus von Hohenheim (1493–1541). His theory was that it is the last possible and most sublime extractive, the *Quinta essentia* (quintessence) which represents the efficient part of every drug, and that the isolation of this extractive should be the goal of pharmacy. This theory undoubtedly laid the basis for research in the preparation of essential oils after his time. The very name "essential" oils recalls the Paracelsian concept—the *Quinta essentia.*

[2] Milwaukee (1900), 22.

II. FROM THE SIXTEENTH TO THE EIGHTEENTH CENTURY

There is still other evidence that the production and use of essential oils did not become general until the second half of the sixteenth century. In 1500 and 1507, there appeared at Strassburg the two volumes of Hieronymus Brunschwig's famous book on distillation, "Liber De Arte Distillandi." The author (1450–1534) was a physician at Strassburg. Although obviously endeavoring to cover the entire field of distillation techniques and products, he mentions only four essential oils, namely, the oil of turpentine (known since antiquity), oil of juniper wood and oils of rosemary and spike. Brunschwig states that oil of spike is produced in "Provinz," meaning undoubtedly the French Provence. This is confirmed in the "New Gross Destillirbuch" of the Strassburg physician, Walter Reiff (Ryff), published in 1556 at Frankfort on the Main, and containing a reference to a French industry of essential oils, especially of oil of spike. "The oil of spike or lavender," writes Reiff, "is commonly brought to us from the French Provence, filled into small bottles and sold at a high price" ("*gemeyncklich aus der Provinz Frankreich zu uns gebracht wird, in kleine Glässlin eingefasst und theuer verkaufft*").

In the part of the book dealing with the appropriate preparation of "some exquisite oils" by means of artificial distillation ("*von rechter Bereytung künstlicher Destillation etlicher furnehmer Ole*"), Reiff mentions, as sources of "precious" oils, clove, mace, nutmeg, anise, spike and cinnamon, as well as many substances that do not contain essential oils, or furnish only traces of them, such as benzoin, sandarac and saffron. The method as described by him, moreover, was by no means apt to produce pure essential oils.

It was the "Kräuterbuch" of Adam Lonicer (1528–1586), the first edition of which appeared at Frankfort on the Main in 1551, which may be regarded as a significant turning point in the understanding of the nature and the importance of essential oils. Lonicer stresses the medicinal value of "many marvelous and efficient oils of spices and seeds" ("*viel herrliche und kräfftige Öhle von Gewurzen und Samen*") and states that "the art of distillation is quite a recent, not an ancient invention, unknown to the old Greek and Latin physicians, and indeed has not been in use at all" ("*Diese Kunst des Destillirens ist fast eine neue, und nicht gar alte Erfindung, den alten griechischen und lateinischen Medicis unbekannt und gar nicht in gebrauch gewesen*").

Further progress in the methods of preparation and the knowledge of the nature of essential oils was made obvious in the "De Artificiosis Extractionibus" written by the German physician, Valerius Cordus (1515–1544), and published in 1561 at Strassburg by the Swiss naturalist, Conrad Gesner (1516–1565). It is significant that Cordus based his reports on the experiments conducted by him in the pharmacy of his uncle, Johannes Ralla, apothecary in Leipzig, by whom his work was supervised. Gesner himself contributed to the progress in the "Thesaurus Euonymi Philiatri," a book published by him at Zürich under the pen name, Euonymus Philiatrus. The most important publication on essential oils during that period, however, came from the pen of one of the most prolific and careful scientific writers of all times, the Neapolitan, Giovanni Battista della Porta (1537–1615). In his "De Destillatione libri IX," written about 1563, he not only differentiates distinctly between expressed fatty and distilled essential oils, but describes their preparation, the ways of separating the volatile oils from water and the apparatus used for this purpose.

In 1607, in his famous "Pharmacopoea Dogmaticorum Restituta" (Frankfort on the Main), the French physician, Joseph Du Chesne, latinized Quercetanus (1544–1609), one of the most ardent Paracelsians, could already state that "the preparation of essential oils is well known to everybody, even to the apprentices" ("*praeparatio omnibus fere, imo ipsis tyronibus, nota et perspecta est*").[3] Quercetanus states that all the *praeparationes chymicae*, among which were included the essential oils, could be obtained in the pharmacies, and he gives enthusiastic praise to the manager of the court pharmacy at Cassel in saying that it was primarily the example set in this pharmacy which inspired parts of his book ("*Officina haec mihi typus primus fuit, ad cuius imitationem meam pharmacopeam conatus sum*").[4] As to the preservation of essential oils in the pharmacies, Quercetanus writes that "15 or 20 different oils were kept in small round boxes and, when asked for, they were delivered by means of a toothpick, i.e., in a minute quantity achieving, nevertheless, the best results" ("*Eiusmodi essentiae conservantur in parvis theculis rotundis, quarum singulae capiunt 15 vel 20 diversa essentiarum genera, quae, cum usus postulat, cum dentiscalpio, hoc est, in minima quantitate exhibebuntur, et effectus nihilominus proferent optatissimos*").[5]

Official pharmacopoeias have always been more or less conservative. Thus, only such drugs as had found general acceptance in contemporary medical science were given a place in these official pharmaceutical standards. Hence, it is not quite as surprising as it may seem at first sight that, in the "Dispensatorium Pharmacopolarum" of Valerius Cordus (published and made official in the Imperial city of Nuremburg in 1546), only three essential oils were listed, in spite of the author's own extensive study of them. These were: oil of turpentine, oils of spike (lavender) and of juniper berries, in general use at least since the end of the fifteenth century. Of interest is the reference to an industry of essential oils, which makes it practicable to buy the oils of juniper berries and of spike, rather than to prepare them in the

[3] P. 245.
[4] P. 246.
[5] P. 246.

laboratories of the pharmacies. As to the oil of juniper berries, the Dispensatorium does not give an explicit formula, because, as it states, the product is bought at a price lower than the cost of preparation by the individual pharmacist ("*quia vero minoris emitur, quam ut ab aliquo pharmacopoea praeparari queat, confectionem eius non indicaminus*").[6] The section dealing with oil of spike states that it is more advantageous to buy the oil from merchants who import it from France and names Narbonne as the seat of the industry ("*apud nos maioribus sumptibus fit quam in Gallia Narbonensi, ideo potius emendum est a mercatoribus qui illud e Gallia afferunt*").[7] The second official Nuremburg edition of the "Dispensatorium Valerii Cordi," issued in 1592, lists not less than 61 distilled essential oils,[8] which fact illustrates the rapid development of the knowledge of essential oils as well as official acceptance.

In the seventeenth and eighteenth centuries, it was chiefly the pharmacists who improved methods of distillation and made valuable investigations into the nature of essential oils. Of special importance was the work of the French apothecaries, M. Charas (1618–1698), N. Lemery (1645–1715), A. J. Geoffroy (1685–1752), G. Fr. Rouelle (1703–1770), J. F. Demachy (1728–1803), and A. Baumé (1728–1804); their German colleagues, Kaspar Neumann (1683–1737), J. Ch. Wiegleb (1732–1800), and F. A. C. Gren (1760–1798); and the German-Russian pharmacist, J. J. Bindheim (1750–1825). Of other investigators of this period, we may mention two famous physicians, the Dutch, H. Boerhave (1668–1738) and the German, Fr. Hoffmann (1660–1743); and finally the man regarded as one of the first great industrial chemists, the German, J. R. Glauber (1604–1670).

III. MODERN DEVELOPMENT

The revolution in the science of chemistry, which began at the end of the eighteenth century with the work of A. Lavoisier (1743–1794), resulted in a new and illuminating approach to the investigation of the nature of essential oils. It is of interest that the first really important modern investigation in the field was devoted to the oldest essential oil known, oil of turpentine. Submitting the oil to elementary analysis, J. J. Houton de la Billardière found the ratio of carbon to hydrogen to be five to eight—the same ratio that was later established for all hemiterpenes, terpenes, sesquiterpenes and polyterpenes. The investigator published his results in a pharmaceutical periodical, the *Journal de Pharmacie* (**4** [1818], 5).

The systematic study of essential oils may be said to have begun with the analysis of a number of stearoptenes by the great French chemist, J. B.

[6] "Dispensatorium Pharmacopolarum Valerii Cordi," Norimbergae, 1546, col. 220.
[7] *Ibid.*, col. 241.
[8] Winkler, "Das Dispensatorium des Valerius Cordus," Mittenwald, 1934, 11.

Dumas (1800–1884), who had started his career as a pharmacist. He published his first treatise devoted to essential oils in *Liebig's Annalen der Pharmacie* (6 [1833], 245).

Of considerable importance in the further development of the chemistry of volatile oils were the investigations of the French chemist, M. Berthelot (1827–1907), devoted primarily to the hydrocarbons contained in these oils. About 1866 the name Terpene was mentioned in a textbook written by Fr. A. Kekulé (1829–1896) who apparently coined this term. In 1875 one of the greatest English chemists emerging from pharmacy, W. Tilden (1842–1926), introduced nitrosyl chloride as a reagent for terpenes, a reaction perfected and used to such an extent and with such excellent results by the German chemist, O. Wallach (1847–1931), that the renowned Swiss pharmacognosist, Fr. A. Flückiger (1828–1894), called Wallach the Messiah of the terpenes.

This very active and far-reaching research was the result, as well as the cause, of the wide expansion in the use of essential oils during the latter half of the nineteenth century; and it is difficult to decide which ranks first, the result or the cause. Gradually the use of essential oils in medicinal drugs became quite subordinate to their employment in the production of perfumes, beverages, foodstuffs, etc. The work of O. Wallach and his pupils, and of F. W. Semmler (1860–1931) and collaborators, on terpenes and terpene derivatives introduced what might properly be called the "Elizabethan Age" of the essential oil industry. Discovery followed discovery; one essential oil after the other was thoroughly investigated and its composition elucidated. Newly identified constituents were synthesized, and many of them manufactured commercially. Our industry of synthetic and isolated aromatics had its origin mainly in the work of these great explorers. Illustrious names, such as O. Aschan, E. Gildemeister, H. Walbaum, S. Bertram, A. Hesse, C. Kleber, E. Kremers, H. Barbier, L. Bouveault and E. Charabot, etc., are connected with these classical investigations, which are still being carried on with ever greater results by some of our greatest contemporary scientists, including L. Ruzicka in Zürich and J. L. Simonsen in London, not to mention other diligent workers in the United States, in the British Commonwealth, the U.S.S.R., Switzerland, France, Germany and in the Far East. It should be emphasized in this connection that Wallach's work also laid the foundation for another important chapter in the chemistry of essential oils, viz., the analysis or assay of products which, because of their high price, are prone to fraudulent manipulation and adulteration by unscrupulous producers or dealers.

IV. PRODUCTION OF ESSENTIAL OILS IN THE UNITED STATES OF AMERICA

It appears that almost every important discovery in the history of the essential oils is connected with the oil of turpentine. The first large-scale production of an essential oil in the United States of America was that of oil of turpentine. There were, naturally, good reasons for this fact: the enormous areas covered by pine forests, especially in North and South Carolina, Georgia and Alabama, and the great and steadily growing demand for the oil at home, as well as abroad.

Although tar, pitch and common turpentine (the oleoresin) were mentioned as products of Virginia in official reports as early as 1610 (D. Hanbury in *Proc. Am. Pharm. Assocn.* **19** [1871], 491), the production of oil of turpentine in North Carolina and Virginia seems not to have started until the second half of the eighteenth century. One of the earliest authentic reports on the production of oil of turpentine in Carolina was given by the German physician and explorer, J. D. Schopf, in his book entitled "Reise durch einige der mittleren und sudlichen Vereinigten Nordamerikanischen Staaten in den Jahren 1783 und 1784," (Erlangen 1788, Vol. 2, 141, 247–252).

In the early nineteenth century, the production of other essential oils was started in the United States and it is generally assumed that the oils of three indigenous American plants, of sassafras, of American wormseed (*Chenopodium anthelminticum* L.) and of wintergreen (a closely related and similar oil can be obtained from the bark of sweet birch) were, in addition to oil of turpentine, the first oils to be produced in the United States of America. The oils of wintergreen and American wormseed have always been held in especially high esteem on the North American continent. It was by their introduction into the first "United States Pharmacopoeia," published in 1820, that, for the first time, both oils were given official recognition.

Of oil of wintergreen, Jacob Bigelow tells in his "American Medical Botany" (Boston 1818, Vol. 2, 31) that it is "kept for use in the apothecaries' shops." No less a person than the apothecary, William Proctor, Jr.—called "The Father of American Pharmacy"—identified the principal constituents of the oils from wintergreen (*Gaultheria procumbens* L.) and from the bark of sweet birch (*Betula lenta* L.) already hinted at by Bigelow. The use of wintergreen oil for medicinal, cosmetic and flavor purposes has been nowhere so popular as in the United States, and yet there is no evidence whatsoever of a production of wintergreen oil on a commercial scale before or until shortly after 1800.

The same holds true as to the essential oil of American wormseed. Benjamin Smith Barton mentions the wormseed plant in his "Collections for an Essay Towards a Materia Medica of the United States" (Philadelphia, 1798,

40 and 49) in the rubric "Anthelmintics." He states that "it is the seeds that are used" and does not mention the oil. James Thatcher in "The American New Dispensatory" (Boston, 1810, 99) tells that "the whole plant may be employed" as an anthelmintic, and that "sometimes the expressed juice is used." In general, however, the seeds "are reduced to a fine powder, and made into an electuary with syrup." He repeats the same statement in the second and third editions of his book (1813 and 1817); and it was not until 1821, i.e., after the issuance of the U.S.P. 1820 listing *Oleum chenopodii*, that Thatcher, in the fourth edition of his dispensatory (pp. 173–174), added to the above cited text the following passage: "The essential oil of chenopodium or wormseed is found to be one of the most efficacious vermifuge medicines ever employed." Any large-scale production of American wormseed oil had in all probability not taken place before the twenties or even thirties of the nineteenth century. In the course of a controversy concerning the quality of the oil prepared from plants grown in Maryland and in "the western states" about 1850, we are told that "about twenty or thirty miles north of Baltimore, some fifty or sixty persons grow the plant in small or large patches on their land" for the production of essential oil (*Am. J. Pharm.* **22** [1850], 303).

Apparently there existed on the North American continent a large-scale production of oil of peppermint prior to any remarkable American preparation of the other essential oils mentioned. Until quite recently it was generally assumed that the distillation of American peppermint oil on a commercial scale had its origin in Wayne County, New York, in 1816. We know now that such an industry must have been in existence at least as early as 1800. In the booklet "150 Years Service to American Health," published by Schieffelin and Company, New York, in 1944, we are told of an offer of "homemade oil of peppermint" made by "Dr. Caleb Hyde, Physician and Druggist, Lenox, Berkshire, Mass." to Jacob Schieffelin, in 1805. The addressee answered as follows:

> "The oils of peppermint and common mint has (*sic*) in consequence of the large quantities made in the United States become a mere drug in our market and no sale for it—I have exported a quantity—it has lain for years in England without a purchaser and I shall eventually become a loser thereby."

The oils of turpentine and of peppermint were not only the first essential oils to be produced on a commercial scale in the United States but they have been up to the present among those that rank first in the quantity produced. Others, such as oil of orange, lemon, grapefruit, etc., will be described elsewhere in this work.

V. GENERAL PRINCIPLES OF PRESENT DAY ESSENTIAL OIL PRODUCTION [9]

Developed, in the course of centuries, from obscure beginnings into an important modern industry, the present-day production of essential oils is based upon principles which vary between two extremes (the first still retaining its original primitive character).

(1) In most instances the aromatic plants grow wild or are cultivated as garden or patch crops by natives of the area concerned. Cultivation of the plants and distillation of the oil represent a family industry—often, indeed, only a "side occupation" of members of the family. By primitive methods, and limiting themselves generally to one oil, the natives produce small quantities of an oil, which they sell through field brokers to village buyers, until the lots finally reach exporters in the shipping ports. The price of these oils depends upon the market, which, in turn, is influenced by supply and demand. The natives are usually well aware of prevailing quotations, and prefer stocking up their output to selling it at unattractive prices. This primitive industry is at a very definite advantage because the native operators never value very highly the work done by themselves or their families, while modern methods of production involve specialized and highly priced labor.

This old-fashioned method of essential oil production is characterized by dispersion rather than by concentration. Lack of roads prevents transport of the plant material to centrally located processing plants. The stills, usually small, portable contraptions, low priced and easy to operate, have to be scattered over the regions concerned, thus following the plant material. Such conditions still exist with respect to numerous oils and in many parts of the world, for example in East India (oil of lemongrass, palmarosa, etc.), in China (oil of star anise, cassia), and in Java (oil of cananga), etc.

(2) Advanced processing methods, based upon modern principles of plant breeding, mechanized agriculture, engineering and mass production, represent the competing counterpart of the primitive methods described above. The oils obtained in regular essential oil factories generally possess a quality superior to those produced by natives in backward districts; but the operating expenses are high. In addition to the higher standard of living and consequent higher wages and salaries involved, the amortization of invested capital, taxes and other general overhead expenses increase the costs of production. Under these conditions, a modern factory trying to

[9] This part of the survey leans on writings by Ernest Guenther, "Essential Oil Production in Latin America," in "Plants and Plant Science in Latin America," p. 205, by Frans Verdoorn, Waltham, Mass., 1945; "Essential Oils and their Production in the Western Hemisphere," New York, 1942, Fritzsche Brothers, Inc.; and "A Fifteen Year Study of Essential Oil Production Throughout the World," New York, 1940, Fritzsche Brothers, Inc.

specialize in the production of only one yearly crop could hardly survive; operation is profitable only if a variety of plants can be processed and thereby the enterprise kept busy during most of the year. Such an organization would have to produce oils from plants grown mainly in the vicinity, or from dried plants which could be shipped from afar at low cost. In other words, a factory of this type would have to be located near large plantations, connected by good roads, and would require conditions of soil, climate and altitude permitting the growth of varied crops of aromatic plants. Theoretically, this would offer the ideal solution for the essential oil industry, but it involves heavy capital investment and necessitates a great deal of experience and lengthy, systematic agricultural research work before the proper location can be found and the proper crops selected and grown. Highly mechanized farming equipment (bulldozers, tractors, planting, cultivating and harvesting machinery, trucks, etc.) must be employed in order to reduce the high cost of American labor (labor is a much smaller item in the "cost calculation" of the native patch croppers abroad). Furthermore, scientific plant breeding would have to aim at strains with a high oil yield.

Only in a few instances has the production of essential oils been placed on a really modern agricultural and technical basis. Previous to World War I, the great essential oil factories in and near New York City, London, Leipzig, and Grasse (Southern France) used to distill essential oils (oils of sandalwood, vetiver, patchouly, etc.) from plant material imported from abroad. The problem of shipping space for bulky raw material which arose during the war forced local growers in various countries abroad to install their own distillation equipment and to process their own plant material for oil. As a result, after World War I, the high cost of transporting raw material prevented manufacturers in Europe and the United States from competing with native producers abroad. Hence the production of essential oils in many instances reverted from a centralized and highly developed system to a primitive and scattered one.

Today only a few essential oils are produced by very modern or "centralized" methods. Among these are the natural flower oils in the Grasse region of Southern France, and the citrus oils of California and Florida. The latter states have succeeded in producing large quantities of high quality oils because they possess a network of good roads and railroads permitting the trucking or hauling of fruit from distant orchards and sections to centrally located, modern processing plants. Because of this feature, the United States has become a large producer and exporter of these oils. In fact, it has achieved independence as regards oils of lemon and orange. In the coming years, a corresponding evolution may take place also in other oils which so far have been distilled in far-off corners of more primitive countries. However, although most essential oils are still imported from regions abroad, where old methods prevail, the American essential oil industry has reached a high standard because of field work carried out abroad and untiring analytical work in the laboratories of private firms and scientific institutions.

The essential oil industry in its present stage is not limited to the production and distribution of essential oils and the improvement of methods, nor to the establishment and maintenance of standards of quality alone, but has come more and more to be concerned with the development, production and testing of synthetic aromatics and mixtures which today find their way into so many products of our advanced civilization.

Botany, agriculture, pharmacy and chemistry, engineering, a knowledge of world markets, commercial ingenuity and responsibility have all contributed to the development of the modern industry of essential oils. It is the maintenance of this combination which will keep up the high standard and the general usefulness of this industry.

CHAPTER 2

THE CHEMISTRY, ORIGIN AND FUNCTION OF ESSENTIAL OILS IN PLANT LIFE

BY

A. J. HAAGEN-SMIT

Note. All temperatures in this book are given in degrees centigrade unless otherwise noted.

CHAPTER 2

THE CHEMISTRY, ORIGIN AND FUNCTION OF ESSENTIAL OILS IN PLANT LIFE

I. THE CHEMISTRY OF ESSENTIAL OILS

Early in his history, man evinced a great deal of interest in the preservation of the fragrant exhalation of plants, and those who were later to be called chemists occupied themselves with separating the essence of the perishable plants. It was probably observed that heating of the plant caused the odoriferous principle to evaporate and that upon condensation and subsequent cooling, droplets united and formed a liquid consisting of two layers—water and oil. While, in such primitive experiments, the water from the plant is used to carry over the oils, additional water or steam was later introduced in "stills" to obtain better yields and quality.

In early work, therefore, we find the term "essential oil" or "ethereal oil" defined as the volatile oil obtained by the steam distillation of plants. With such a definition, it is clearly intended to make a distinction between the fatty oils and the oils which are easily volatile. Their volatility and plant origin are the characteristic properties of these oils, and it is for this reason more satisfactory to include in our definition volatile plant oils obtained by other means than by direct steam distillation.[1,2] Bitter almond and mustard oil, obtained by enzymatic action, followed by steam distillation; lemon and orange oil isolated by simple pressing, and certain volatile oils obtained by extraction are, therefore, included among the essential oils.

In the early stages of development of organic chemistry, the chemical investigation of oils was limited to the distillation of a great number of plants, and the oils which were obtained in this way were used to compose perfumes according to recipes, some of which are still used at the present time; e.g., the eau de Cologne prepared in 1725 by Johann Maria Farina in Cologne.

Gradually with the advance of science came improvements in the methods of preparing the oils, and parallel with this development a better knowledge of the constituents of the oils was gained. It was found that the oils contain chiefly liquid and more or less volatile compounds of many

[1] Thomas, "Ätherische Öle," in Klein, "*Handbuch der Pflanzenanalyse*," Vol. III, 1 (1932), 454.

[2] Rosenthaler, *Pharm. Acta Helv.* 19 (1944), 213.

classes of organic substances. Thus, we find acyclic and isocyclic hydrocarbons and their oxygenated derivatives. Some of the compounds contain nitrogen and sulfur. Although a list of all the known oil components would include a variety of chemically unrelated compounds, it is possible to classify a large number of these into four main groups, which are characteristic of the majority of the essential oils, i.e.:

1. Terpenes, related to isoprene or isopentene;
2. Straight-chain compounds, not containing any side branches;
3. Benzene derivatives;
4. Miscellaneous.

Representatives of this last group are incidental and often rather specific for a few species (or genera) and they contain compounds other than those belonging to the three first groups (Fig. 2.1).

$CH_2{=}CH{-}CH_2{-}N{=}C{=}S$	allyl isothiocyanate
$CH_2{=}CH{-}CH_2{-}S{-}CH_2{-}CH{=}CH_2$	diallyl sulfide
$CH_3{-}CH_2{-}\underset{\displaystyle CH_3}{\underset{\vert}{CH}}{-}S{-}S{-}CH{=}CH{-}CH_3$	sec.-butyl propenyl disulfide
$CH_3{-}CH_2{-}CH_2{-}CH_2{-}SH$	*n*-butyl mercaptan
$CH_3{-}CH{=}CH{-}CH_2{-}S{-}CH_2{-}CH{=}CH{-}CH_3$	dicrotyl sulfide

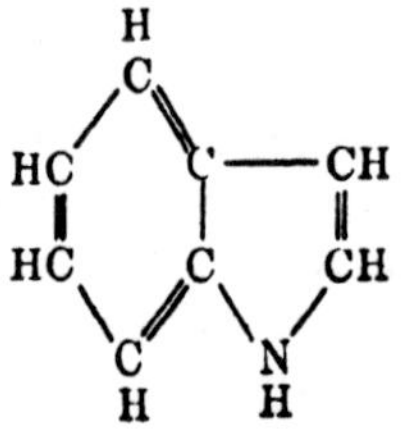

Indole

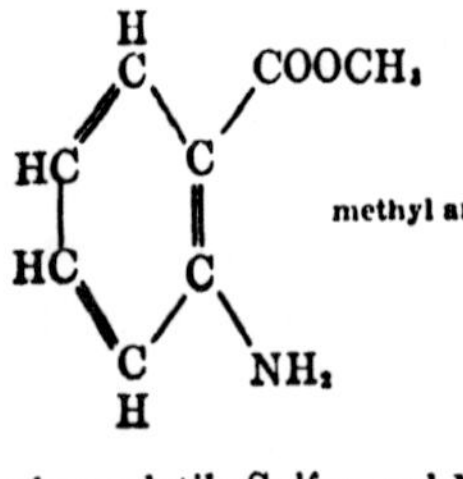

methyl anthranilate

FIG. 2.1. Natural occurring volatile Sulfur and Nitrogen containing compounds.

For example, the mustard oils, containing allyl isothiocyanate, are found in the family of the *Cruciferae*; allyl sulfides in the oil of garlic. The oil from *Ferula asafoetida* L., belonging to the family of the *Umbelliferae*, gained reputation from its active component, secondary butyl propenyl disulfide,[3] a

[3] Mannich and Fresenius, *Arch. Pharm.* **274** (1936), 461.

competitor of the odoriferous principles of the skunk, primary *n*-butyl mercaptan and dicrotyl sulfide.[4] The more pleasant smelling orange blossom and jasmine perfume betrays the presence of small amounts of anthranilates and indole, both compounds related to the amino acid, tryptophane.

Although it is possible to list a considerable number of such singular cases, the most characteristic group present in many essential oils contains hydrocarbons, as a rule of the formula $C_{10}H_{16}$ and a group of oxygen-containing compounds with the empirical formula $C_{10}H_{16}O$ and $C_{10}H_{18}O$. The classical book of Wallach indicates the names of these two types of compounds in its title "Terpene und Campher." The English word "terpene" and the German "Terpen" are derived from the German word "Terpentin," English "turpentine" and French "térebenthine." The name "Terpen" is commonly attributed to Kékulé, who is said to have introduced it as a generic term for hydrocarbons $C_{10}H_{16}$ to take the place of such words as Terebene, Camphene, etc.[5,6] The name "camphor" formerly was used to indicate the crystalline oxygen compounds, such as thyme camphor (C. Neumann, 1719) and peppermint camphor (Gaubius, 1770); these are now known respectively as thymol and menthol. The name "camphor" is at present limited to a specific compound and its more general meaning, covering the oxygenated derivatives, has been taken over by the term "terpene." With an increase in our knowledge, this broadened definition in its turn became too narrow and had to be modified to cover new and more distantly related compounds. Not all terpenes are represented by the formula C_5H_8; there exist compounds which contain less hydrogen, still others which are more saturated. We also find terpenes, like santene (C_9H_{14}), which have only 9 carbon atoms. The close resemblance to and probable connection with the C_{10} compounds through the terpene acid, santalic acid, make it impractical to omit such a compound from the terpene literature.

At the present time, therefore, we use the term *terpene* both in its broadest sense to designate all compounds which have a distinct architectural and chemical relation to the simple C_5H_8 molecule, and in a more restricted sense to designate compounds with 10 carbon atoms derived from $C_{10}H_{16}$. When confusion with the general designation is possible, members of the C_{10} group are often referred to as *monoterpenes*. Compounds having a more distant connection with the terpenes, but still containing features which link them with terpene structures, are sometimes called terpenoids or iso-

[4] Stevens, *J. Am. Chem. Soc.* 67 (1945), 407.
[5] Gildemeister and Hoffmann, "Die Ätherischen Öle," 2d Ed., Vol. I (1910), 90.
[6] Kremers and collaborators, "Phytochemical Terminology," *J. Am. Pharm. Assocn.* 22 (1933), 227.

prenoids in analogy with the term steroids, which includes not only sterols, but many more remotely connected relatives.[7, 8, 9, 10]

Characteristic for many of these oil constituents is their instability and the ease with which intramolecular rearrangements occur. These properties have been a great hindrance to the study of these compounds. Another drawback in the analysis of these oils is that most of the compounds are liquids so that thorough fractionation is necessary to separate the constituents which boil within a restricted temperature range. Since in the early stages of research it was difficult to define sharply the isolated fraction, a great number of terpenes were named after the plant from which they were obtained.

Order was brought into this chaos by Wallach, who saw clearly that the first task in the study of the oils was the identification of the terpenes with the help of crystalline derivatives, this being the only practical way we possess at present to identify chemical substances with certainty. Based on Wallach's investigation, about 500 compounds have since been isolated and characterized in the essential oils. After a general idea was obtained of the great number of distinct chemical compounds in oils, Wallach started the second part of his working program, i.e., studies of the relationship between the terpenes and the camphors. By reason of their fundamental nature and the clear presentation of the problems they involved, these studies provided great stimulus not only to his contemporaries—Semmler, Harries, Tilden and others—but had a pronounced influence on the development of chemistry as a whole. The establishment of the constitution and the relationship of the terpenes revealed a certain regularity in their structures. As early as 1869 Berthelot had discovered how the hydrocarbons $C_{10}H_{16}$, $C_{15}H_{24}$, and $C_{20}H_{32}$ are related to the hydrocarbon isoprene (C_5H_8) isolated by Williams[11, 12] a few years before. However, it was through the combined work of the aforementioned investigators that this hypothesis was established on a firm basis.

The compounds which we find in the monoterpene series can be figuratively divided into 2 isopentene chains; such a hypothetical combination gives substances of the empirical formula $C_{10}H_{16}$. If three of these isopentene units can be recognized in the molecule, the name sesquiterpene is given. In the course of time there have been added diterpenes derived from $C_{20}H_{32}$, triterpenes, $C_{30}H_{48}$, and tetraterpenes, $C_{40}H_{64}$, and finally polyterpenes with an indefinitely large number of these units (Fig. 2.2).

[7] Kremers, *ibid.*

[8] Gildemeister and Hoffmann, "Die Ätherischen Öle," 3d Ed., Vol. I, 15.

[9] Ruzicka, *Ann. Review Biochem.* 1 (1932), 581.

[10] Fieser and Fieser, "Organic Chemistry," Heath Co. (1944).

[11] Kremers and collaborators, "Phytochemical Terminology," *J. Am. Pharm. Assocn.* 22 (1933), 227.

[12] Williams, *Jahresber.* (1860), 495.

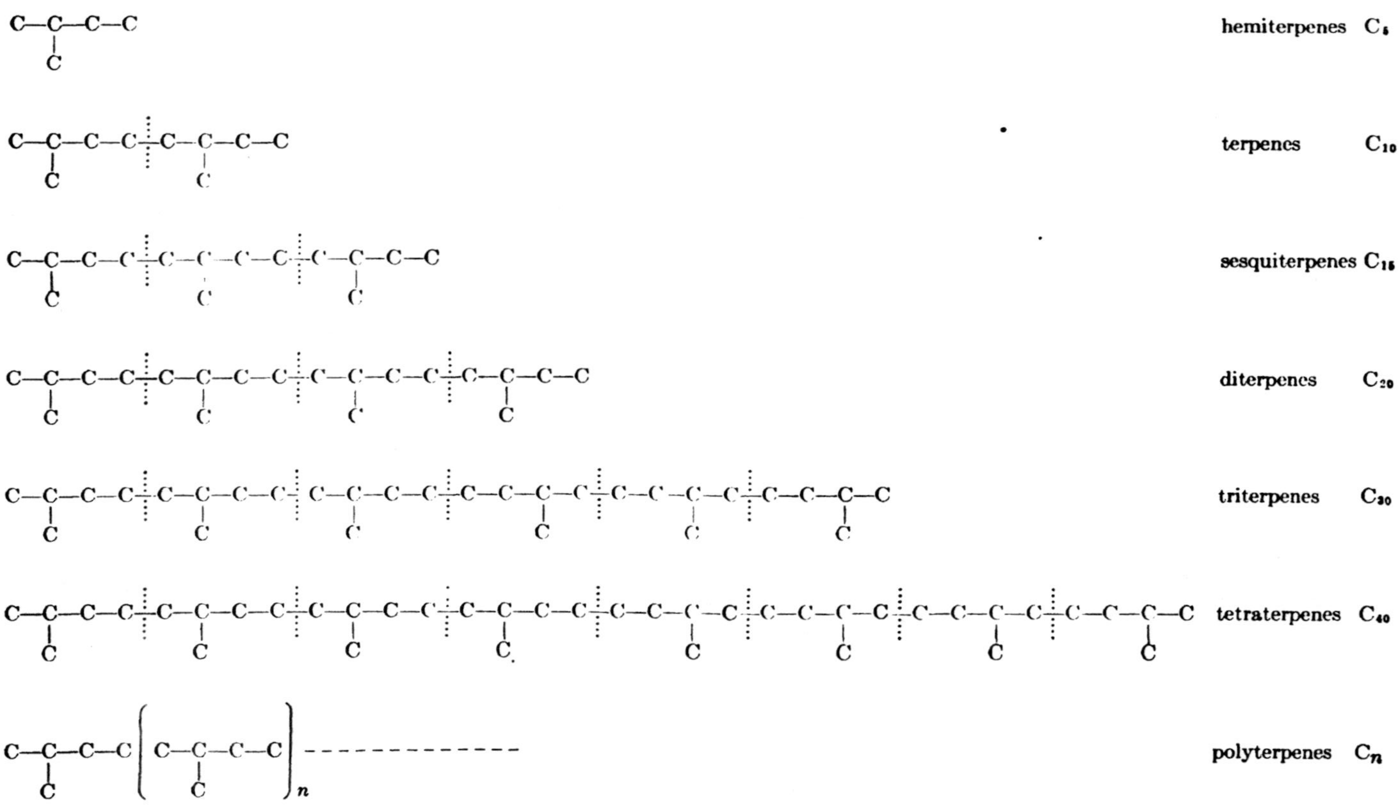

```
C—C—C—C                                                    hemiterpenes   C5
  |
  C

C—C—C—C⋮C—C—C—C                                            terpenes       C10
  |       |
  C       C

C—C—C—C⋮C—C—C—C⋮C—C—C—C                                    sesquiterpenes C15
  |       |       |
  C       C       C

C—C—C—C⋮C—C—C—C⋮C—C—C—C⋮C—C—C—C                            diterpenes     C20
  |       |       |       |
  C       C       C       C

C—C—C—C⋮C—C—C—C⋮C—C—C—C⋮C—C—C—C⋮C—C—C—C⋮C—C—C—C            triterpenes    C30
  |       |       |         |       |         |
  C       C       C         C       C         C

C—C—C—C⋮C—C—C—C⋮C—C—C—C⋮C—C—C—C⋮C—C—C—C⋮C—C—C—C⋮C—C—C—C⋮C—C—C—C   tetraterpenes C40
  |       |       |       |         |       |       |         |
  C       C       C       C         C       C       C         C

C—C—C—C ( C—C—C—C )n - - - - - - - - - - - -               polyterpenes   Cn
  |         |
  C         C
```

Fig. 2.2. Carbon Skeletons of Terpenes.

A saturated acyclic hydrocarbon with 10 carbon atoms would have the formula $C_{10}H_{22}$, possessing 6 H atoms more than a compound $C_{10}H_{16}$. This lower hydrogen content may be caused by the occurrence of double bonds, by ring structure, or by both, giving rise to acyclic, monocyclic and bicyclic representatives, with 3, 2 and 1 double bond, respectively. We have, therefore, the following possibilities for a molecule with the formula $C_{10}H_{16}$ (monoterpene):

Acyclic	No ring	3 double bonds
Monocyclic	One ring	2 double bonds
Bicyclic	Two rings	1 double bond
Tricyclic	Three rings	No double bonds

All these structural variations of the same empirical formula are found in the constituents of volatile plant oils. A chemical shorthand, developed by terpene chemists, has been introduced to show more clearly the principal structural details. This greatly simplified way of writing formulas consists in assuming a carbon atom at a place where valency lines end, or form an angle. As many C's and H's as feasible are omitted and only double bonds and substituents, such as hydroxyl and amino groups, are written in full (Fig. 2.3). Others prefer to indicate all end groups such as methyl and methylene groups in full.

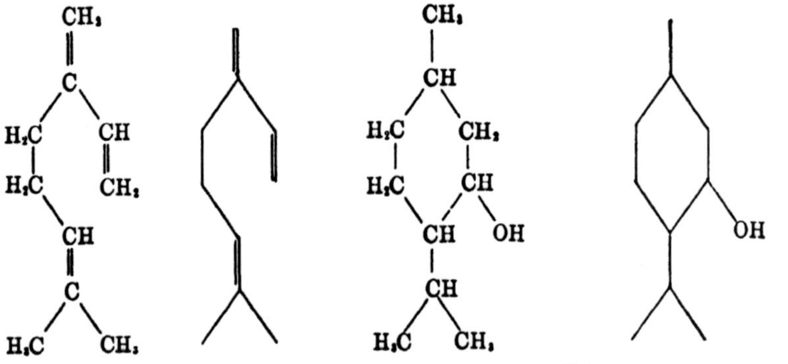

myrcene

menthol

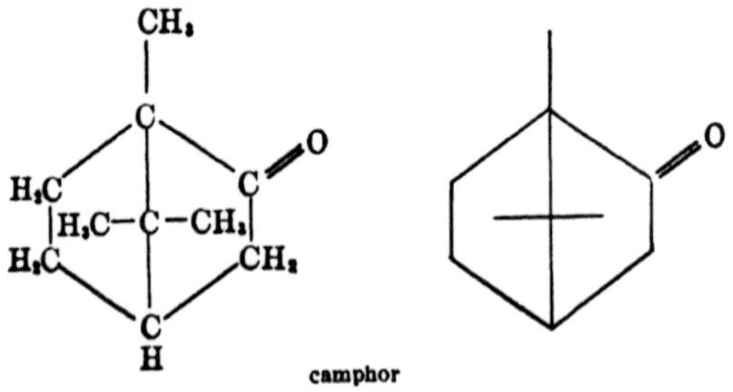

camphor

FIG. 2.3. Abbreviated formulas of Terpenes.

As examples of the acyclic terpenes with 3 double bonds, we find ocimene and myrcene. In the frequently occurring acyclic alcohols geraniol and linaloöl, in the aldehydes citronellal and citral, and in dehydrogeranic acid we see several stages of oxidation and reduction of this type of terpene hydrocarbons (Fig. 2.4). Many of these compounds can be converted into

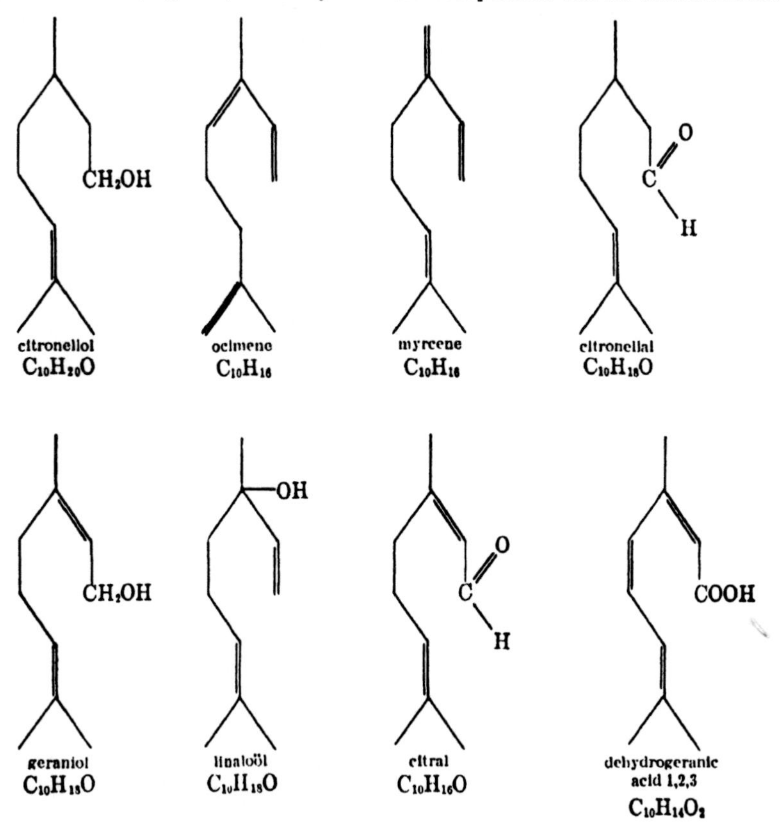

FIG. 2.4. Oxidation Stages of Acyclic Terpenes.

[1] Cahn, Penfold and Simonsen, *J. Chem. Soc.* (1931), 3134.
[2] Kuhn and Hoffer, *Ber.* 65 (1932), 651.
[3] Fischer and Lowenberg, *Liebigs Ann.* **494** (1932), 263.

each other with great ease. Geraniol, the chief constituent of rose and geranium oil, is easily converted into the monocyclic alcohol α-terpineol, the chief constituent of the oil of hyacinth, and into linaloöl, which as acetate constitutes the characteristic component of lavender oil.

Geraniols of variant origin have variant constants and odors, due to the presence of isomers. The double bond between carbon atoms 2 and 3 makes the existence of *cis-* and *trans-*isomers possible, and the relative ease of ring formation permits one to distinguish between these forms, which have been called nerol and geraniol according to their origin. The double bond near the terminal carbons is another source of isomerism. Thus geraniol, nerol and other compounds with similar structure, such as citronellol and rhodinol, and citronellal and rhodinal, consist of varying quantities of isomers containing the double bond, between either carbon atoms 7 and 8, or 6 and 7, resulting in a further source of variation in the constants of the oil constituents (Fig. 2.5).

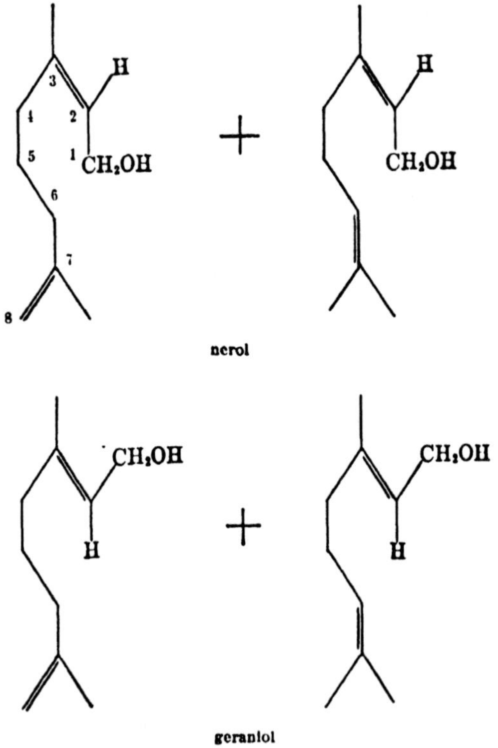

nerol

geraniol

Fig. 2.5. Isomerism of Geraniol.

Most of these compounds easily form cyclic derivatives under the influence of acids, and the formulas are usually written intentionally in such a way as to indicate where the ring closure takes place. A saturated monocyclic terpene has the formula $C_{10}H_{20}$ and is called *menthane*. If the compound has the empirical formula $C_{10}H_{16}$, there must be 2 double bonds, since the ring occurs in the place of one of the 3 double bonds present in aliphatic terpenes. Such hydrocarbons are called menthadienes, and the method of indicating the position of the double bond given by Baeyer makes use of the Greek capital letter Δ (delta), and an index number indicating the carbon atom from which the double bond starts. If the double bond is in the side chain, then it will be necessary to indicate toward which carbon atom the double bond goes. This number is placed in brackets behind the number of the first carbon atom, as is indicated in Fig. 2.6.[13]

We find many representatives of this class of menthadienes among the terpene fractions in essential oils. For example, dipentene, formed by the polymerization of isoprene under the influence of acids, is such a compound. The official name of this compound would be $\Delta^{1,8(9)}$-menthadiene.

Carbon atoms, indicated by an asterisk in Fig. 2.6 have four substituents, each of different nature, and these substituents can be arranged in two different ways around the carbon atom, thereby forming mirror images. These two forms show similar chemical properties and have the same melting and boiling points, but differ in their behavior toward light. When plane polarized light is passed through this type of compound, the plane in which the light vibrates is rotated and the amount of this rotation is determined in a polarimeter. The two forms give a rotation of equal magnitude, but in opposite directions. Often equal amounts of the two forms crystallize together and this combination acts in many ways as a third isomer. These so-called "racemates" or their derivatives have melting points, solubilities, etc., different from the two components, but do not show any optical rotation.

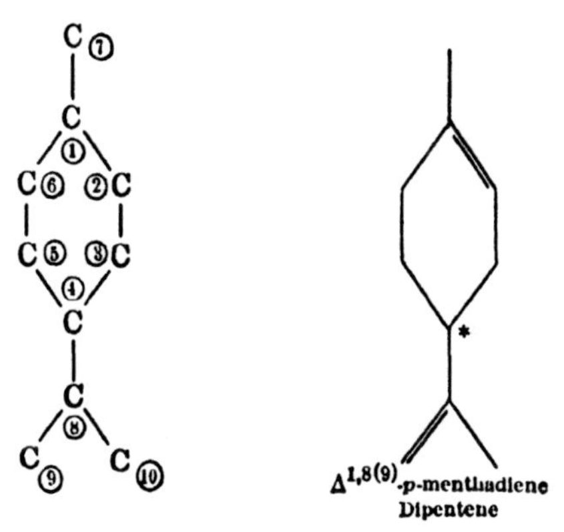

Fig. 2.6.

In the $\Delta^{1,8(9)}$-menthadienes all forms and mixtures of these optical isomers and racemates occur in nature. In pine needle and lemon oil we find a laevorotatory isomer called *l*-limonene; the *d*-form we find in oil of lemon and caraway, whereas the racemate, viz., dipentene, occurs in the oil of turpentine.

When we arrange the double bonds in a different way we can make a total of 32 isomers, which include possible optical antipodes and their racemic mixtures and *cis- trans-* forms. We see, therefore, that for this one

[13] It is customary to indicate a double bond from C_8 to one of the atoms 9 or 10 as $\Delta^{8(9)}$, although with unsubstituted end groups no confusion could arise using the index Δ^8. In the modern American literature the Δ sign is generally no longer used with monocyclic terpenes, but it is still employed in the nomenclature of the bicyclic terpenes and derivatives.

type of terpene alone, a great number of possibilities exists. Some of these structures occur in nature, as is indicated in Fig. 2.7.

The number of possibilities is further increased if the methyl and the isopropyl group occupy positions other than 1,4 on the cyclohexane ring. We find representatives of this structure in sylvestrene, a terpene derived from the commercial oil of *Pinus sylvestris.*

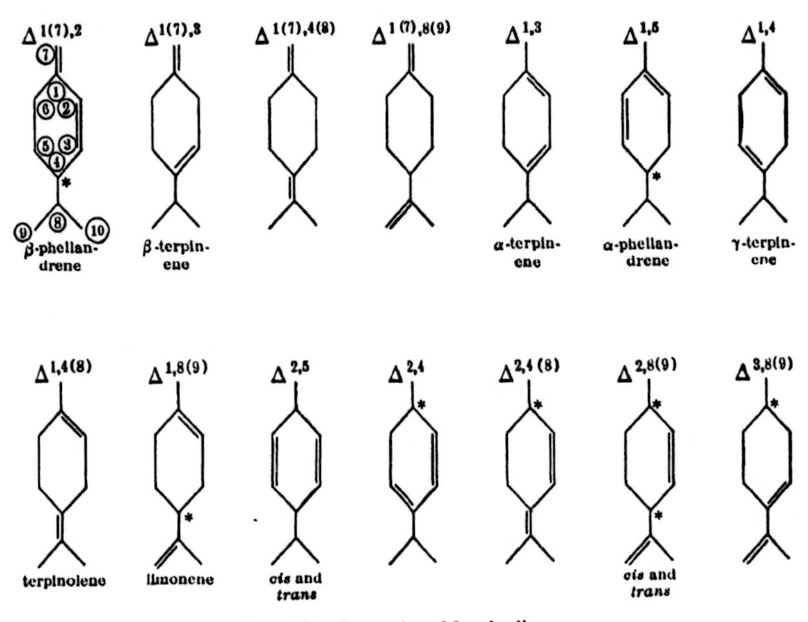

FIG. 2.7. Isomeric *p*-Menthadienes.

The monocyclic terpenes also occur in many stages of oxidation, and after we have seen the diversity in the aliphatic series, it is not surprising to find compounds with less and with more hydrogen than the general formula $C_{10}H_{16}$ requires, i.e., 1-methyl-4-isopropenylbenzene $C_{10}H_{12}$, in hashish oil;[14] *p*-cymene $C_{10}H_{14}$, in eucalyptus oil; and Δ^3-menthene $C_{10}H_{18}$, with only one double bond, in thyme oil (Fig. 2.8). Oxygen-containing derivatives (alcohols and carbonyl compounds) of these hydrocarbons also belong to the monocyclic terpene group, and many of these structures can be converted into each other by relatively simple chemical reactions (Fig. 2.9).

In all these derivatives we see a so-called head-to-tail union of the branched C_5 chains. The first discovered deviation from this structural scheme was looked upon with a great deal of suspicion. However, after the synthesis of a derivative of one of these compounds (i.e., tetrahydroar-

[14] Simonsen and Todd, *J. Chem. Soc.* (1942), 188.

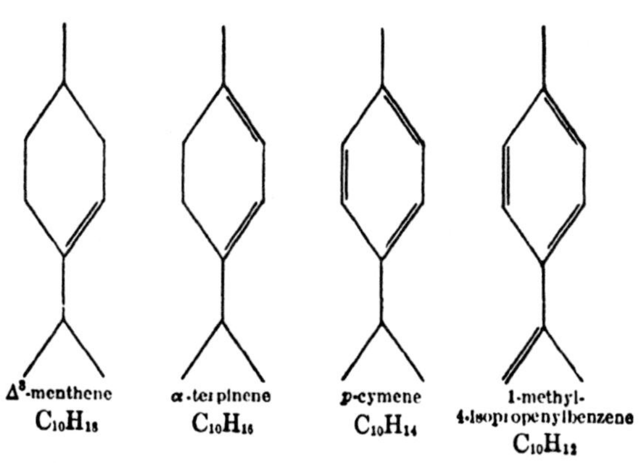

FIG. 2.8. Oxidation stages of Monocyclic Terpenes.

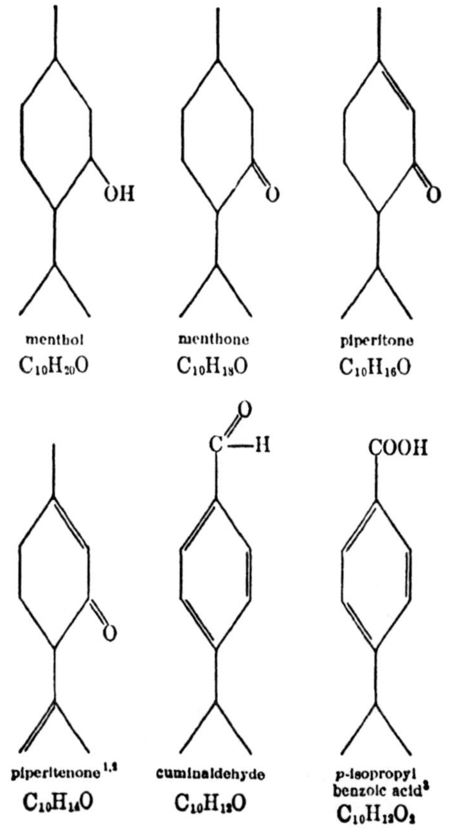

FIG. 2.9. Oxidation stages of Monocyclic Terpenes.

[1] Naves and Papazian, *Helv. Chim. Acta* **25** (1942), 984.
[2] Penfold, Ramage and Simonsen, *J. Chem. Soc.* (1939), 1496.
[3] Malavya and Dutt, *Proc. Indian Acad. Sci.* **16A** (1942), 157.

temesia ketone) was achieved, there could no longer be any doubt about the possibility of irregular build-up in the terpene series. The formula of artemesia ketone can still be divided into branched C_5 chains, but no head-to-tail union is found. Recently two similar cases have become known, senecic acid, which occurs bound in senecio alkaloids in some *Compositae*, and lavandulol in oil of lavender, which is accompanied by the ester of the regularly built linaloöl (Fig. 2.10).[15]

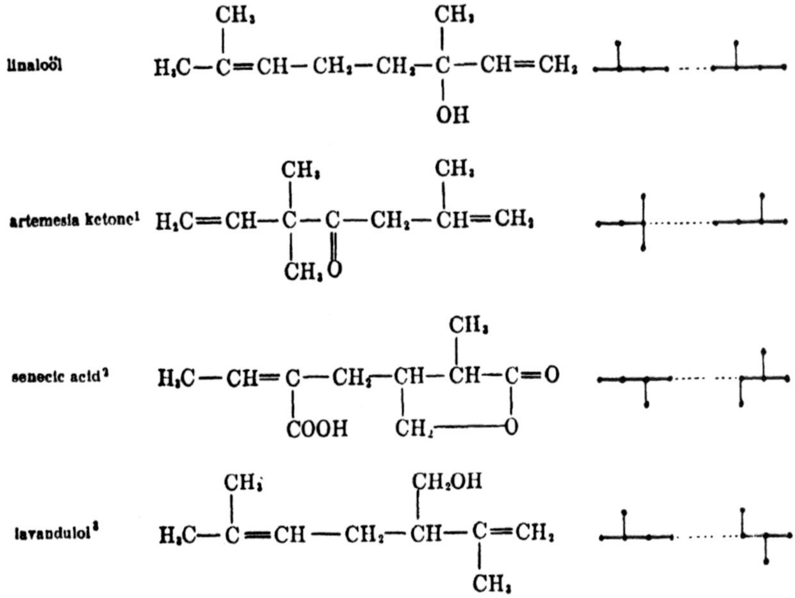

linaloöl $H_3C-C(CH_3)=CH-CH_2-CH_2-C(CH_3)(OH)-CH=CH_2$

artemesia ketone[1] $H_2C=CH-C(CH_3)(CH_3)-C(=O)-CH_2-CH(CH_3)=CH_2$

senecic acid[2] $H_3C-CH=C(COOH)-CH_2-CH(CH_2-O)-CH(CH_3)-C=O$ (ring closed: CH_2—O—C=O)

lavandulol[3] $H_3C-C(CH_3)=CH-CH_2-CH(CH_2OH)-C(CH_3)=CH_2$

Fig. 2.10.

[1] Ruzicka, Reichstein and Pulver, *Helv. Chim. Acta* 19 (1936), 646.
[2] Manske, *Can. J. Research* **17B** (1939), 1. Barger and Blackie, *J. Chem. Soc.* (1936), 743.
[3] Schinz and Seidel, *Helv. Chim. Acta* **25** (1942), 1572.

Still another way of connecting the two branched C_5 chains is observed in chrysanthemum acid which, esterified with pyrethrolone, is a part of the ester pyrethrin, the active component of insect powder made from *Chrysanthemum cinerarefolium*. The origin from a regular built carane-like bicyclic terpene through oxidation is indicated (Fig. 2.11).

The third group of (bicyclic) monoterpenes has two rings, leaving room for only one double bond for a formula $C_{10}H_{16}$. This type of structure we find in compounds such as carane, pinane, camphane, bornylane, isobornylane and fenchane. Through rearrangements of these different structures

[15] For a discussion on theoretically possible monoterpene structures see: Schinz and Bourquin, *Helv. Chim. Acta* **25** (1942), 1599; and Schinz and Simon, *Helv. Chim. Acta* **28** (1945), 774.

(Fig. 2.12), other ring systems can be derived. Molecular rearrangements, shifting of double bonds to other places in the molecule, oxidation or dehydrogenation and hydrogenation occur more or less readily, whereas treatment with acids may open these rings.

The fourth group of (tricyclic) monoterpenes has as its only natural occurring representative teresantalic acid.

Molecular rearrangements, leading to a shifting of the connections between the carbon atoms, made the investigations of these compounds very complicated. On the other hand, through these many rearrangements, a considerable number of conversions of great importance occur. The oxidation of α-pinene to camphor includes a simultaneous reattachment of the carbon atom carrying the two methyl groups to the carbon atom carrying one methyl group as indicated in Fig. 2.13. Comparable is the conversion of α-pinene into derivatives of the fenchyl series, i.e., fenchyl alcohol and fenchone. A most ingenious application of this versatility is the complete conversion of *l*-camphor into *d*-camphor and vice versa by Houben and Pfankuch.[16]

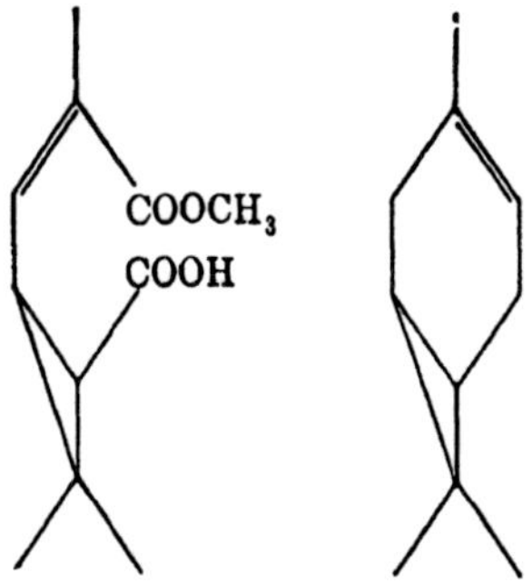

COOCH$_3$
COOH

chrysanthemum dicarboxylic ester

Δ^3-carene

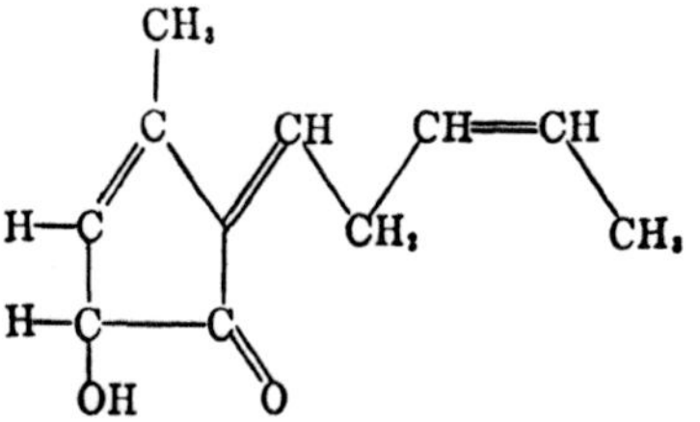

CH$_3$ C CH CH═CH H—C CH$_2$ CH$_3$ H—C—C OH O

pyrethrolone[1]

FIG. 2.11. Pyrethrin in its relation to Terpenes.

[1] Staudinger and Ruzicka, *Helv. Chim. Acta* 7 (1924), 201, 212. West, *J. Chem. Soc.* (1944), 239.

The next class of terpene compounds contains three of our building stones and its members are called sesquiterpenes, since they contain one and a half times as many carbon atoms as the monoterpenes. Here, as in the terpene series, we can predict what type of compound we might expect from their general formula, $C_{15}H_{24}$, i.e.:

4 double bonds, no rings, aliphatic
3 double bonds, one ring, monocyclic
2 double bonds, two rings, bicyclic
1 double bond, three rings, tricyclic
0 double bonds, four rings, tetracyclic.

[16] *Ber.* **64B** (1931), 2719. *Liebigs Ann.* **501** (1933), 235; **507** (1933), 37.

Representatives of the first four groups are known, and here, as in the C_{10} series, we see representatives of different stages of oxidation and reduction. We count among these some valuable perfume constituents like the aliphatic sesquiterpene alcohols, farnesol and nerolidol. These C_{15} alcohols have the same relation to each other as geraniol and linaloöl in the mono-

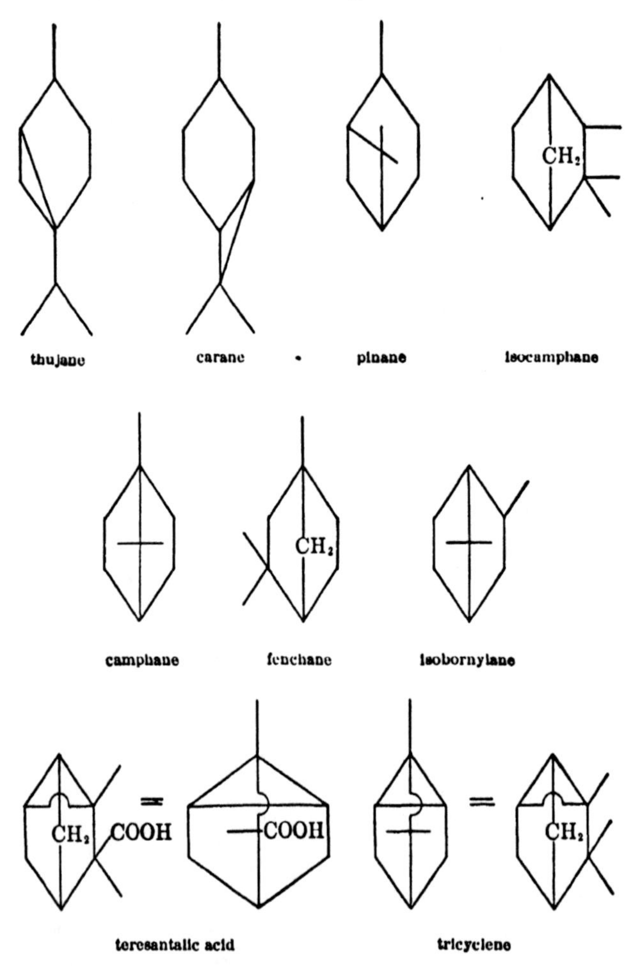

Fig. 2.12. Carbon skeletons of Bicyclic and Tricyclic Terpenes.

terpene series. Ruzicka has, therefore, called them the geraniol and linaloöl of the sesquiterpene group. This similarity is expressed by their interconversions and ring-closures to monocyclic sesquiterpenes of the bisabolene type, analogous to the formation of terpineol and dipentene in the terpene series (Fig. 2.14).

By the addition of the unsaturated C_5 chain, the number of possibilities for secondary ring-closure has been increased, and the formation of a bicyclic sesquiterpene takes place easily. These ring-closures can often be effected by treatment with acids or by dehydrogenation with sulfur,[17] selenium[18] and catalysts, such as platinum and palladium.[19]. By applying

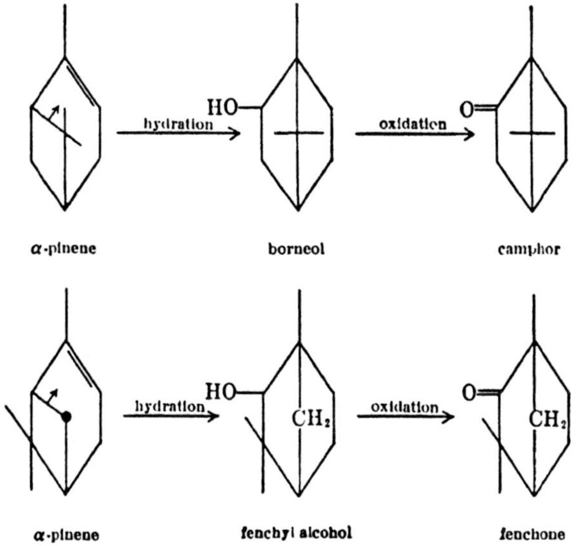

α-pinene —hydration→ borneol —oxidation→ camphor

α-pinene —hydration→ fenchyl alcohol —oxidation→ fenchone

FIG. 2.13. Conversion in the Bicyclic Terpene series.

these methods, Ruzicka has done much of the ground work in the sesquiterpene group. It has been found that three chief classes of compounds were formed, one belonging to the ring structure characteristic for a group of blue hydrocarbons sometimes found in essential oils, and two to that characteristic of the naphthalene group, i.e., cadalene and eudalene. Although the cadalene formation is easily explained on the basis of a ring-closure, as described for bisabolene, eudalene, a C_{14} compound, could only have been formed from a C_{15} compound by loss of an angular CH_3 group. The structure of the sesquiterpene belonging to this eudalene group shows an architecture which we find quite often in the higher terpenes, resin acids and carotenoids, i.e., a cyclogeraniol or cyclocitral structure. These ring compounds can readily be prepared from aliphatic terpenes under the influence of concentrated acids, such as phosphoric and sulfuric. Derivatives of this type of cyclization are found among the ionones. The bicyclic

[17] Vesterberg, *Ber.* **36** (1903), 4200.
[18] Diels and Karstens, *Ber.* **60** (1927), 2323.
[19] Zelinsky, *Ber.* **44** (1911), 3121. *J. Russ. Phys. Chem. Soc.* **43** (1911), 1220.

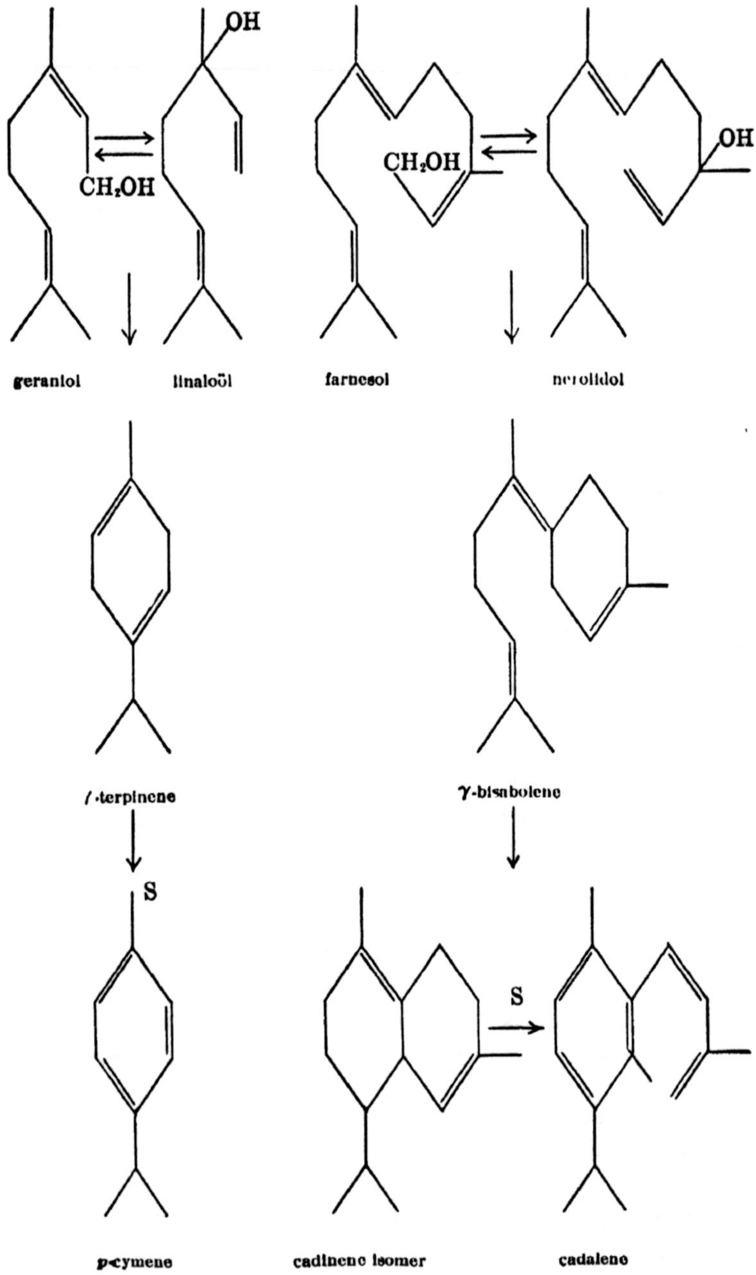

OH

CH_2OH CH_2OH OH

geraniol linaloöl farnesol nerolidol

γ-terpinene γ-bisabolene

S

S

p-cymene cadinene isomer cadalene

Fig. 2.14. Cyclization in the Sesquiterpene series, Cadalene formation.

sesquiterpenes of the eudalene type may be described as being of this cyclogeraniol ring-closure structure followed by a ring-closure of the linaloöl——→ terpinene type (Fig. 2.15). To this group belong the alcohol, eudesmol,[20] and the lactone, santonine[21] (the active principle of Levant wormseed [*Flores cinae*], well known for its anthelmintic properties).

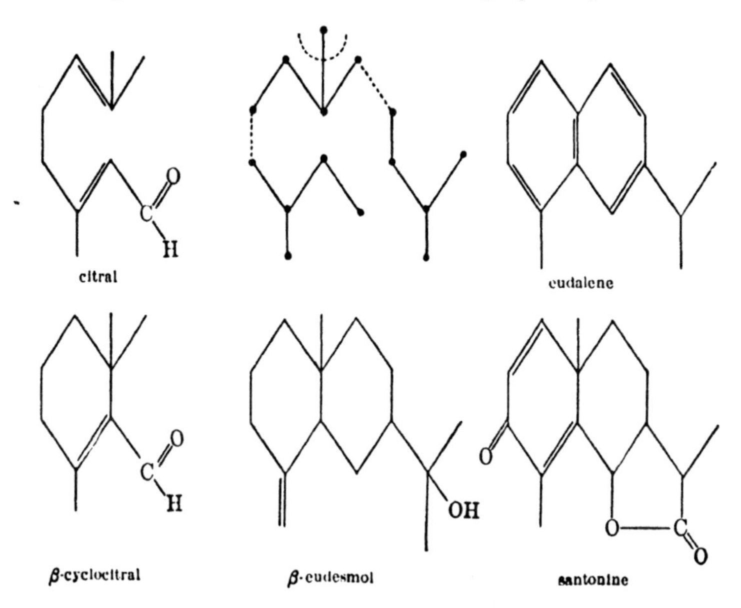

Fig. 2.15. Cyclization of the Sesquiterpene series, Eudalene formation.

The third method by which we may derive certain natural sesquiterpenes from an aliphatic chain results in compounds which are present in some oils and which give a blue color on dehydrogenation. Such bicyclic terpene derivatives, guaiol and vetivone for example, when treated with sulfur or selenium, are oxidized by removal of hydrogen; a stable system of conjugated and cross-conjugated double bonds in the rings is established and intensely blue hydrocarbons are formed. These so-called *azulenes*, mixed with yellow components, are responsible for the green color of many oils. This color is not caused by the presence of copper compounds from the stills as was formerly believed.

It is worthy of remark here that the blue color which appears on the freshly cut surfaces of some mushrooms is due to the formation of the same azulene as that obtained from the guaiol in the oil of guaiac wood, and from

[20] Ruzicka, Plattner and Fürst, *Helv. Chim. Acta* 25 (1942), 1364.

[21] Clemo, Haworth and Walton, *J. Chem. Soc.* (1929), 2368; (1930), 1110, 2579. Ruzicka and Eichenberger, *Helv. Chim. Acta* 13 (1930), 1117.

many sesquiterpenes in other oils, i.e., guaiazulene.[22] The investigations of Ruzicka and of Pfau and Plattner have shown that these azulenes or their hydrogenated precursors consist of a five- and a seven-membered ring fused together, wherein the methyl and isopropyl side chains are placed in such a manner that we may describe the compound as being formed from an aliphatic sesquiterpene such as farnesol. This unrolling and connecting of different carbon atoms of a C_{15} chain can be done in several ways. At

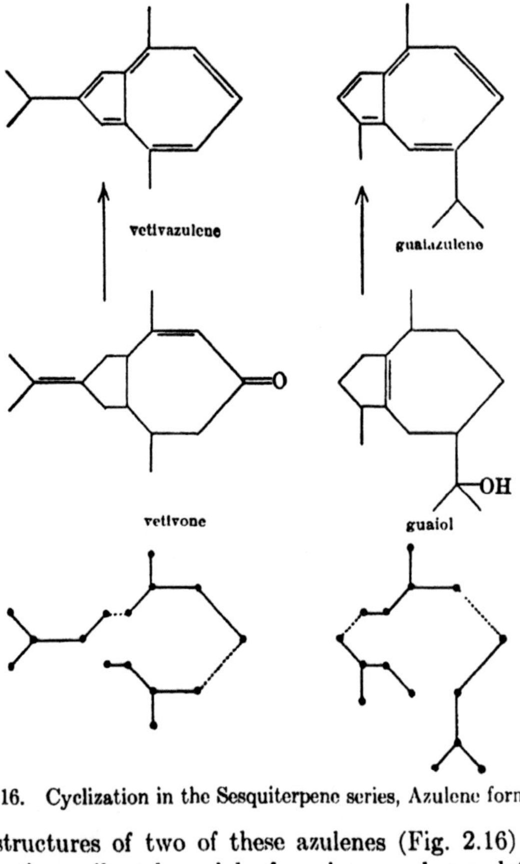

vetivazulene guaiazulene

vetivone guaiol

Fig. 2.16. Cyclization in the Sesquiterpene series, Azulene formation.

present, the structures of two of these azulenes (Fig. 2.16) derived from vetivone of vetiver oil and guaiol of guaiac wood are definitely established,[23, 24, 25, 26, 27, 28, 29] and syntheses of many azulenes have made this type

[22] Willstaedt, *Ber.* 69 (1936), 997.
[23] Naves and Perrottet, *Helv. Chim. Acta* 24 (1941), 1.
[24] Plattner and Lemay, *Helv. Chim. Acta* 23 (1940), 897.
[25] Plattner and Magyar, *Helv. Chim. Acta* 24 (1941), 191.
[26] Plattner and Magyar, *Helv. Chim. Acta* 25 (1942), 581.
[27] Pfau and Plattner, *Helv. Chim. Acta* 19 (1936), 865; 22 (1939), 640.
[28] Ruzicka and Rudolph, *Helv. Chim. Acta* 9 (1926), 118.
[29] Ruzicka and Haagen-Smit, *Helv. Chim. Acta* 14 (1931), 1104.

of compound easily available. Through the formation of well-characterized molecular addition products with picric acid and trinitrobenzene, the azulenes, in addition to cadalene and eudalene, have become a welcome tool for identification of the carbon skeletons of a number of unknown sesquiterpenes.

If the rule of the regular head-to-tail union of the C_5 units fails, and if no known dehydrogenation product betrays the general structure, the difficult road of gradual degradation has to be followed. Such has been the case, for many years, with the investigations of the structure of caryophyllene and cedrene.

After the general structure has been established, important details such as position of double bonds and substituents have to be settled. Excellent examples of this type of work can be found in Ruzicka's publications. By the use of certain methods (chiefly oxidation) on original, dehydrated, or on partially hydrogenated products, we obtain compounds for the most part complex and unknown. They may, however, indicate combinations of groups such as carbonyl and carboxyl and thus facilitate the choice among a number of proposed formulas. This elimination procedure has been very fruitful in the sesqui- and higher terpene groups. Its obvious limitation makes us welcome new direct methods of attack, such as that employed in determining the position of the nuclear double bonds in dextro pimaric acid,[30] which consists in marking the position of the double bond by oxide formation followed by substitution with a methyl group and dehydrogenation to a methyl substituted aromatic compound.

Campbell and Soffer[31] used this method to revise the position of the double bonds in the cadinene and isozingiberene formulas of Ruzicka. Ruzicka's degradation acids obtained from cadinene agree with the new formula as well as with the old one, but the new formula does not agree with the oxidation results on the tricyclic sesquiterpene copaene which gives the same dihydrochloride as cadinene. In such cases doubt arises about the homogeneity of the copaene and the cadinene, since it is possible that cadinene hydrochloride obtained from fractionated copaene actually belongs to cadinene, the latter being present as an impurity (Fig. 2.17).

This example emphasizes the great need for care in the purification of the substances under investigation. Crystalline derivatives rarely form quantitatively, and hence we cannot be sure that we are dealing with a homogeneous compound. It is to be expected that the application of chromatographic adsorption will contribute a great deal to the clarification of these problems.

[30] Ruzicka and Sternbach, *Helv. Chim. Acta* **23** (1940), 124.
[31] *J. Am. Chem. Soc.* **64** (1942), 417; **66** (1944), 1520.

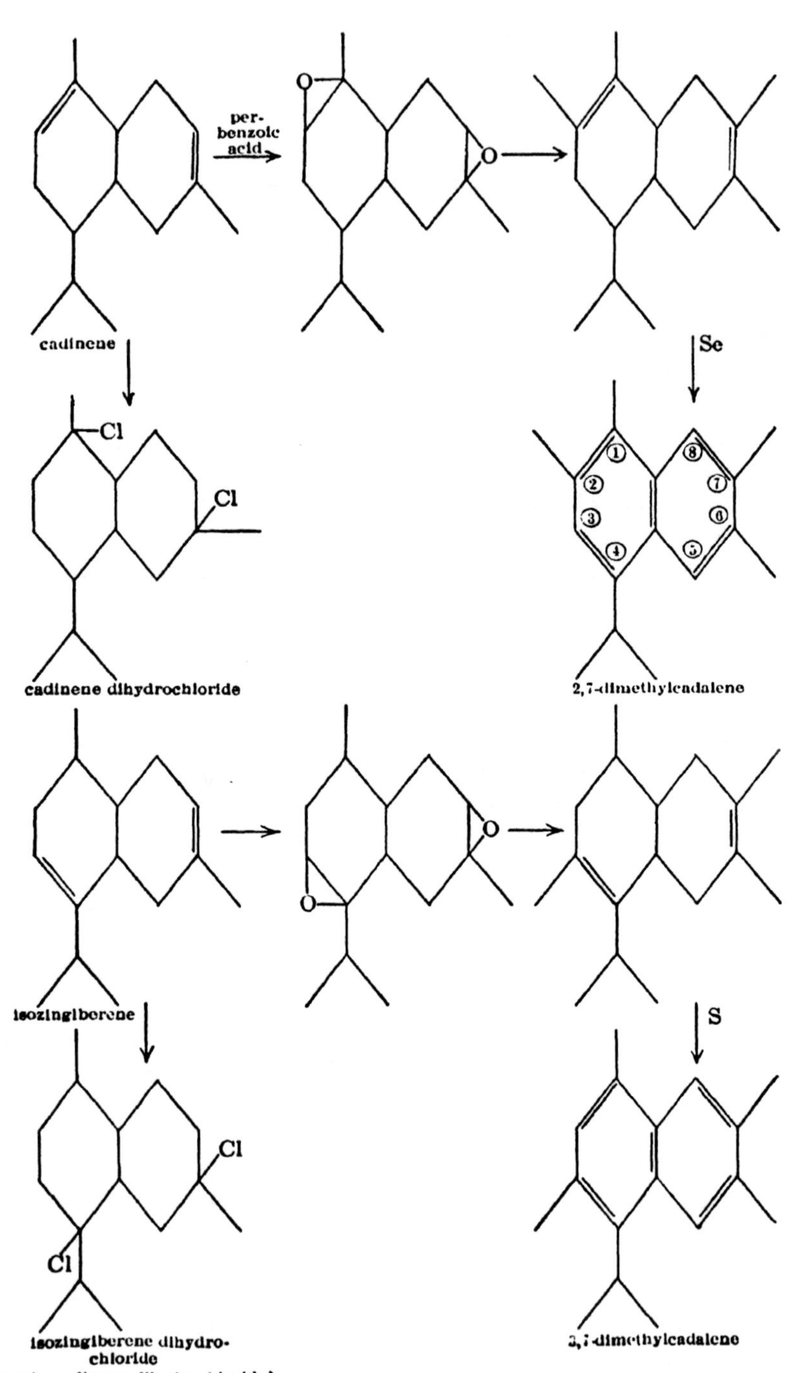

FIG. 2.17. Structure of Cadinene.

The members of the next group of compounds contain 4 isopentane units; hence the possibilities of coupling such units are much more numerous than they were in the previous groups. These compounds are difficult to study, since the separation techniques are limited through the high boiling points, the increased chances for isomerizations, and the similarity of their physical properties. Fully established structures in the diterpene series are, therefore, few and belong chiefly to some commercially important products—among them the resin acids from rosin,[32] galipot[33] and kaurikopal,[34] and compounds like vitamin A and the chlorophyll alcohol phytol (Fig. 2.18).

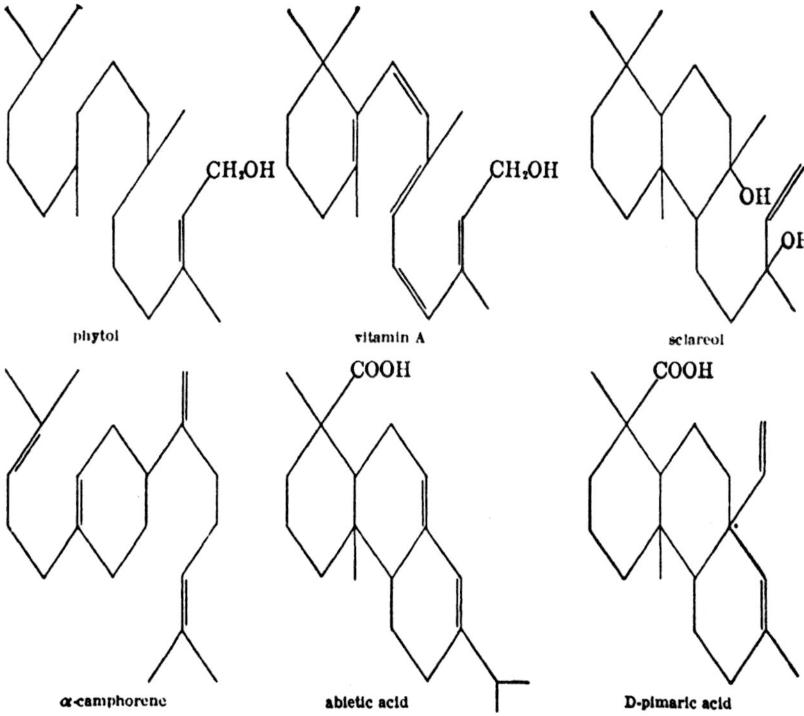

phytol vitamin A sclareol

α-camphorene abietic acid D-pimaric acid

FIG. 2.18. Diterpenes.

With the exception of α-camphorene,[35] sclareol,[36] and related manoyl oxide and manoöl, the structures of the diterpenes which occur in the highest boiling fractions of the essential oils are not yet satisfactorily established. This group is at present going through the same early stages of develop-

[32] Ruzicka, Sternbach and Jeger, *Helv. Chim. Acta* **24** (1941), 504.
[33] Ruzicka and Sternbach, *ibid.* **23** (1940), 124.
[34] Ruzicka, Bernold and Tallichet, *Helv. Chim. Acta* **24** (1941), 223.
[35] Ruzicka and Stoll, *Helv. Chim. Acta* **7** (1924), 271.
[36] Ruzicka, Seidel and Engel, *Helv. Chim. Acta* **25** (1942), 621.

ment, i.e., a sharper characterization, as the monoterpenes fifty years ago. A crystalline diterpene phyllocladene has been found, which appears to be identical with sciadopitene and dacrene isolated from other sources and isomeric with podocarpene and isophyllocladene.[37] Dihydrophyllocladene is found to be identical with the lignites (fossil resins): hartite, bombiccite and hofmannite.[38,39] To find still higher isoprene homologues, the separation technique with steam distillation is unsatisfactory and we have to resort to reduced pressure distillation or solvent extraction techniques. Through the use of these methods, products have been isolated which show a continuation of the branched C_5 building plan and consist of six C_5 units. Some of these we find in the nonvolatile part of the resins—like β-amyrin; other examples are boswellic acid from incense and betulin, the white pigment of the bark of birch. Interesting members are found in woolfat as lanosterol,[40] in quinine bark as chinovic acid,[41] in cloves and olive leaves as oleanolic acid (Fig. 2.19). This acid is also known to be present as glycoside in several plants. These types of glycosides, as a result of their detergent reaction, have received the name of saponins. Many drug plants, such as sarsaparilla and smilax, owe their pharmacological action to the presence of these C_{30} compounds.

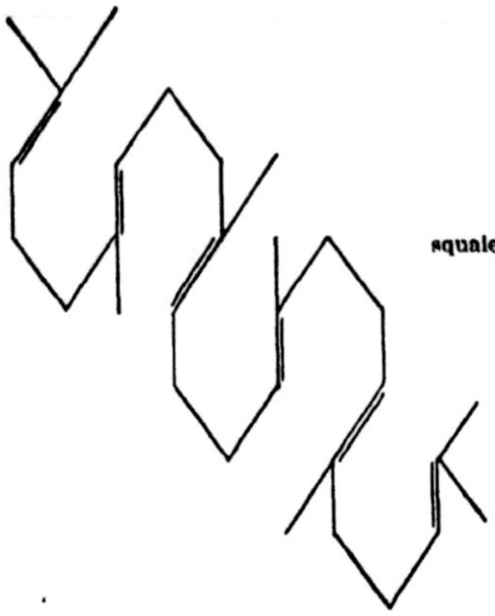

squalene

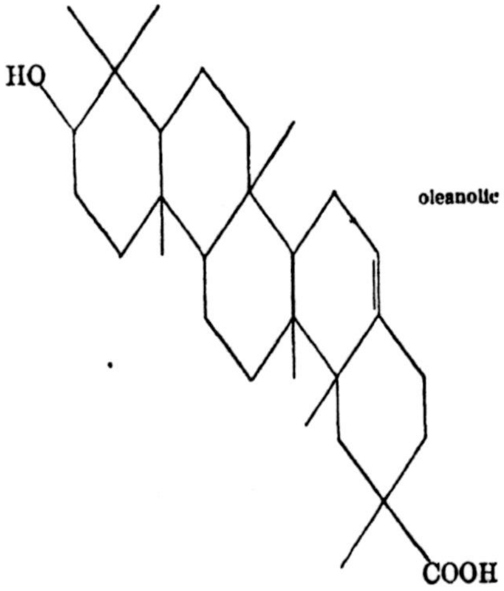

HO

oleanolic acid[1]

COOH

FIG. 2.19. Triterpenes.

[1] Bilham, Kon and Ross, *J. Chem. Soc.* (1942), 532. Cf. Ruzicka, et al., *Helv. Chim. Acta* **26** (1943), 227, 280. Noller, *Ann. Review Biochem.*, **14** (1945), 381.

[37] Briggs, "Review of the Diterpenes," *Rep. Meeting Australian New Zealand Assocn. Adv. Sc.* **23** (1937), 45.

[38] Soltys, *Monatsh.* **53–54** (1929), 175.

[39] Briggs, *J. Soc. Chem. Ind.* **60** (1941), 226T.

[40] Bellamy and Doree, *J. Chem. Soc.* (1941), 172.

[41] Ruzicka and Prelog, *Helv. Chim. Acta* **20** (1937), 1570.

The structure determination of these tetracyclic triterpenes relies heavily on the formation of picene derivatives through dehydrogenation. Since all angular methyl groups are removed in this process, special degradation reactions should provide proof of their position. This difficult structural detail has not yet been accomplished for all methyl groups, and their present fluctuating position in the proposed formulas rests largely on the application of the "isoprene hypothesis."

The simple symmetrical constitution of the triterpene squalene (Fig. 2.19) present in liver of sharks and many vegetable oils[42] permitted rapid progress in the elucidation of its structure, and even its synthesis has been accomplished by combining 2 mols of the acyclic sesquiterpene alcohol farnesol.[43, 44]

This symmetrical build up, which we see for the first time in squalene, is quite common in the next group, viz., the tetraterpenes. The known members of this group belong to the yellow and red crystalline carotenoid pigments from plants and animals. The carbon skeleton of these compounds can be described as a doubling of a regularly built diterpene. In lycopene (Fig. 2.20), the red pigment of the tomato, we find a chain of 32 carbon atoms with 8 methyl side chains. In most carotenoids the ends of the long chain have formed a cyclogeraniol type of cyclization, as in β-carotene and xanthophyll (Fig. 2.20).

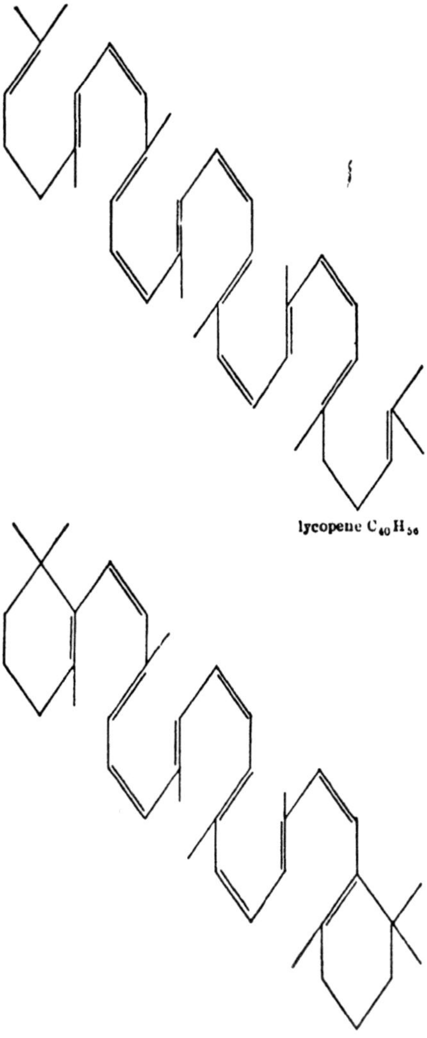

lycopene $C_{40}H_{56}$

β-carotene $C_{40}H_{56}$

Fig. 2.20. Tetraterpenes.

[42] Täufel and Heiman, *Biochem. Z.* **306** (1940), 123.
[43] Schmitt, *Liebigs Ann.* **547** (1941), 115.
[44] Karrer and Helfenstein, *Helv. Chim. Acta* **14** (1931), 78.

The presence of a large number of conjugated double bonds (14 in lycopene) distinguishes this group from other terpenoids, and is responsible for the red and yellow colors which are so characteristic of this group of substances.

As a result of our lack of knowledge of representatives belonging to the intermediate groups, we must look for the next higher terpenes in a number

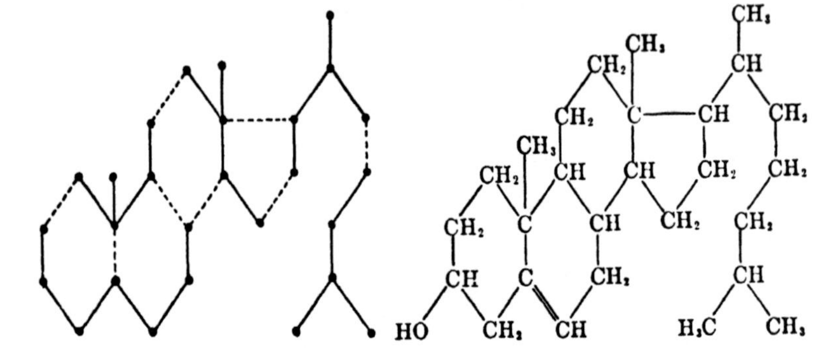

cholesterol

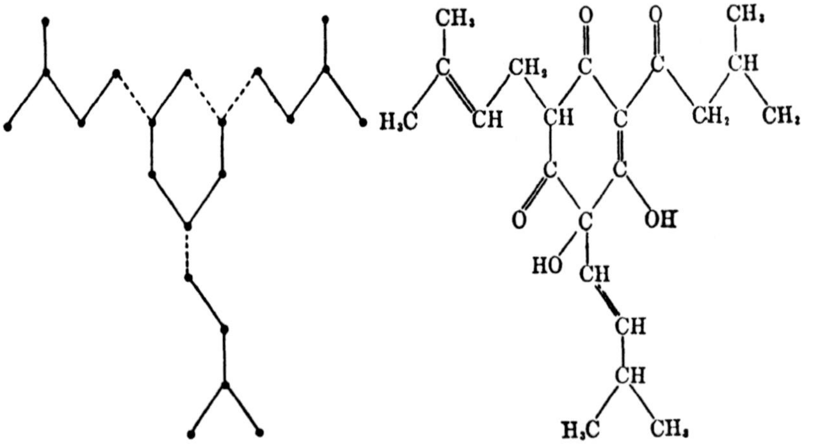

humulone[1]

FIG. 2.21. Mixed building schemes.

[1] H. Wieland and E. Martz, *Ber.* **59B** (1926), 2352.

of compounds which have attracted the attention of investigators on account of their elastic properties. The different rubbers from Hevea, Guayule, etc., belong to this polyterpene group and contain up to several thousand C_5 units.

While all of these compounds can be divided completely into branched C_5 chains, a number of natural products contain structures in which we can recognize one or more of these units, but which we cannot fully describe in this way. In such cases some of the carbon atoms are left over; these form often a straight chain, as in cholic acid and cholesterol. Also in humulone (Fig. 2.21) the connection with the terpene compounds is unmistakable; nevertheless, we cannot fully divide the molecule into branched C_5 chains.

In some cases the recognition of such building principles serves as a guide in the laboratory synthesis, and also indicates a possible biogenesis. An interesting illustration is furnished by the structure of cannabidiol, one of the constituents of Egyptian hashish. We can easily recognize in this molecule the structures of cymene or 1-methyl-4-isopropenylbenzene, the first of which form the main part of the essential oil of hashish[45] (Fig. 2.22).

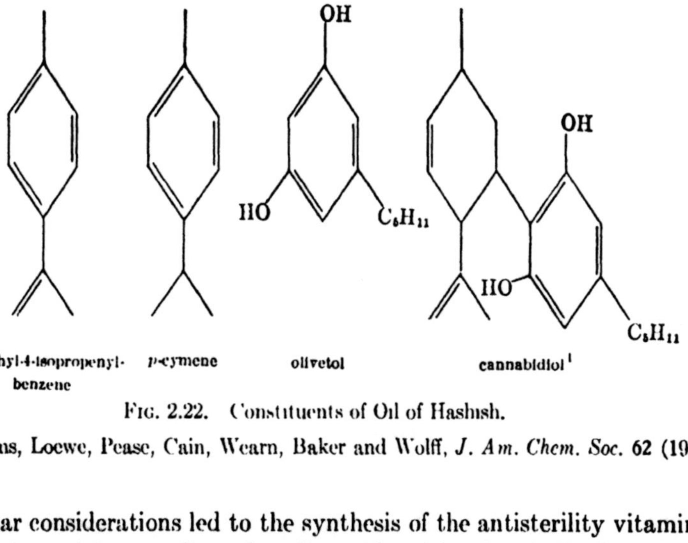

1-methyl-4-isopropenyl-benzene p-cymene olivetol cannabidiol[1]

FIG. 2.22. Constituents of Oil of Hashish.

[1] Adams, Loewe, Pease, Cain, Wearn, Baker and Wolff, *J. Am. Chem. Soc.* 62 (1940), 2566.

Similar considerations led to the synthesis of the antisterility vitamin *E* which is formed by coupling phytylbromide with trimethylhydroquinone, whereas the antihemorrhagic vitamin K_1 is synthesized by condensing the same bromide with the sodium salt of 2-methyl-1,4-naphthoquinone.[46]

The second major group of oil components contains only straight chain hydrocarbons and their oxygen derivatives: alcohols, aldehydes, ketones, acids, ethers and esters. These essential oil hydrocarbons range from *n*-heptane, which forms 90 per cent of the oil of *Pinus sabiniana* and *P. jeffreyi*, to compounds with 15–35 carbon atoms.[47] The higher paraffin-like materials may crystallize out during cooling and storage of the oils and are called "stearoptenes." The number of carbon atoms in some of these hydrocarbons suggests a connection with the natural occurring fatty acids through de-

[45] Simonsen and Todd, *J. Chem. Soc.* (1942), 188.
[46] *Dam. Ann. Review* IX(1940), 362.
[47] Schorger, *Trans. Wisconsin Acad. Sci.* 19 (1919), 739, 752.

carboxylation or ketone formation. The formation from the wax alcohols through dehydration and reduction is also held possible.[48]

The alcohols, aldehydes and ketones are quite often contained in the low boiling fraction of the volatile oil. A typical example is found in the so-called leaf alcohol (*cis*- or *trans*-hexen-3-ol-1), carrier of the odor of grasses, green leaves, etc. Oxidation of the alcohol with chromic acid furnishes a hexenal which has been recognized in many green plants, including tea, ivy, clover, oak, beech, wheat, robinia, black radish, violet leaves and cucumbers. However, the volatile oil of cucumber consists largely of nonadiene-2,6-ol-1 with some nonadiene-2,6-al-1[49] (Fig. 2.23). This aldehyde has also been recognized in the leaf oil of violets.[50]

$$CH_3—CH_2—CH{=}CH—CH_2—CH_2{\cdot}OH$$

leaf alcohol

$$\underset{⑨}{CH_3}—\underset{⑧}{CH_2}—\underset{⑦}{CH}{=}\underset{⑥}{CH}—\underset{⑤}{CH_2}—\underset{④}{CH_2}—\underset{③}{CH}{=}\underset{②}{CH}—\underset{①}{CH_2}{\cdot}OH$$

nonadiene-2,6-ol-1

$$CH_3—CH_2—CH{=}CH—CH_2—CH_2—CH{=}CH—C\begin{matrix}{/\!\!/}O \\ \backslash H\end{matrix}$$

nonadiene-2,6-al-1

Fig. 2.23.

In this group are also included the many fatty acids which occur free or esterified with alcohols of different chain length and different degrees of saturation. This group is present in a number of volatile oils from fruit.

The third major group of essential oil components comprises a number of important flavor and perfume constituents derived from benzene and more specifically from *n*-propyl benzene. As in the preceding groups, we find these compounds in many stages of oxidation. The aromatic ring may carry hydroxy, methoxy and methylene dioxy groups; the propyl side chain may contain hydroxyl or carboxyl groups, or form a part of a lactone group, as in coumarin and its many derivatives[51] (Fig. 2.24). Many members of this group are related through simple chemical reactions. For example, on isomerization followed by oxidation, eugenol is converted to the corresponding vanillin (Fig. 2.25), the flavoring principle of the vanilla bean.

[48] *Bull. Univ. Wisc.*, Serial No. 1919, Gen. Series No. 1703 (1934). "Phytochemistry," III, Kremers and collaborators. The methane series of hydrocarbons.

[49] Takei and Ono, *J. Agr. Chem. Soc. Japan* 15 (1939), 193.

[50] Ruzicka and Schinz, *Helv. Chim. Acta* **25** (1942), 760.

[51] Sethna and Shah, *Chem. Rev.* **36** (1945), 40.

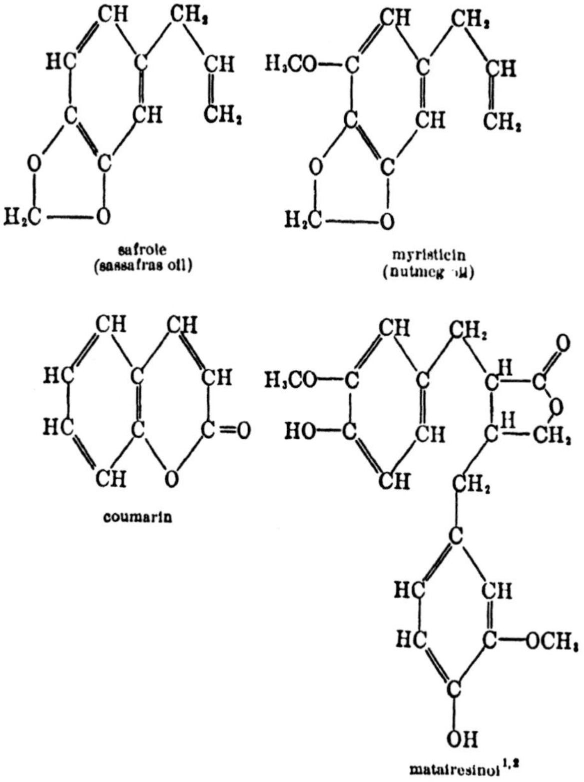

safrole (sassafras oil)

myristicin (nutmeg oil)

coumarin

matairesinol[1,2]

Fig. 2.24. Aromatic oil constituents.

[1] Haworth and Kelly, *J. Chem. Soc.* (1937), 384.
[2] Ishiguro, *J. Pharm. Soc. Japan* **56** (1936), (Abstracts in German), 68.

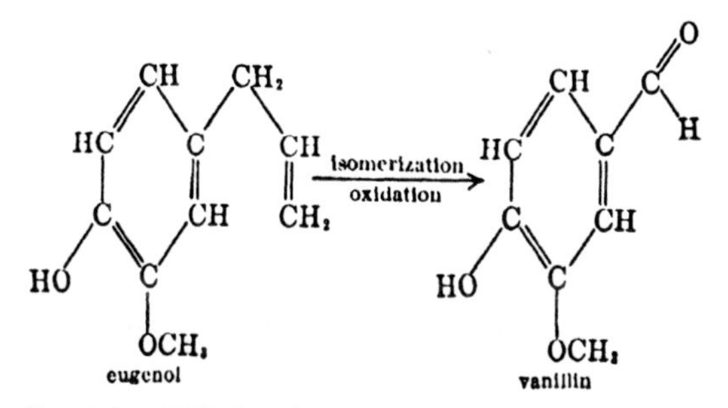

Fig. 2.25. Relations between aromatic oil constituents.

This group of compounds shows a definite relationship to some of the resins with aromatic structures, like matairesinol,[52] which represent a doubling

[52] Haworth and Slinger, *J. Chem. Soc.* (1940), 1098.

of the propyl benzene structure. This dimerization has been demonstrated *in vitro* for isoeugenol methyl ether which is doubled into *bis*isoeugenol methyl ether[53] (Fig. 2.26). It is probable that a condensation of a large

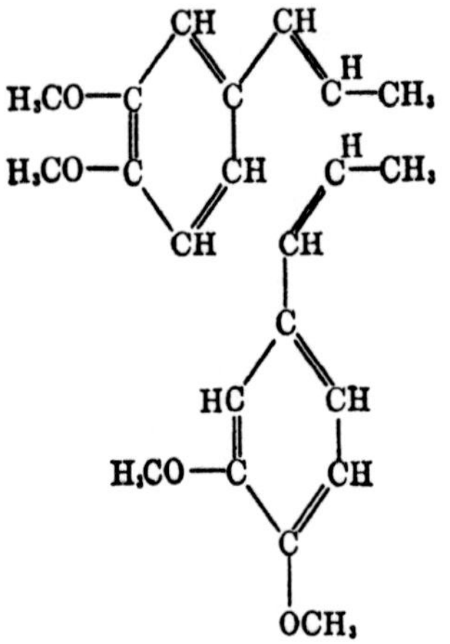

isoeugenol methyl ether

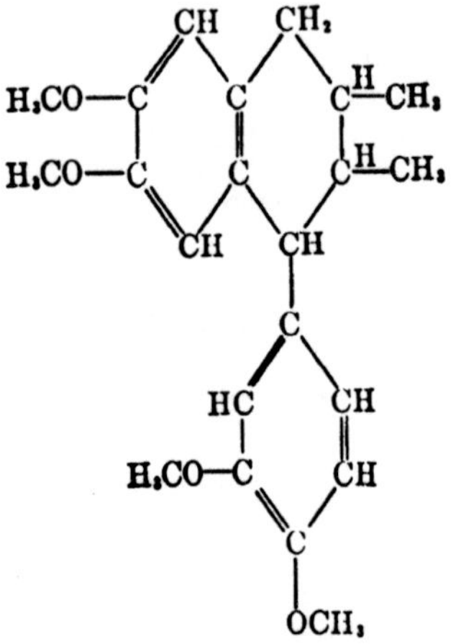

bis-isoeugenol methyl ether

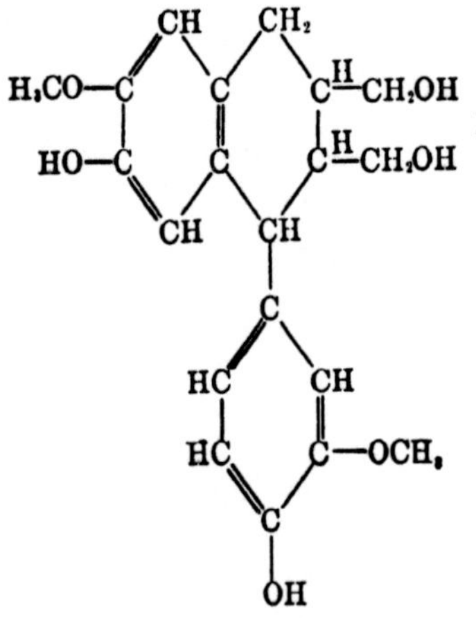

isolariciresinol

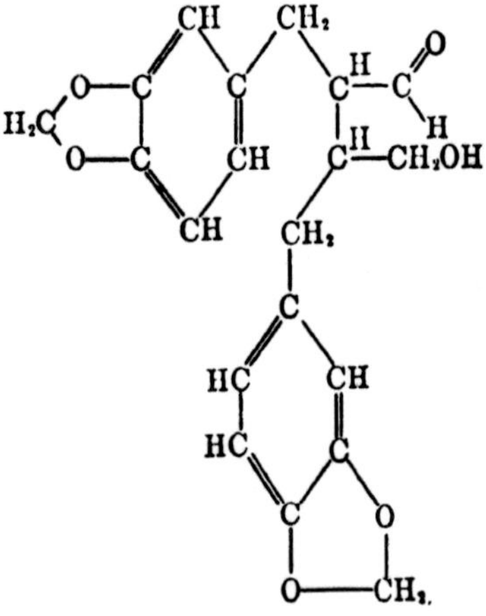

cubebin

FIG. 2.26. Relation between aromatic oil constituents and resinols.

number of analogous units has led to the formation of lignin, the widely distributed component of woody tissues. Such a relation is comparable to the formation of rubber from the smaller terpene building units. A somewhat more distant connection can be seen in the formation of the anthocyans

[53] Muller and Hartai, *Ber.* 75B (1942), 891.

and flavones, since propyl benzene derivatives have been postulated as taking part in the biosynthesis of these plant pigments.

From this short account of our chemical knowledge of the essential oil components and their near relatives, it is clear that their study must have occupied the minds of a large number of chemists interested in natural products. In numerous cases the purification, characterization, structural determination and synthesis of a single terpene has been the lifetime work of many of the terpene chemists.

The difficulties are immediately apparent when the starting material arrives in the laboratory. Separation has to be accomplished on a large diversity of compounds, since the only links between them are their plant origin and their volatility. In addition, many of the constituents are easily converted into other compounds with similar properties. The investigation of the essential oils has, therefore, served as a hard schooling in chemical separation techniques, in which none of the existing methods, physical or chemical, can be neglected. When the oils are obtained from the plant by steam distillation, the steam carries over the volatile component at a temperature of somewhat less than 100°.* This, however, does not represent the actual boiling point of the oil components. At the boiling point of a mixture of oil and water, the sum of the partial pressures of oil and water is equal to the atmospheric pressure. The boiling temperature of the steam and vapor mixture is, therefore, lower than the boiling temperature of water alone.

In a mixture of oil of turpentine and water, which boils at 95.6° at 760 mm. pressure, the vapor pressure of the oil contributes 113 mm., the water 647 mm. Without the help of the water vapor, the bulk of most oils would distill at 150°–300°, at which temperature labile substances would be destroyed and a strong resinification would occur. With the aid of steam distillation, the majority of these compounds are carried over a few degrees below the boiling point of water. Vapor pressure data[54, 55, 56, 57] of single components make it possible to calculate their boiling points by steam distillation and the proportion of oil and water which is distilled at different pressures.

While steam distillation is a simple procedure, we cannot *per se* assume that a steam distilled oil is identical with the oils as occurring in the plants. Several cases are known where certain compounds are formed by the action of the steam. These are for the most part degradation products of carbohydrates, like furfural. Loss of water from alcohols and hydrolysis of esters

* Note. All temperatures in this book are given in degrees centigrade unless otherwise noted.

[54] Charabot and Rocherolles, *Compt. rend.* 135 (1902), 175.

[55] Pickett and Peterson, *Ind. Eng. Chem.* 21 (1929), 325.

[56] Linder, *J. Phys. Chem.* 35 (1931), 531.

[57] Schoorl, *Rec. trav. chim.* 62 (1943), 341, 350, 354, 358, 363, 366, 375.

will result in the formation of new hydrocarbons and acids. Likewise, nitrogen compounds often have a secondary origin. This destruction usually runs parallel with the loss of the delicate nuances in smell and is a certain indication that changes in the original composition of the oil have taken place, a matter of concern for both production and research departments.

Since the oils contain chemical compounds of many classes, it is often desirable to remove at least those groups of substances that contain more reactive groupings than the hydrocarbons, among them acids, bases, phenols, ketones and aldehydes. The oils are, therefore, treated with dilute aqueous alkali solutions to remove the acidic substances, or with bases to remove the acids, with sulfite, bisulfite or Girard's reagent to isolate ketones and aldehydes, and sometimes with phthalic anhydride to remove the alcohols.

In the oil layer or in the solvent extract of the steam distillate (including that of the distillation water), compounds boiling lower and higher than water are found, and the desirability of fractionating the original, or the treated oil, into fractions which preferably contain only one component is indicated. When we assemble the fractionation data in a graph and plot the quantity of oil distilled within a certain temperature interval along the ordinate, although the abscissa shows the boiling points, we notice in the fractionation curve of different oils maxima indicating the presence of distinct components of the oils. In the fractionation of the American oil of peppermint (Illustration 2.1), a typical terpene oil, the first volatile components which distill are small quantities of two compounds which were postulated by some investigators as the building stones of the branched C_5 chains, viz., acetone and acetaldehyde, accompanied by dimethyl sulfide, a compound containing sulfur. These are followed by the hemiterpenes, isovaleraldehyde and isoamyl alcohol; and these in turn are followed, at a temperature of approximately 150°, by a number of compounds, which by analysis are shown to consist of carbon and hydrogen only. Several small maxima in the boiling point curve indicate the presence of several of these terpenes. At this stage the determination of physical constants, such as specific gravity, refractive index and optical rotation, may aid considerably in indicating the nature of the terpenes.

With the exception of camphene and bornylene, all the terpene hydrocarbons are liquid at ordinary temperatures, and their tendency to crystallize at lower temperatures is negligible. The tendency to form a distinct crystalline pattern can be greatly increased when polar groups are introduced into the molecule. The first crystalline derivative of a hydrocarbon to be prepared in this way is the so-called "artificial camphor" obtained by Kindt[58] in 1803, when he passed hydrogen chloride into oil of turpentine.

[58] *Bull. Univ. Wisc.*, Serial No. 1813, Gen. Series No. 1597 (1932). "Phytochemistry," II, Kremers and collaborators. Chemical properties of hydrocarbons.

The systematic study of these methods is, however, due to Wallach,[59] who about fifty years ago introduced several more of these procedures. Since the double bonds are the only reactive points in the terpenes, these groups are introduced into the molecule by simple addition reactions to the double bond. In this way, halides, dihalides, nitrosohalides, nitrosites and nitrosates are formed. Through these reactions, the hydrocarbon derivatives may crystallize even from impure fractions, and their identification is

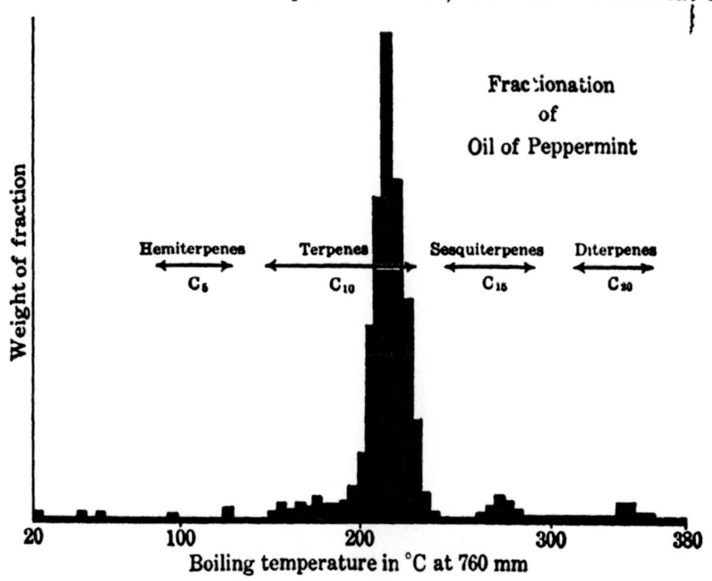

ILL. 2.1.

possible through comparisons of the melting points of the same derivatives of known terpenes. Modern chemistry has added only a few more reagents to this list, among them, maleic anhydride for the characterization of compounds with conjugated double bonds[60, 61, 62] (Fig. 2.27).

When the distillation is continued, the next large group of fractions boils at about 200°–230°. They consist mainly of the oxygen derivatives of the terpenes, $C_{10}H_{18}O$, and in our fractionation example of peppermint oil, a fraction of menthone is obtained, followed by a larger fraction of menthol. While some oxygen derivatives in peppermint oil are crystalline, in many other oils these have to be characterized by reaction products with certain identifying reagents, such as phenylisocyanate and nitrobenzoylchlorides for alcohols, and nitrophenylhydrazines for ketones and aldehydes.

[59] Ruzicka, "The Life and Work of Otto Wallach," *J. Chem. Soc.* (1932), 1582.
[60] Diels and Alder, *Liebigs Ann.* 460 (1928), 98.
[61] Birch, *J. Proc. Roy. Soc. N. S. Wales* 71 (1937), 54.
[62] Goodway and West, *J. Soc. Chem. Ind.* 56 (1937), 472T.

The next group of compounds we encounter in the fractionation is again of a hydrocarbon nature. We have come into the region of sesquiterpenes C_{15}, and these C_{15} compounds are again followed by their oxygen derivatives which in turn are followed by diterpenes. If the oils contain a number of benzene and aliphatic compounds their fractions will be superimposed on this general scheme.

The boiling point regularities observed in the fractionation of the oils are clearly expressed in a graph showing the boiling points of the normal

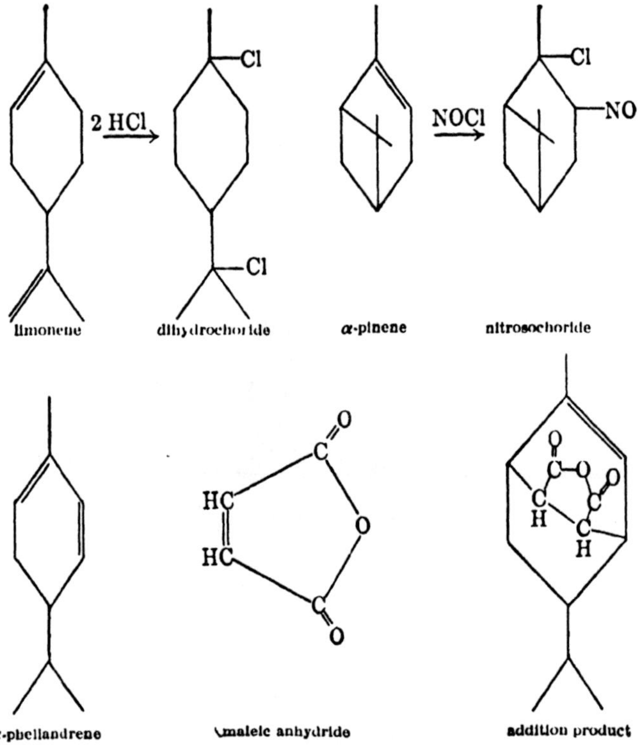

FIG. 2.27. Crystalline reaction products of Monoterpenes.

hydrocarbons of different chain length (Illustration 2.2). If the ordinate is plotted on a logarithmic scale nearly straight lines are obtained.[63] The boiling temperatures of the essential oil hydrocarbons are usually lower than those of the straight chain compounds, since branching of the chain tends to lower the boiling point. This counterbalances the possible rise in boiling point through the introduction of unsaturation. The net result is a boiling

[63] *Bull. Univ. Wisc.*, Serial No. 1919, Gen. Series No. 1703 (1934). "Phytochemistry," III. The methane series of hydrocarbons (Foote, Relationship between chemical constitution and boiling points of hydrocarbons of the methane series).

interval, for the majority of monoterpenes, of from 155° to 185° (representing the range of boiling points from pinene to terpinolene), whereas the straight chain normal decane boils at 175°.

A further look at the graph shows some irregularities at 300°, which indicate that even the normal saturated hydrocarbons are destroyed. To avoid this disagreeable behavior, a reduction in pressure, resulting in a lowering of the boiling point, is resorted to when we intend to continue the fractionation beyond this point. In the case of the much more sensitive terpenes, this phenomenon will appear much earlier, and it is usually not safe to raise the

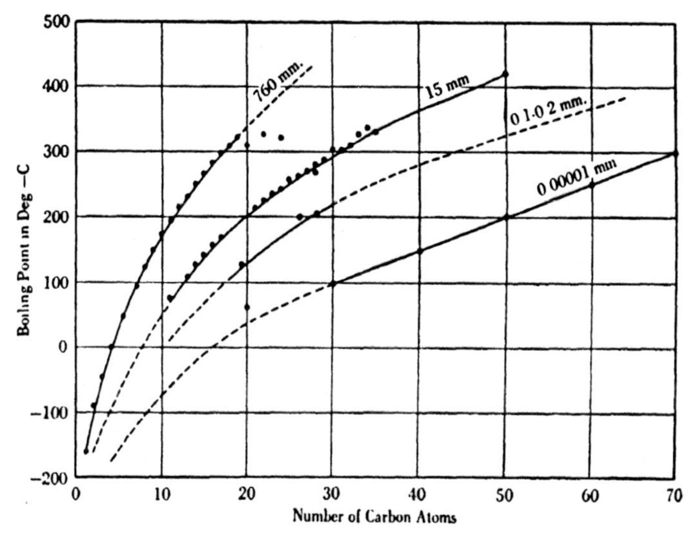

ILL. 2.2. Boiling points of straight-chain hydrocarbons.

outside bath temperature to higher than about 180°. Thus a fraction containing sesquiterpenes (C_{15}) boiling at 250° at ordinary pressure can be investigated by distilling at aspirator vacuum of 15 mm. at about 120°, or at a still lower pressure of 0.1 mm. at about 60°.

In this way and by applying ultra high-vacuum, it is possible to study still higher terpene homologues. But gradually these compounds become too complicated and too fragile to give satisfactory fractionation data. For this reason, it is desirable to obtain these compounds, not by steam distillation, but by extraction with solvents, such as alcohol, acetone or petroleum ether, or by chromatographic adsorption analysis. Following these procedures we will also find the compounds of the isoprene structure which have taken part in reactions which made them nonvolatile, such as the combination of the diterpene alcohol, phytol, with the complex phorbine ring system of chlorophyll.

When the purified fractions are characterized by preparation of crystalline derivatives, and when these are compared by melting point and mixed melting point with the known derivatives, there remain always some fractions which cannot be characterized in this way. In such cases, chemical degradations of the molecules have to be applied. The principle involved in these degradations consists of breaking up the molecule into smaller parts until the pieces have become so simple that they can be recognized. For this purpose, oxidation with ozone, potassium permanganate and chromium trioxide is often used. Sometimes several of these degradations may be necessary before the pieces obtained are small enough to be identified. On the basis of these degradations, a possible structure is postulated and attempts are made to confirm this structure by synthesis. This work has been carried out on about 500 constituents of essential oils. One-fifth of this number is made up of monoterpenes, and only a start has been made on the investigations of sesqui- and higher terpenes. In view of the greatly increased possibility of structural isomerism, every time 5 carbons are added to a molecule we may look forward to the addition to our present knowledge of a great number of the higher terpene homologues when a more extensive survey is made.

II. THE ORIGIN OF ESSENTIAL OILS

In the foregoing discussion of the components of the volatile oils, we saw that they consist of a variety of compounds which belong to all chemical classes. We cannot expect to find a common history for such varied substances. We do observe, however, certain chemical relations between a number of the components. Indeed, it was this similarity that led us to discuss the results of chemical research in terms of four groups, i.e., straight chain hydrocarbons, benzene derivatives, terpenes and miscellaneous compounds. In view of their structural similarity straight chain hydrocarbons are generally considered as connected with fatty acid metabolism, while benzene and propyl benzene derivatives are connected with carbohydrate metabolism. The group which gives rise to most of the speculation, however, comprises the terpenes.

We have seen that members of this series could conveniently be described as divisible into branched C_5 chains. This statement refers to an established fact; but we enter the field of speculation and hypothesis in assuming that such a structure as a C_5 chain actually represents the basic unit in the formation of the terpenes in the plants.

Many terpene investigators have risked guesses as to the nature of this basic unit, but few have tried to support their hypothesis by experiments. One of the oft-mentioned precursors (as we may call them) is isoprene (C_5H_8), belonging to the group of hemiterpenes. This compound in its turn is postulated to arise from the condensation of acetone or derivatives like dihydroxyacetone and acetaldehyde.[64] Through polymerization and addition of isoprene to higher terpenes, terpene homologues can be prepared.[65] Among several condensation products dipentene and a bisabolene-like sesquiterpene can be identified[66] (Fig. 2.28). When such reactions are

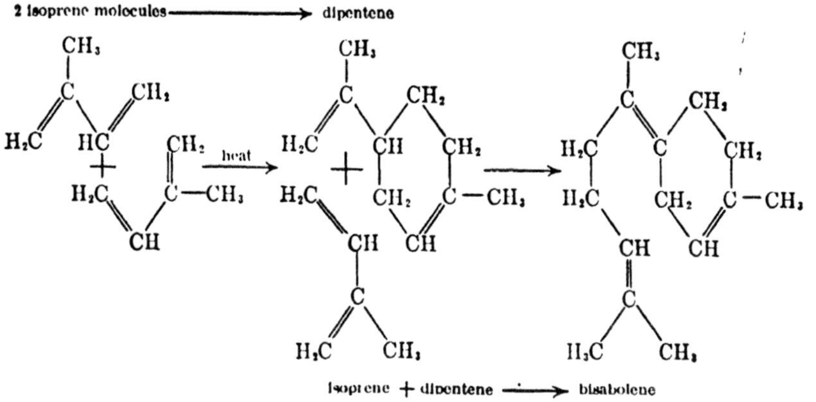

2 isoprene molecules ⟶ dipentene

isoprene + dipentene ⟶ bisabolene

FIG. 2.28. Polymerization of Isoprene.

carried out under simultaneous hydrogenation or hydration, the reactive ends of the molecules are saturated and further condensation and resinification are thereby largely prevented. Following this principle, Midgley et al.[67] carried out the condensation of isoprene under reducing conditions with sodium amalgam and obtained the terpene hydrocarbon 2,6-dimethyloctane. Wagner-Jauregg[68] condensed two mols of isoprene in the presence of sulfuric and acetic acids. Under these conditions water is added to the double bonds and geraniol can be isolated from the condensation mixture (Fig. 2.29). Ingenious as these experiments are, they do not furnish proof of the isoprene hypothesis.

The same can be said for the hypothetical precursor 3-methylbutenal (Fig. 2.30). This compound would very well satisfy the demands for a reactive precursor. *In vitro* experiments with 3-methylbutenal, with its conjugated carbonyl group and double bond, clearly demonstrate great reactivity and readiness to react with many other molecules. Fischer[69]

[64] Aschan, "Naphtenverbindungen, Terpene und Campherarten," Walter De Gruyter & Co., Berlin and Leipzig (1929), 127.

[65] *Fortschritte Chem. Org. Naturstoffe* **3** (1939), 1. Bedeutung der Diensynthese für Bildung, Aufbau und Erforschung von Naturstoffen, Diels.

[66] Egloff, "Reactions of Pure Hydrocarbons," Reinhold Publishing Co. (1937), 759.

[67] *J. Am. Chem. Soc.* **51** (1929), 1215; **53** (1931), 203; **54** (1932), 381.

[68] *Liebigs Ann.* **496** (1932), 52.

[69] Fischer and Löwenberg, *Liebigs Ann.* **494** (1932), 263.

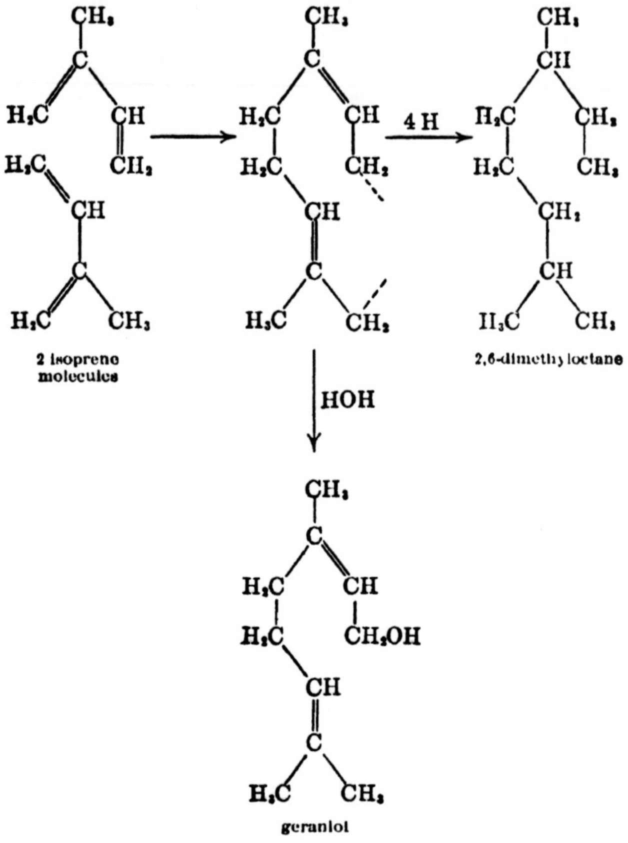

2 isoprene molecules

4 H

2,6-dimethyloctane

HOH

geraniol

FIG. 2.29. Dimerization of Isoprene.

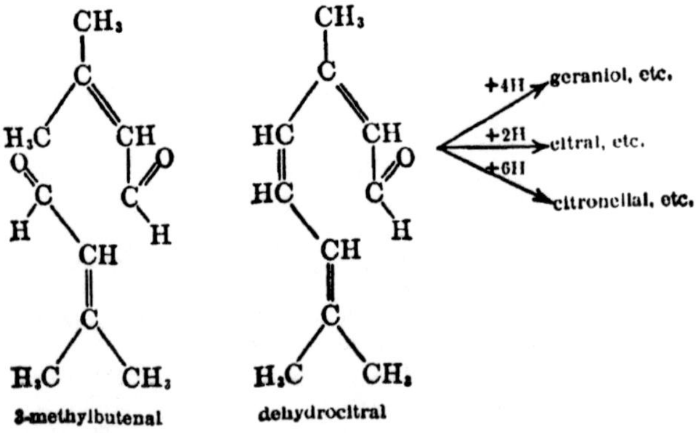

3-methylbutenal

dehydrocitral

+4H geraniol, etc.

+2H citral, etc.

+6H citronellal, etc.

FIG. 2.30. Terpene synthesis from 3-Methylbutenal.

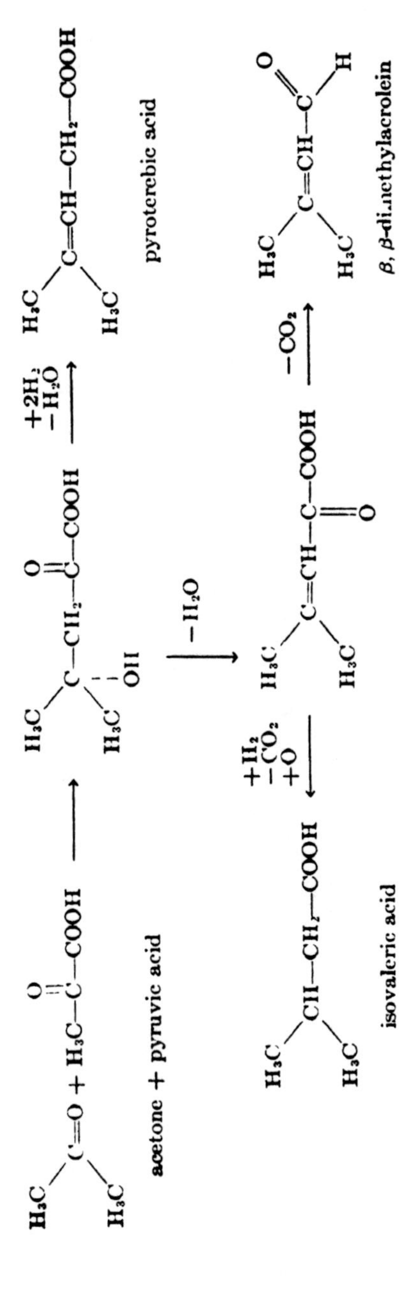

FIG. 2.31. Hypothetical Terpene Formation from Acetone and Pyruvic Acid.

succeeded in this way in building up dehydrocitral which might easily serve as the basic substance for aliphatic, as well as cyclic, terpenes.

An added proof would be the synthesis of 3-methylbutenal from acetaldehyde and acetone. Unfortunately, this follows a different addition scheme *in vitro*; and others have, therefore, suggested the formation of 3-methylbutenal by condensation of acetone with pyruvic acid, followed by decarboxylation. This would also furnish an explanation of the presence of

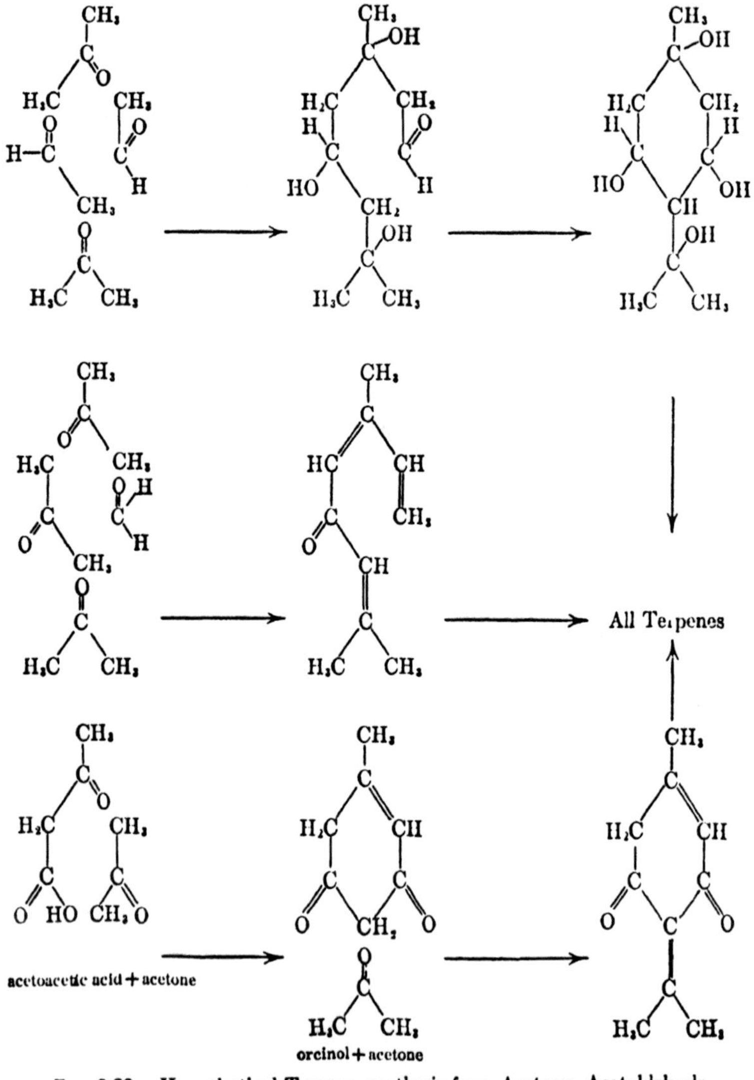

All Terpenes

acetoacetic acid + acetone

orcinol + acetone

Fig. 2.32. Hypothetical Terpene synthesis from Acetone, Acetaldehyde and Acetoacetic acid.

isovaleric acid and pyroterebic acid in the oil of *Calotropis procera*, where the latter acid occurs esterified with a diterpene alcohol[70] (Fig. 2.31).

However, Francesconi[71] can claim these compounds for his scheme in which isoamyl alcohol has a prominent place. This alcohol is obtained through degradation of carbohydrates, proteins or amino acids like leucine. From leucine, pyroterebic acid and isovaleric acid can be derived with great ease. Huzita[72] follows Ostengo in considering isovaleraldehyde to take a prominent place among the number of proposed precursors. Still another possibility is mentioned by Simpson,[73] who couples acetoacetic acid with 2 mols of acetone to obtain the monocyclic terpenes. The aliphatic terpenes are constructed on paper by linking 3 mols of acetone with one of formaldehyde (Fig. 2.32). Similar hypothetical schemes, using 2 acetone and 2 acetaldehyde molecules, are published by Singleton,[74] and Smedley-MacLean.[75] Available experimental data on these reactions speak against these types of condensation and special factors and conditions have to be postulated in order to account for the directive nature of the plant processes (Fig. 2.32).

Since none of these theories can be definitely rejected or accepted, it is clear that the presence of the branched chain represents a weak foundation on which to build hypotheses on the formation of the terpenes. We also

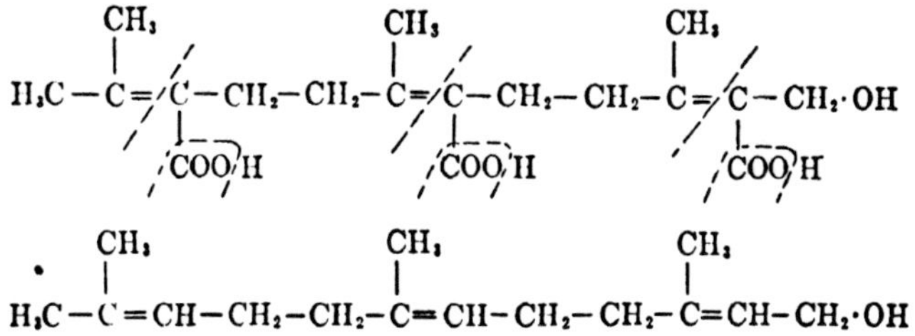

$$H_3C-\overset{CH_3}{\overset{|}{C}}=\underset{COOH}{C}-CH_2-CH_2-\overset{CH_3}{\overset{|}{C}}=\underset{COOH}{C}-CH_2-CH_2-\overset{CH_3}{\overset{|}{C}}=\underset{COOH}{C}-CH_2\cdot OH$$

$$H_3C-\overset{CH_3}{\overset{|}{C}}=CH-CH_2-CH_2-\overset{CH_3}{\overset{|}{C}}=CH-CH_2-CH_2-\overset{CH_3}{\overset{|}{C}}=CH-CH_2\cdot OH$$

farnesol formation

FIG. 2.33. Terpene synthesis according to Emde.

have to admit the possibility that the 5 carbon units into which we can divide the molecules of the terpenes may have their origin in larger units. This suggestion was made by Emde,[76] who postulated a physiological synthesis from sugars, through a coupling of levulinic acid-like molecules, followed by loss of CO_2 and the addition of smaller fragments of sugar meta-

[70] Hesse, in "Organic Chemistry" by Fieser and Fieser (1944), 981.
[71] *Rivista ital. essenze profumi* 10 (1928), 33.
[72] *J. Chem. Soc. Japan* 60 (1939), 1025.
[73] *Perfumery Essential Oil Record* 14 (1923), 113.
[74] *Chemistry Industry* (1931), 989.
[75] *J. Chem. Soc.* 99 (1911), 1627.
[76] *Helv. Chim. Acta* 14 (1921), 881.

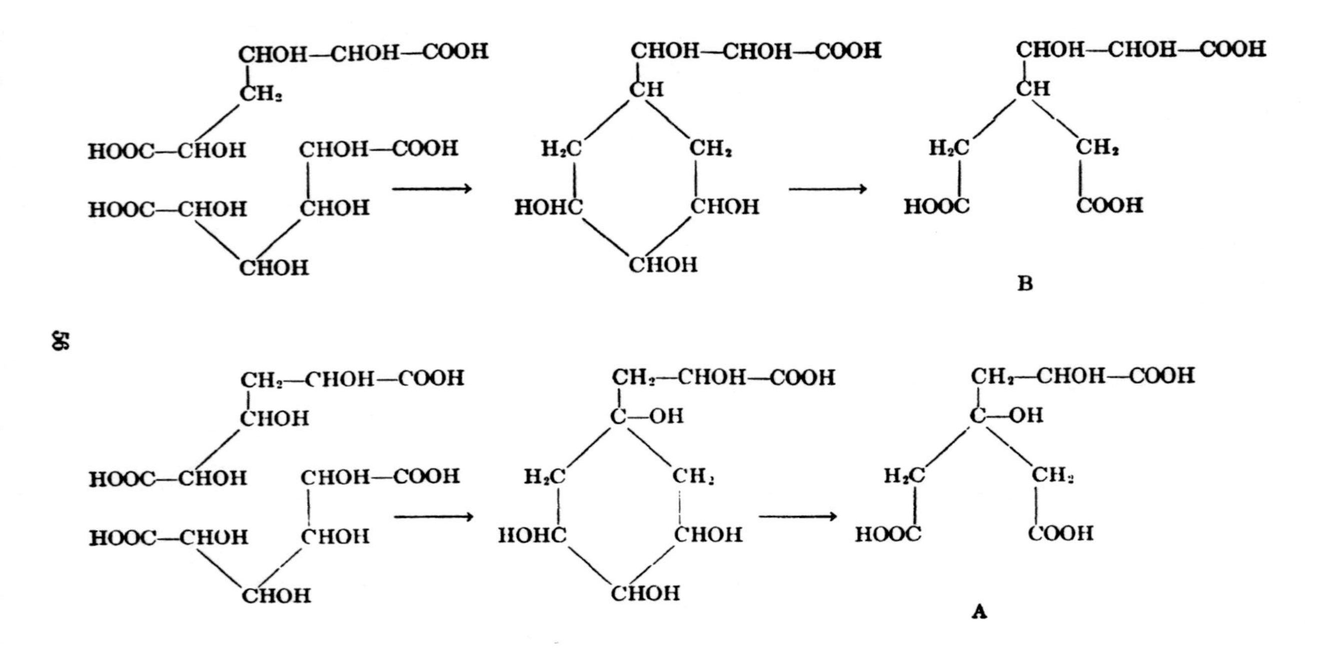
CHOH—CHOH—COOH
CH₂
HOOC—CHOH
CHOH—COOH
HOOC—CHOH
CHOH
CHOH
CHOH—CHOH—COOH
CH
H₂C
CH₂
HOHC
CHOH
CHOH
CHOH—CHOH—COOH
CH
H₂C
CH₂
HOOC
COOH
B
CH₂—CHOH—COOH
CHOH
HOOC—CHOH
CHOH—COOH
HOOC—CHOH
CHOH
CHOH
CH₂—CHOH—COOH
C—OH
H₂C
CH₂
HOHC
CHOH
CHOH
CH₂—CHOH—COOH
C—OH
H₂C
CH₂
HOOC
COOH
A

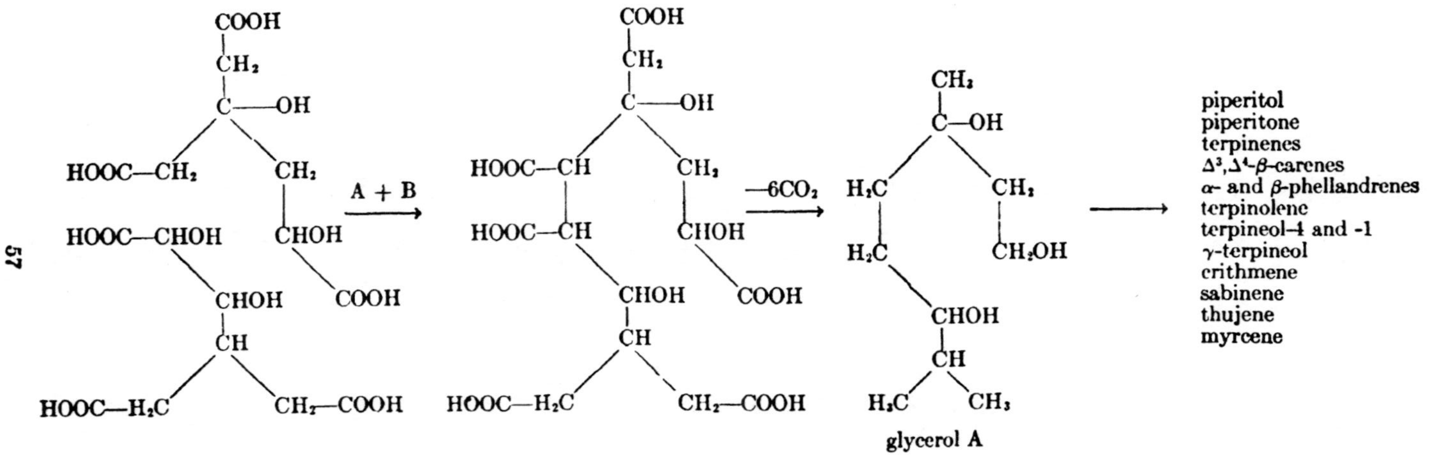

Fig. 2.34. Terpene Synthesis according to Hall.

bolism when necessary (Fig. 2.33). The chief value of this clever hypothesis is probably that it points to other ways of constructing branched molecules. This applies especially to the theory of Hall,[77] who attributes the formation of terpenes and benzene derivatives to the condensation and degradation of sugar derivatives. In this way different hypothetical "half molecules" were postulated which are finally combined to give the desired structures. An example of the proposed formations of a terpene precursor is pictured in Fig. 2.34.

Extensive schemes for the derivation of other terpenes and the further synthesis of higher terpenes can hardly contribute to the acceptance of any one of these theories, because once a terpene-like compound is synthesized on paper it is not difficult to explain the many combinations of terpenes we encounter in nature. Oxydases, reductases, esterases and even special ring-closing enzymes ("Kyklokleiasen" of Tschirch) are therefore welcome instruments in the hands of theorists. *In vitro* many of the terpenes have been converted one into the other by simple chemical reactions, which take place under physiologically possible conditions. Under the influence of light, air and water, we can expect reactions to take place which we observe *in vitro* in improperly stored essential oils, i.e., oxidation and polymerization. Free acids, if present, may cause loss of water, cyclization and esterification.

Considering the long storage of these oils in the plant, it is not astonishing that analyses of the oils indicate a gradual change in the expected direction with the maturing of the plant. Experiments on peppermint show an increase in the menthone content with an accompanying decrease in menthol content due to oxidative processes. At the same time, the percentage of compounds other than menthol and menthone increases, indicating a splitting off of water and polymerization.

It is very probable that, in a number of cases, especially in oxidation and reduction reactions, enzymes play an important role. Neuberg succeeded in the reduction of citronellal[78] to *d*-citronellol, and of citral[79] to geraniol with yeast. These experiments, extended by Fischer,[80] disclosed certain laws which govern the enzymatic hydrogenation of double bonds between carbon and carbon, and carbon and oxygen. The double bond conjugated with the aldehyde group in citral is slower in its hydrogen uptake than the carbonyl group; and we see, therefore, that the formation of geraniol takes precedence over the formation of citronellal. When geraniol is subjected to further hydrogenation, citronellol is formed, leaving the double bond at C_6 untouched. Citronellol produced in this way from optically inactive

[77] "Relationships in Phytochemistry," *Chem. Rev.* **20** (1937), 305.
[78] Mayer and Neuberg, *Biochem. Z.* **71** (1915), 174.
[79] Neuberg and Kerb, *Biochem. Z.* **92** (1918), 111.
[80] *Fortschritte Chem. Org. Naturstoffe* **3** (1939), 30.

geraniol is optically active-dextrorotatory (specific rotation $[\alpha_D] = +6$) as in citronella oil. No further hydrogenation of the isolated double bond can be effected in this way, and it is interesting to note that in plants also, the hydrogenation has come to a halt at the citronellol stage (Fig. 2.35).

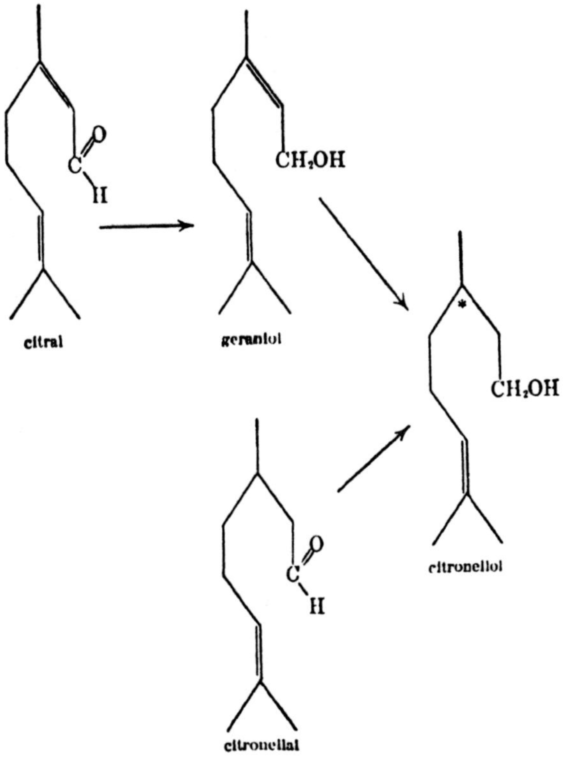

citral

geraniol

citronellal

citronellol

FIG. 2.35. Enzymatic reductions.

Substituents greatly influence the speed of the enzymatic hydrogenations, as seen in the slower hydrogen addition to keto groups, and to double bonds on tertiary carbon atoms. Carvone, main constituent of caraway oil, when subjected to enzymatic treatment, is reduced with difficulty to dihydrocarvone, another constituent of this oil (Fig. 2.36). The absence of the totally hydrogenated carvomenthol suggests that similar laws are followed in the production of these terpenes in the plant.

These biological reductions can also be followed by studying the excretion products in urine during feeding or injection experiments. While in general, advanced oxidative degradations outweigh hydrogenation processes, a careful analysis of the excretion product shows similar reactions, as in the more simple experiments with yeast or enzyme-systems. Perhaps due to the branching of the chains, the reaction products of terpenes, such as citral, geraniol and geranic acid,[81] can be recognized in the urine of rabbits after feeding or injection experiments. Fig. 2.37 shows that, notwithstanding the simultaneous oxidation in other parts of the molecule, the double bond in α,β-position to alcohol, aldehyde or acid groups is hydrogenated. Similar experiments on citronellol[82] confirm these observations; no reduction of the double bond in the isopropylidene group can be observed, but further oxidation produces dihydro Hildebrandt acid and hydroxydihydrogeranic acid (Fig. 2.37).

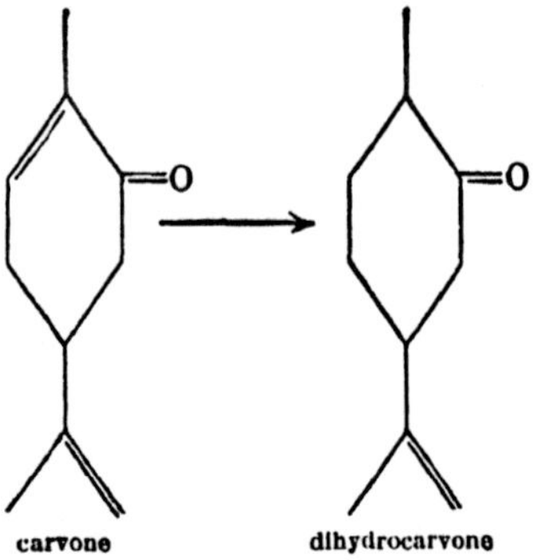

FIG. 2.36. Enzymatic reductions.

In β-ionone,[83] however, where the double bonds are conjugated, reduction of the carbonyl group and its neighboring double bond takes place, leaving the double bond between the two tertiary C atoms unchanged. Further oxidation introduces a hydroxyl group at one of the methyl groups (Fig. 2.38). The agents responsible for similar oxidations in the plant are suspected to be of enzymatic nature, but this has not been established experimentally.

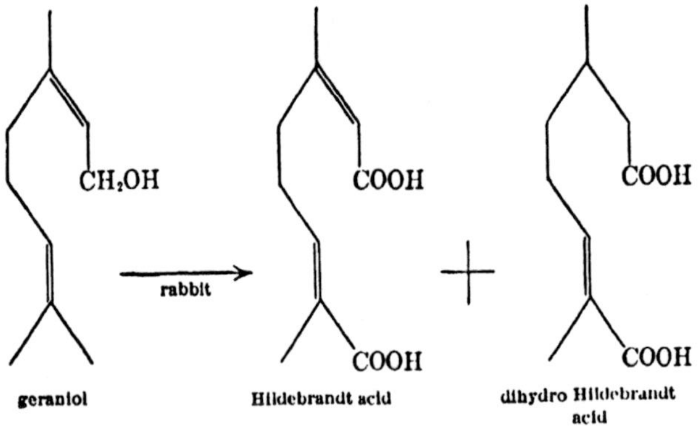

FIG. 2.37. Oxidation and reductions of Geraniol in animals.

Based on the not too improbable assumption that the terpenes present in a specific oil are interrelated, several building schemes were developed involving a stepwise conversion of the components, starting with a common

[81] Hildebrandt, *Z. Physiol. Chem.* 36 (1902), 441.
[82] *Ibid.*
[83] Fischer, *Fortschritte Chem. Org. Naturstoffe* 3 (1939), 30.

precursor. In this way, Francesconi[84] explained the simultaneous presence of citral, citronellal, linaloöl, dipentene, methyl heptenone and acetaldehyde in lemongrass oil. Likewise, Kremers[85] correlated the components of American peppermint oil, acetone, acetaldehyde, citral, citronellal, isopulegol, menthol and menthone.

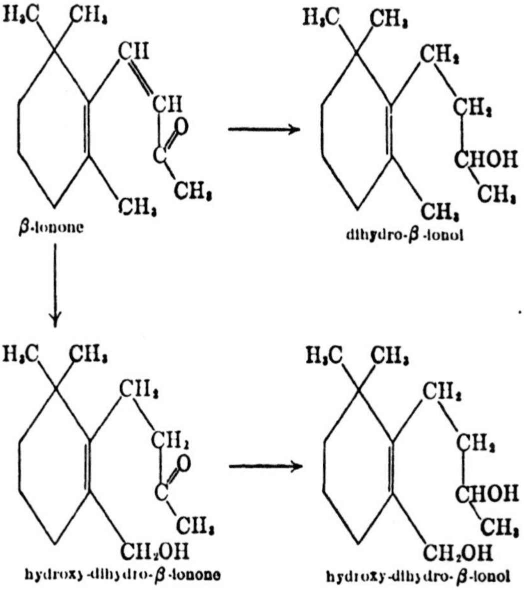

β-Ionone → dihydro-β-ionol

hydroxy-dihydro-β-ionone → hydroxy-dihydro-β-ionol

FIG. 2.38. Biological oxidations and reductions of β-Ionone.

The following biogenesis of the two groups of substances found in the oils of American black mint and spearmint was suggested by Kremers.[86] The names of substances actually found in the oils are italicized, while the two reducible groups in the citral molecule are underlined(Fig. 2.39).

Structural relationship and frequent occurrence in mint and eucalyptus oils has been noticed by Read[87] for the terpenes, piperitone, piperitol, α-phellandrene and Δ^4-carene. Piperitone is always accompanied by geranyl acetate, from which many cyclic terpenes can be formed. Read, therefore, has expressed the opinion that the geraniol is a possible intermediate precursor of a number of terpenes. In *Eucalyptus macarthuri* the chain of reaction apparently stopped at the formation of geraniol, since the oil contains 77 per cent geranylacetate, while in most other species (under different conditions in the plant), more advanced transformations take place.

[84] *Rivista ital. essenze profumi* 10 (1928), 33.
[85] *J. Biol. Chem.* 50 (1922), 31.
[86] *Ibid.*
[87] *J. Soc. Chem. Ind.* 48 (1929), 786.

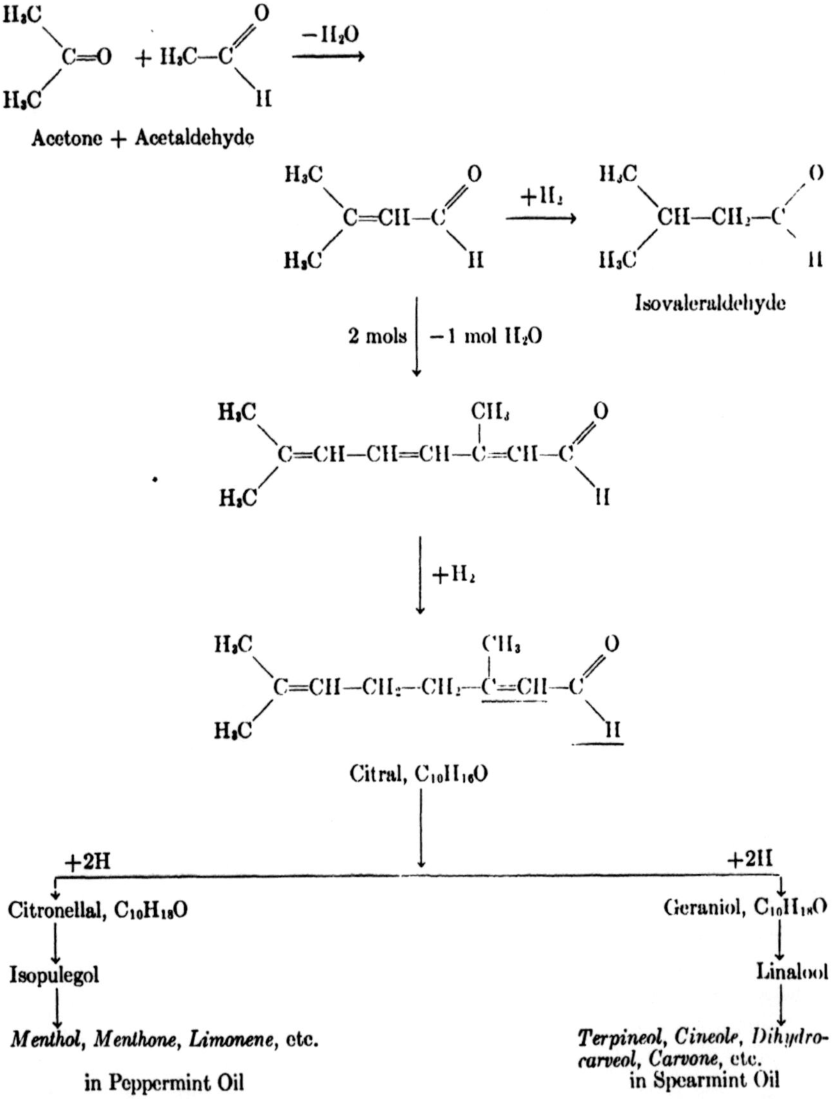

$$(H_3C)_2C{=}O + H_3C{-}CHO \xrightarrow{-H_2O}$$

Acetone + Acetaldehyde

$$(H_3C)_2C{=}CH{-}CHO \xrightarrow{+H_2} (H_3C)_2CH{-}CH_2{-}CHO$$

Isovaleraldehyde

2 mols ↓ −1 mol H_2O

$$(H_3C)_2C{=}CH{-}CH{=}CH{-}C(CH_3){=}CH{-}CHO$$

↓ $+H_2$

$$(H_3C)_2C{=}CH{-}CH_2{-}CH_2{-}C(CH_3){=}CH{-}CHO$$

Citral, $C_{10}H_{16}O$

+2H	+2H
Citronellal, $C_{10}H_{18}O$	Geraniol, $C_{10}H_{18}O$
↓	↓
Isopulegol	Linaloöl
↓	↓
Menthol, Menthone, Limonene, etc.	*Terpineol, Cineole, Dihydrocarveol, Carvone*, etc.
in Peppermint Oil	in Spearmint Oil

Fig. 2.39. Biogenesis of Terpenes in Oil of Peppermint and of Spearmint.

Similar relations are discussed in the genus *Orthodon* (fam. *Labiatae*). These oils mostly contain major quantities of thymol, carvacrol, cymene, cineole, thujene and thujyl alcohol.[88] However, one species, *Orthodon linaloöliferum* Fujita, contains 82 per cent linaloöl. This compound can be

[88] Naves, "The Formation of the Terpenes in the Labiates," *Tech. Ind. Schweiz. Chem. Ztg.* 25 (1942), 203.

converted into many oil components of other species of the same genus. Huzita,[89] therefore, considers this linaloöliferum plant as the parent species of the genus *Orthodon*. It is, however, equally well possible that the reactions become blocked at the linaloöl stage through a mutation process.

Although the tendency has been to explain the formation of the terpene compounds from a C_{10} precursor like geraniol or citral, it is quite feasible that the condensation of the units takes a different and individual path for a number of terpenes. We are naturally forced to accept this for irregularly built compounds such as artemesia ketone and lavandulol, but it might also be equally true for a number of the regularly built terpenes, e.g., pinene. α-Pinene is one of the most frequently occurring oil constituents,[90] and, although the preparation of this ring structure from an aliphatic terpene is unknown, easy roads lead from pinene to a number of mono- and bicyclic compounds, such as terpineol, borneol, camphene, camphor, fenchone, fenchyl alcohol, dipentene, 1,4-cineole, terpin, pinol, myrtenol, dihydromyrtenol and verbenone (Fig. 2.40). Laboratory experiments may indicate groups of compounds which can easily be converted into each other,[91] but we have always to refer to the composition of the natural oils to give these groups a physiological meaning. It appears likely that in different oils the synthesis of specific compounds (such as limonene) might have taken place in several ways—such as by ring opening from pinenes or ring closure of citral, geraniol or other cyclic terpenes, or by even direct synthesis.

This individuality of many couplings is further supported by our experience in the higher terpenes, where often, as in abietic acid, one unit is in an irregular position. For an explanation of the different groups of higher terpenes, we have to accept formations from single units, single and double units, doubling of double units, and doubling of triple and quadruple units.

Having reviewed all of these theories, let us summarize the established facts, in order to draw a conservative conclusion regarding the possible synthesis in the plant. We know that:

1. The structural formula of a large number of the compounds in plants can be divided up into branched C_5 chains.

2. The arrangement of the branched C_5 units is in most cases a head-to-tail union, but exceptions occur in the monoterpene group, and are common in sesqui-, di- and triterpenes.

3. Ring compounds are easily formed from aliphatic terpenes, whereas the reverse can only be accomplished with difficulty.

4. Oxidation, reduction, shifting of double bonds and polymerization take place readily.

[89] *J. Chem. Soc. Japan* 61 (1940), 424.

[90] α-Pinene occurs in 375 oils, according to Ganapathi, *Current Sci.* 6 (1937), 19.

[91] Okuda, *J. Chem. Soc. Japan* 61 (1940), 161.

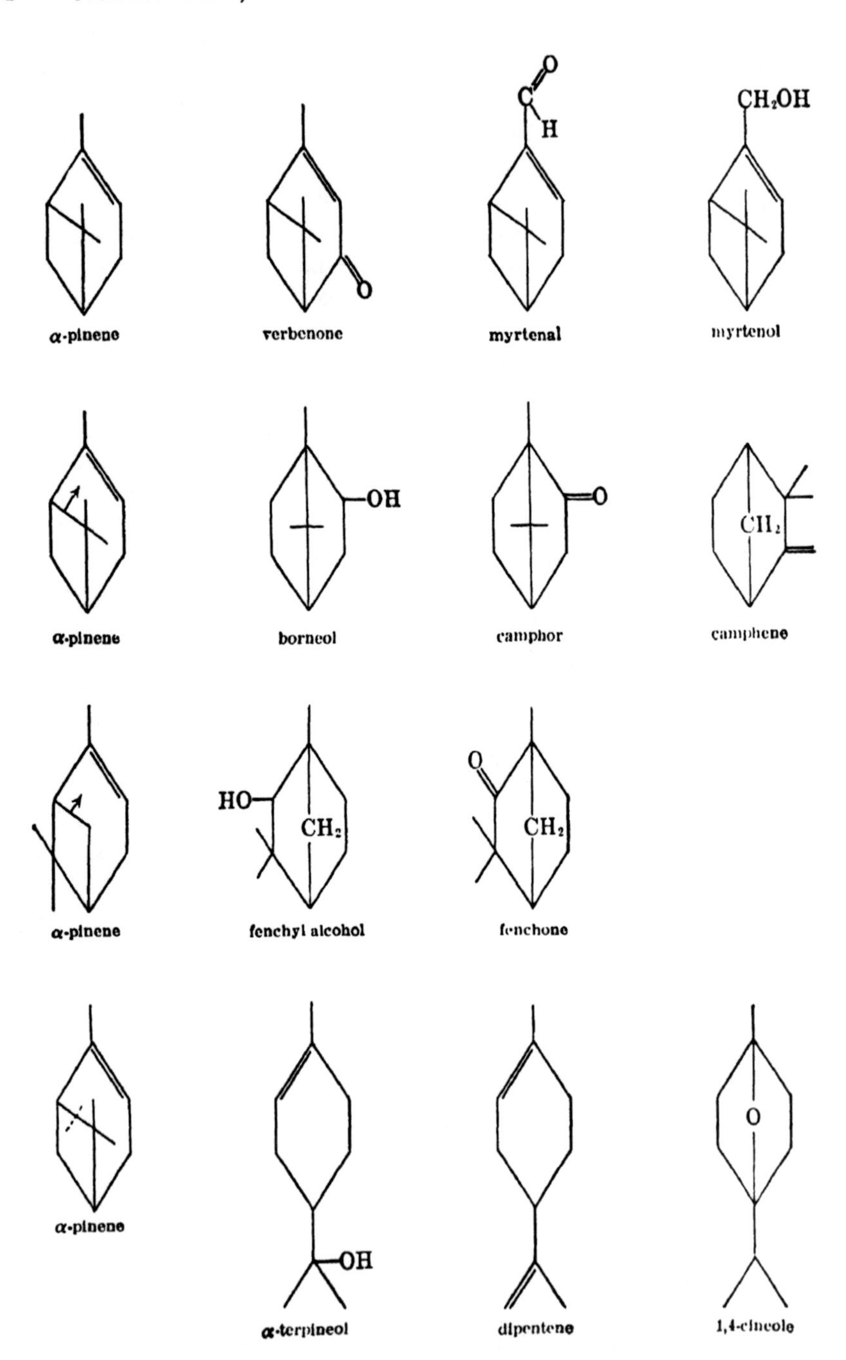

FIG. 2.40. Terpene family.

5. The branched C_5 unit is distinguishable in the formulas of a number of nonterpenes coupled with nonbranched structures.

6. The terpenes are often accompanied with propyl benzene derivatives and straight chain hydrocarbons.

On the basis of these facts, we may safely conclude that a number of terpenes are formed from a unit which can give rise to one or more branched C_5 chains before or after the condensations. It is possible that the C_5 unit is not the actual structure undergoing condensation, and that more complex compounds are involved, which split off certain groups after condensation has taken place. This would include the precursors as described by Hall and Emde, viz., phosphoric acid esters as the sugar precursor, and their degradation products, and protein complexes carrying the condensing structures which release the terpene compounds when formed. The regular head-to-tail union may be predetermined in the compound from which the terpene is formed, or the mechanism of the condensation may be such that this type of union occurs.

The terpenes already formed readily undergo secondary changes, such as reduction, oxidation, esterification and cyclization, and this fact may explain the large variety of derivatives of the same pattern. These families of terpenes may have their origin in independently formed key terpenes, such as geraniol, citral, pinene, etc. Higher terpenes may have been formed through a condensation of lower terpenes of the same or different chain length, whereby quite often derivatives from the regular and symmetrical architecture can be observed. No indications are available that would justify connecting the terpenes directly with other essential oil components, such as straight chain hydrocarbons or propyl benzene derivatives. Although the majority opinion favors a connection through the carbohydrate metabolism in the plant, there is no reason to assume that these products are formed in the same phase of these processes.[92, 93] Other essential oil components show structural features strongly suggesting connections with fat and nitrogen metabolism. From chemical evidence we can draw the conclusion that the complexity of the oil composition is caused by excretion or secretion of products formed in many metabolic processes taking place in the plant.

Since the volatile oils are intimately connected with vital processes in the plant, the presence of these specific components has been used also in the determination of the evolutionary status of plant families.[94]

A continued, thorough chemical study of the volatile, and especially of the nonvolatile, components will undoubtedly give us a more complete

[92] Simpson, *Perfumery Essential Oil Record* 14 (1923), 113.
[93] Hall, "Relationships in Phytochemistry," *Chem. Rev.* 20 (1937), 305.
[94] McNair, *Am. J. Botony* 21 (1934), 427. *Bull. Torrey Botan. Club* 62 (1935), 219.

picture of the processes which take place and of the structures which are formed in the metabolic activities of the plant. Although this knowledge must be the basis for any speculation on the mechanism involved, we have to turn our attention again to the living plant itself in order to collect experimental support for our theory of what actually happens. One of the ways in which the plant physiologist tries to solve these problems is to study the cells in which the oils are deposited, and the circumstances under which oil formation takes place

The observation has been made that some of the cells or spaces in plant tissue are filled with oily droplets, difficult to distinguish from fatty oils.

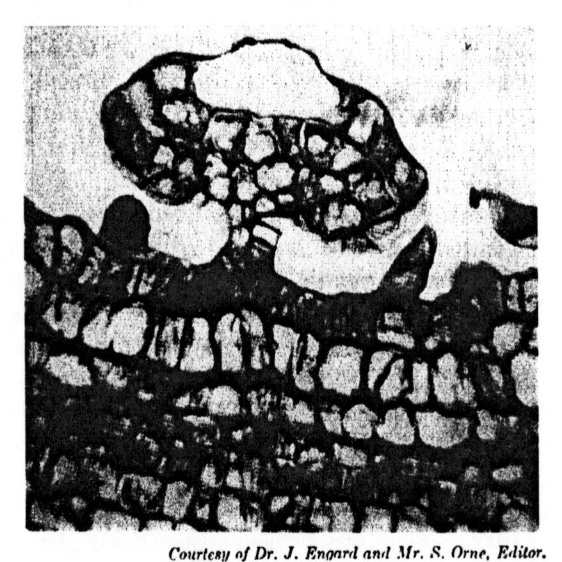

Courtesy of Dr. J. Engard and Mr. S. Orne, Editor.

ILL. 2.3. Lysigenous oil sac in *Rubus rosaefolius* Smith.[1]

[1] Engard, *Univ. Hawaii, Research Publication,* No. 21 (1944).

These oils can be detected by staining with sudan and osmic acid, and a distinction from fatty oils is best made by taking advantage of the presence of substances with a chemically more active character than the unsaturated hydrocarbons and alcohols, i.e., aldehydes and phenols. For example, droplets containing phenols can sometimes be stained with phloroglucinol hydrochloride. The presence of aldehydes is shown with fuchsin and sodium bisulfite reagents.[95]

The oil secretion appears in different cell groups (Illustrations 2.3 and 2.4), and distinctions have been made between external and internal gland

[95] Czapek, "Biochemie der Pflanzen," Vol. III, 593, Dritte Auflage (1925), Verlag G. Fischer, Jena.

cells.[96] The external glands are epidermal cells or modifications of these, such as the *excretion hairs*. The secretion product is usually accumulated outside the cell between the cuticle and the rest of the cell wall. The cuticle is a thin skin covering the secretions and a slight touch suffices to break this thin piece of skin. Thus, on touching the plant, we observe immediately its well-known scent.

The internal glands are located throughout the plant; they are formed by the deposition of the oils between the walls of the cells. This schism of cells has been called a schizogenous formation. If this is followed by dissolution

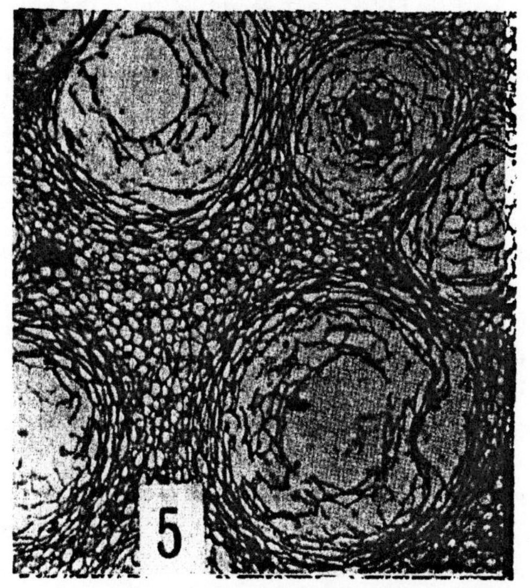

Courtesy of Dr. F. M. Turrell and Dr. L. J. Klotz, The Botan. Gaz., Vol. 101 (1940), 862.

ILL. 2.4. Tangential section showing oil glands of Washington navel orange fruit.

of the surrounding cells, morphologists speak of a schizolysogenous gland formation. Often these intracellular glands have grown to form long canals, coated on the inside with a layer of thin-walled cells. This coating is said to have a double function, viz., the separation of other tissues from the oils and the formation of oils and resins. The secretion forms in the epithelial cells or in the membranes and passes through the cell wall into the interior of the gland. The secretion crosses a mucilagenous material produced by the outer membranes of the secretion cells which has been called the resinogenous layer by Tschirch. This layer does not possess any of the secretory functions ascribed to it, and the designation "resinogenous layer" is in-

[96] Haberlandt, "Physiologische Pflanzen Anatomie," 4776, Aufl. 1924, Verlag Engelmann, Leipzig. Tschirch and Stock, "Die Harze," W35, I, 20 (1933), Verlag Bornträgen, Berlin.

applicable, at least in the cases of the *Umbelliferae* and *Rutaceae* studied by Gilg and collaborators.[97]

Studies on the number and distribution of the glands show unequal distribution. The count of the glandular scales in *Mentha* species shows that the lower surface contains 10–25 scales per sq. mm., the upper surface 1–6 per sq. mm. Dimensions and number of the scales increased near the large vein.[98]

If we search the literature[99] regarding the exact place of formation of substances like terpenes, we find that a few disputed observations are available, wherein it has been noted that secretion vacuoles suddenly appear in the cell, then increase in number and size, while cytoplasm and nucleus degenerate. These oil globules appear to be surrounded by a membrane. Some observers have seen small droplets of oil, formed in or near the chloroplast, which unite later and form the large oil drops. Others have not observed any oil drops at all in the cells, but found the oil in the membrane layers adjoining the secretion pockets.

Certain observations along these lines seem to point toward the region of photosynthetic activity, where carbon dioxide is reduced and synthesized to carbohydrates. Some support is lent to this thesis by experiments which attempt to establish correlations between oil secretion and known metabolic processes in the plant. Examples of this angle of research are to be found in studies on the effect of climatic and growth conditions on oil content.

A typical example of such investigations is contained in a report on the oil content and composition of Japanese mint (*Mentha arvensis*) grown in the United States, in which it was established that conditions in southeastern states do not favor the formation of menthol to the same extent as those in the northern and western states. The average differences in large sections of America are of the order of 74.5–81.0 per cent for combined menthol. Data on the individual oils obtained in the different regions show a spread for total menthol of 65.2–88.7 per cent and for combined menthol of 1.7–11.1 per cent. Sievers and Lowman[100] rightly stress, therefore, the importance of a critical attitude toward the evaluation of results obtained in such surveys. More reliable evidence is obtained when the handling and oil determinations are carried out under strictly controlled conditions.

Although such statistical experiments are important from a commercial and agricultural point of view, it is difficult to draw any theoretical conclu-

[97] *Arch. Pharm.* 268 (1930), 7.

[98] Hocking and Edwards, *J. Am. Pharm. Assocn.* 32 (1943), 225.

[99] Tünmann, *Ber. deut. pharm. Ges.* 18 (1908), 491. Czapek, "Biochemie der Pflanzen," III (1925), 585.

[100] "Commercial possibilities of Japanese Mint in the United States as a source of natural menthol," *U. S. Dept. Agr. Tech. Bull.* 378 (1933), Washington, D. C.

sions as to the physiological effect of climate, soil and other variables. These data, moreover, give an overall picture of the oil content and composition of young and old leaves, branches and flowers alike. We know, however, that different parts of the plant contain oils which are often of very different chemical composition. As an extreme and almost classical example, the composition of the oil of Ceylon cinnamon might be given. The bark yields oil with a high cinnamic aldehyde content, the leaf oil consists chiefly of eugenol, and the root oil contains a high percentage of camphor. Orange and lemon in flowers and fruit contain oils of different composition, and numerous are the examples where only certain parts of the plant contain oil: oil of iris, valerian and calamus occur only in the roots; sweet birch and cinnamon oils are found in the bark; whereas in the case of *santalum album* and cedar, the core wood contains the valuable oils.

Better controlled experiments on the influence of climatological conditions, such as sunlight on the oil formation, are found in a series of articles by Charabot and others.[101] Experiments on shaded and unshaded plants indicate that light favors formation of oil.[102,103] These observations cover a period of several weeks. We possess at least one observation on the daily fluctuations recorded on the oil content of nutmeg sage, the oil yield being 1.5 per cent during the night and in the afternoon only 0.6 per cent. The content of esters is highest toward the evening and least at night. The yield is lower during windy, dry weather.[104]

To study oil formation as affected by plant development, it is necessary to select one type of organ and carry out the experiments under rigidly controlled and nonvariable external circumstances. Since this is usually not feasible, the next best results may be obtained in experiments during a stable weather period on fast growing plants, or through the other extreme of very long periods on slow growing plants, thereby averaging the effect of climatic changes.

Although no experimental data exist which will satisfy the most rigid requirements, the second type of experiment is represented by the analysis of oil from the peppermint plant during different stages of growth. Bauer[105] analyzed the oils of *Mentha piperita* at four stages—before, and during, bud formation; and during, and after, flowering. His findings are recalculated and summarized in Illustration 2.5, in such a way that the curves represent the percentage of the components relative to the fresh weight of the plant. The different corresponding growth stages are indicated, I, II, III and IV

[101] Charabot and Hébert, *Bull. soc. chim.* [3] 31 (1904), 402.
[102] Lubimenko and Norvikoff, *Bull. Appl. Bot.* 7 (1914), 697.
[103] Rabak, *U. S. Dept. Agr., Bur. Plant Ind. Bull.* No. 454 (1916).
[104] Gaponenkov and Aleshin, *J. Applied Chem. U.S.S.R.* 8 (1935), 1049.
[105] *Pharm. Zentralhalle* 80 (1939), 353. Relation between the composition of peppermint oil and the vegetative development and variety of the plant.

representing the period before budding, during bud formation, flowering stage, and after flowering stage.

The percentage of oil increases until flowering, when it either drops or remains constant. This is due chiefly to a decrease in free menthol formation, although the ester menthol continues to increase slowly, but steadily, probably at the cost of the free menthol. The constitution of the oil of a related mint, "Pfälzer mint," shows the same behavior during development in regard to the increase of ester content. Typical for this mint, however,

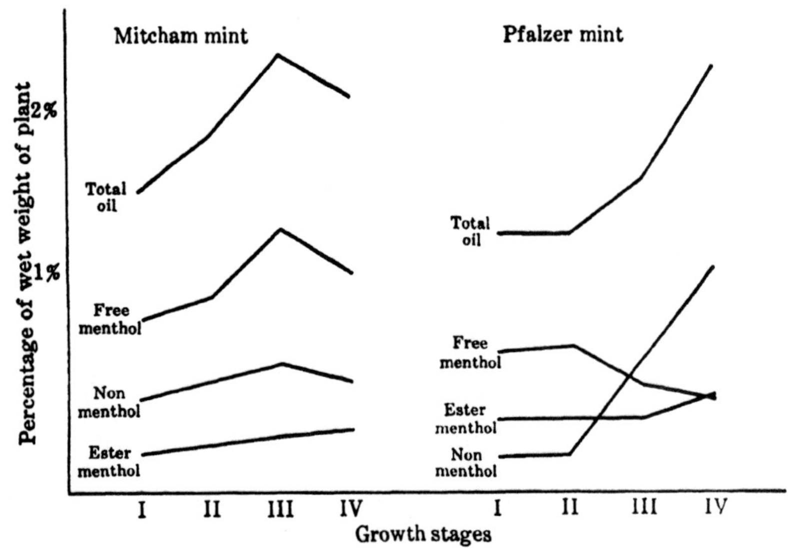

ILL. 2.5. Percentages of mint oils and their components at various stages of development.

is the increase in compounds other than the alcohol, probably menthone or dehydration products. Similar conclusions can be drawn from the investigations of Charabot[106] on leaves of *Lavandula, Mentha piperita, Ocimum*

[106] Charabot and Laloue, *Compt. rend.* **147** (1908), 144. Charabot and Gatin, "Le Parfum chez la Plante," Paris (1908). Charabot, "Les Principes Odorants des Vegetaux," *Encycl. Scient.*, Paris (1912). Charabot, *Am. J. Pharm.* **85** (1913), 550. Charabot, *Compt. rend.* **129** (1899), 728; **130** (1900), 257, 518, 923. Charabot, *Bull. soc. chim.* [3] **23** (1900), 189. Charabot, *Ann. chim. phys.* [7] **21** (1900), 207. Charabot and Hébert, *Compt. rend.* **132** (1901), 159; **133** (1901), 390. Charabot and Hébert, *Bull. soc. chim.* [3] **25** (1901), 884, 955. Charabot and Hébert, *Compt. rend.* **134** (1902), 181; **136** (1903), 1678. Charabot and Laloue, *Ibid.* **136** (1903), 1467. Charabot and Hébert, *Bull. soc. chim.* [3] **29** (1903), 838. Charabot and Hébert, *Compt. rend.* **138** (1904), 380. Charabot and Laloue, *ibid.*, 1513. Charabot and Hébert, *Ann. chim. phys.* [8] **1** (1904), 362. Charabot and Hébert, *Compt. rend.* **139** (1904), 608. Charabot and Laloue, *ibid.* **139** (1904), 928; **140** (1905), 667. Charabot and Hébert, *ibid.* **141** (1905), 772. Charabot and Laloue, *ibid.* **144** (1907), 152. Charabot and Laloue, *Bull. soc. chim.* [4] **1** (1907), 1032. Charabot and Laloue, *Compt. rend.* **144** (1907), 152, 435. Charabot and Laloue, *ibid.* **142** (1906), 798. Charabot and

basilicum, *Verbena tryphylla*, *Artemisia absinthium* and *Pelargonium*. In the later stages of growth the alcohols decrease probably at least partly through ester formation and dehydration to hydrocarbons. This process in turn is followed by oxidative reactions wherein aldehydes and ketones are formed. A decrease in oil content of the leaves during flowering has been observed by Charabot et al. on *Verbena tryphylla*.[107] In Table 2.1

TABLE 2.1. MG. OF OIL IN PARTS OF *Verbena tryphylla* PER WHOLE PLANT

	Flowering	After Flowering	
Root	10	16	increase 6
Stem	8	16	increase 8
Leaves	242	192	loss 50
Flowers	77	56	loss 21
Weight of Plant	366 g.	259 g.	

is listed the mg. oil present in different parts of the plant, during the flowering, and after the flowering period. In this period the leaves lost a considerable amount of oil, as compared with other parts of the plant. Analysis of the flower oil showed that the material lost from the flower consisted chiefly of citral. Charabot attributed this decrease in oil content of the leaves in *Verbena* and *Artemisia absinthium*[108] to a consumption of the oil constituents by the flowers, and postulated, therefore, a flow of oil from the leaves to the flowering parts.

When we take into account the way the oils are stored in the plant, and their toxic action when released, this transfer seems unlikely. It is, however, possible that material which otherwise would have contributed to the formation of the oils is used up in the flowering stage, and that the reduced formation of oil is unable to compensate for the constant loss through evaporation. The same explanations can be made for Charabot's experiment in which it was shown that *Mentha piperita*[109] and *Ocimum basilicum*[110] plants,

Laloue, *Bull. soc. chim.* [3] **35** (1906), 912.

Similar results are recorded by Rabak, *J. Am. Chem. Soc.* **33** (1911), 1242. Nylov, *J. Gov. Bot. Garden Nikita Yalta Crimea* **20** (1929), 3. *Repts. Schimmel & Co.*, 1926, 141, 142, 143. Spiridonova, *J. Gen. Chem. U.S.S.R.* **6** (1936), 1536.

Experiments on salvia seedlings are recorded by Wyslling and Blank, *Verh. Schweizer Naturf. Ges. Locarno* (1940), 163.

Data on oil content at different stages recorded by Francesconi, *Gazz. chim. ital.* **49**, I (1911), 395. Francesconi and Sernagiotto, *Atti accad. Lincei* **20**, II (1911), 111, 190, 230, 249, 255, 318, 383.

Data on camphor tree recorded by Hood, *J. Ind. Eng. Chem.* **9** (1917), 552.

[107] Charabot and Laloue, *Bull. soc. chim.* [4] **1** (1907), 640, 1032.

[108] Charabot and Laloue, *Compt. rend.* **144** (1907), 152, 435.

[109] Charabot and Hébert, *Bull. soc. chim.* [3] **31** (1904), 402.

[110] Charabot and Hébert, *ibid.* [3] **33** (1905), 1121.

after debudding, contain more oil in the leaves than under ordinary circumstances.

Long-term experiments stretching over two years, and averaging the climatic influences, have been carried out by Charabot and Laloue on *Citrus aurantium*. From their extensive data, the total oil present in a twig with an attached leaf can be followed through its development. Illustration 2.6 shows clearly the large increase in absolute weight of the oil during the early

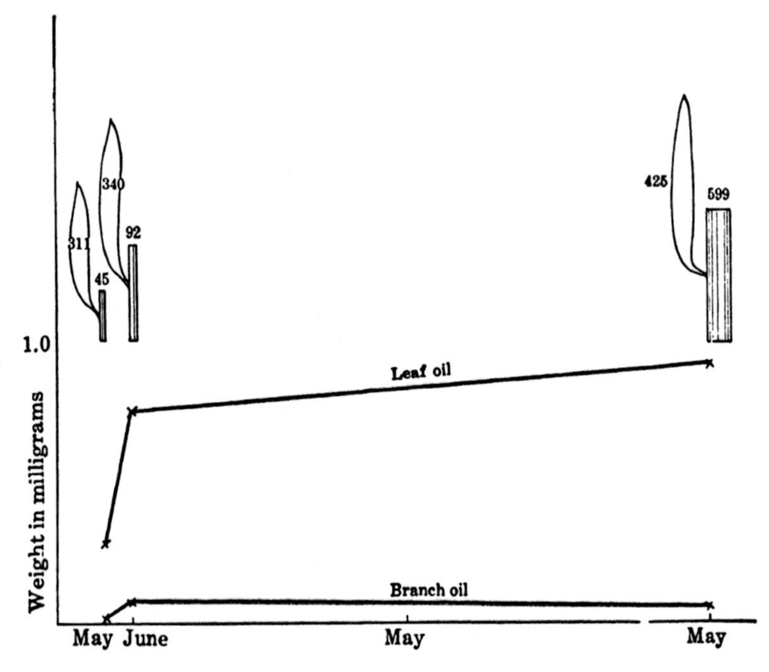

Ill. 2.6. Total oil content in growing leaf and branch of *Citrus Aurantium*.

period of growth. During the later period the formation in the branches is not even intense enough to compensate for the losses, due to consumption, transportation to other parts[111] and evaporation. An increased production of limonene is observed. This is probably formed by the dehydration of the initially present, free and esterified linaloöl and geraniol. Similar experiments on the oil content at different stages of development were carried out on the oil of bergamot. A tendency in the expected direction was actually observed, i.e., an increase of esters and an increase of terpenes, through the loss of water and through cyclization.

[111] Charabot and Laloue, *Compt. rend.* **142** (1906), 798. *Bull. soc. chim.* **35** (1906), 912. Hood, *J. Ind. Eng. Chem.* **8** (1916), 709; **9** (1917), 552. Laloue, *Bull. soc. chim.* [4] **7** (1910), 1101, 1107.

The essential oils extracted from the trunk of the young *Chamaecyparis formosana* tree contain a large amount of *d*-myrtenol and smaller amounts of *d*-α- and β-pinene. *d*-Dihydromyrtenol, which is only present in very small amounts in the young tree, is a major constituent of the older tree. The simultaneous disappearance of the pinenes and myrtenol strongly suggests that the tree converts these substances into the characteristic and rare alcohol, dihydromyrtenol, by oxidation and reduction processes[112] (Fig. 2 41).

A fourth method for the determination of the effect of growth on the oil content consists of the comparison of the analysis of the oils of leaves harvested at the same time, but representing different developmental states. The influence of preceding variations in weather conditions on the older leaves has to be reduced in a way similar to that mentioned previously in describing experimental methods. It has been shown of peppermint oil that the percentage of yield decreases from the upper to the lower leaves. In agreement with the findings of Charabot and others, Nilov and Ponta[113] found the ester and oxygen content higher in the older leaves. The ester content is also increased through the effect of hydrogen sulfide or ethylene.

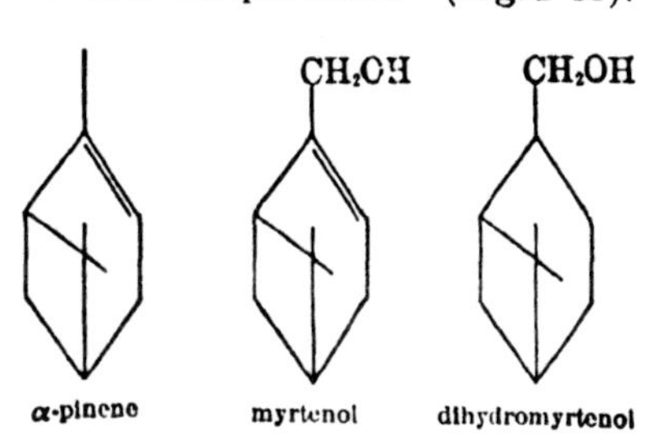

FIG. 2.41. Related bicyclic Terpenes in *Chamaecyparis formosana.*

The same gradient in oil content is seen in *Pogostemon patchouli*,[114] where the oil is chiefly located in the upper three pairs of leaves, confirming the general rule that the production of the oil coincides with the most active growth.

If we want to study the influence of different environmental factors on oil formation, it appears from the preceding discussion that we have to choose plant material of the same physiological age. It is also advisable to study the oil composition of young tissues which, due to their intense synthesis, are better suited to reflect any effects of the environment.

Careful studies in this direction have been made by Berry et al.[115] on the oil of *Eucalyptus cneorifolia.* The oil of this eucalyptus consists chiefly of cineole, the hydrocarbons pinene and *l*-β-phellandrene, the carbonyl compounds *l*-phellandral, cuminal, cryptal, *l*-4-isopropyl-Δ^2-cyclohexene-1-one; also present are *l*-α-phellandrene and some alcohols such as australol.

[112] Sebe, *J. Chem. Soc. Japan* 62 (1941), 22.
[113] *Trudy Vsesoyuz, Nauch.-Issledovatel. Inst. Efirno Masl. Prom. Sbornik Rabot Perechnoi Myate*, No. 5 (1939), 104.
[114] de Jong, *Rec. trav. chim.* 30 (1911), 211.
[115] *J. Chem. Soc.* (1937), 1443.

Although no marked change in the composition or in the amount of oil can be noted in the mature leaves, the case is quite different in the younger stages of development. The total oil content and the amount of the different components from the growing tips of the branches at different times of the year are shown in Illustration 2.7, expressed in percentage of the wet weight of the plant.

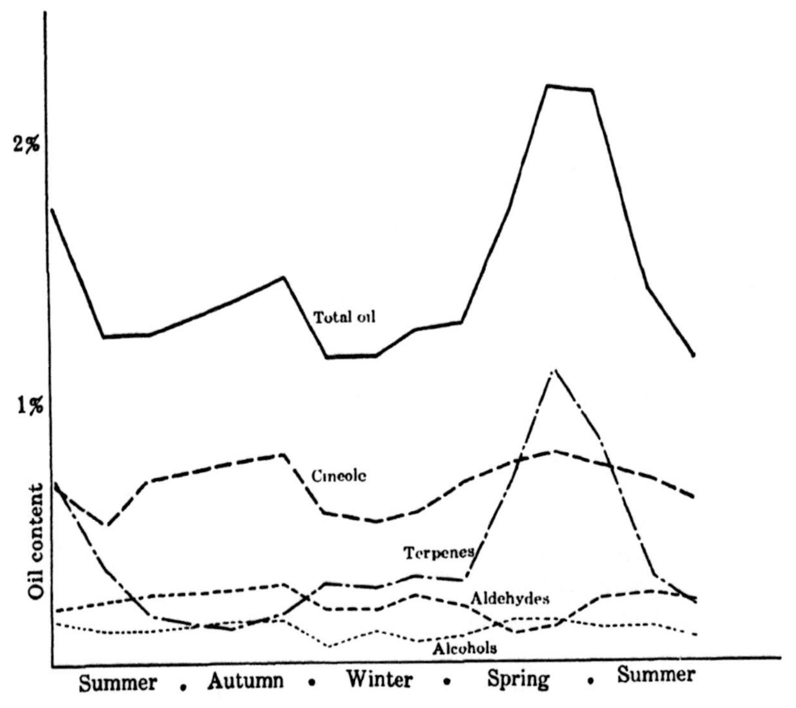

Ill. 2.7. Composition of oil from growing tips of *Eucalyptus cneorifolia.*

It is thus possible to show the absolute formation of each group of terpenes per unit weight of the plant, uninfluenced by an increased synthesis of one of the components, as would be the case if we expressed the composition as a percentage of the oil. The amount of alcohols formed at different periods does not seem to be greatly affected. More aldehydes are formed in autumn than in spring. During the period of maximum growth in spring and summer the formation of oil is highest. This increase cannot be attributed to a greater production of alcohols and aldehydes, because the alcohol content at different periods is not greatly affected; and in the case of the aldehydes we notice even the opposite effect: a decrease during spring. The real contributors toward the increased oil production are the terpene hydrocarbons, viz., β-phellandrene and cymene, and in a lesser degree the terpene-oxide cineole. From analytical data on mature leaves, it is known that the phellandrene content of the oil is greatly reduced, and that the cymene content was only 3–4 per cent, as compared to 19 per cent in the young leaves. From these analyses Berry concludes that α- and β-phellan-

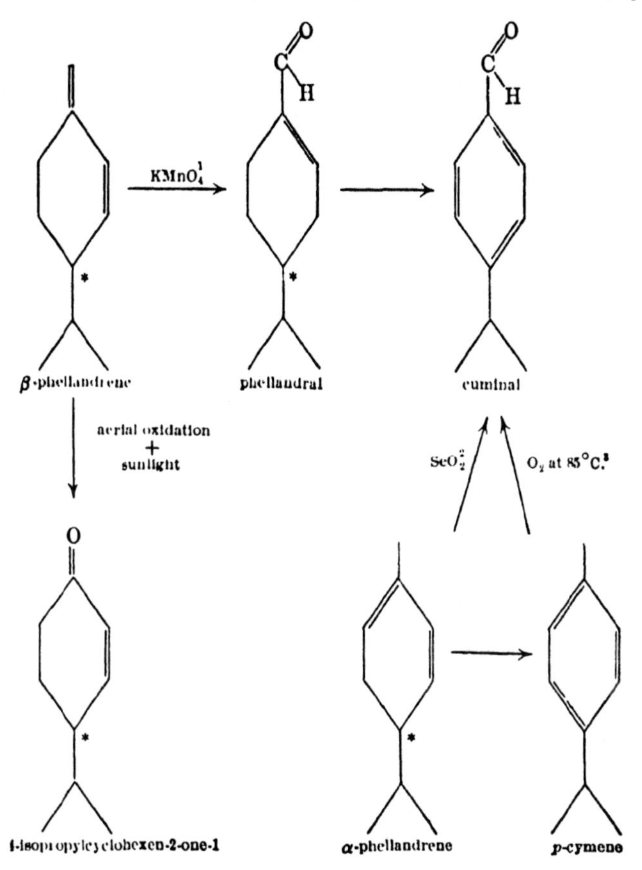

FIG. 2.42. Relations between Terpenes in *Eucalyptus cneorifolia.*

[1] Wallach, *Liebigs Ann.* **340** (1905), 1; **343** (1905), 35.
[2] Borgwardt and Schwenk, *J. Am. Chem. Soc.* **56** (1934), 1185.
[3] Stephens, *J. Am. Chem. Soc.* **48** (1926), 1824.

drene might be the precursor of certain terpenes, such as *p*-cymene, phellandral, cuminal and 4-isopropylcyclohexene-2-one-1. In the laboratory these conversions can be carried out with great ease[116, 117, 118] (Fig. 2.42). The optically active compounds present in this oil are stereochemically

[116] Borgwardt and Schwenk, *J. Am. Chem. Soc.* **56** (1934), 1185.
[117] Stephens, *J. Am. Chem. Soc.* **48** (1926), 1824.
[118] Wallach, *Liebigs Ann.* **340** (1905), 1; **343** (1905), 35.

related and belong to the laevorotatory series, constituting additional evidence of their common genesis. A similar relationship for *d*-phellandrene has been noted by Berry[119] in *Phellandria aquatica*, viz., *d*-α- and *d*-β-phellandrene and the corresponding *d*-ketone.

Many more observations made on the yield and composition of plants grown under different conditions of soil, climate and treatment, and in different stages of development could be added, but most of these are of such a specific and often experimentally vague nature that they can justify only the general conclusion that the more actively the plant grows, the larger the quantity of oil formed.

To gain a deeper insight into the physiological processes involved in formation of essential oils, we have to limit our experimental subjects to well-defined organs of well-defined species of plants. The experimental work on the composition of the eucalyptus group is a warning that the oils from closely related species may be widely different. Even species indistinguishable by ordinary morphological techniques can be distinguished on the basis of the production of oils of different chemical composition.[120]

In many cases, the abnormal behavior is due to hybridization of different species. Extensive genetic work has been carried out by Russian workers, and has led to the conclusion that considerable changes in the synthetic activity of the plants can be observed under the influence of hybridization, so that compounds may appear in the oil which were not present in the parent plants.[121] On the other hand, Mirov in his investigations on the turpentine from the genus *Pinus* describes a Ponderosa-Jeffrey hybrid which contains terpenes inherited from the Ponderosa parent, and heptane from the Jeffrey parent.[122, 123]

Polyploidy and other types of mutations, such as heteroploidy and chromosome aberration, may cause changes in the quantity and composition of the oils, as has been demonstrated in *Pelargonium roseum*.[124]

Many factors are, therefore, involved which change the composition of the oils; and for a successful study of these effects and the solutions of problems of oil formation it is imperative not to add further complications, such

[119] Berry, Killen, Macbeth and Swanson, *J. Chem. Soc.* (1937), 1448.

[120] Foote and Matthews, *J. Am. Pharm. Assocn.* **31** (1942), 65. Penfold and Morrison, *J. Roy. Soc. N. S. Wales* [I] **69** (1935), 111; [II] **71** (1938), 375; [III] **74** (1941), 277.

[121] Snegirev, *Bull. Appl. Bot., Genetics, Plant Breeding U.S.S.R.*, Ser. III, no. 15 (1936), 245. Nilov, Nesterenko and Mikhel'son, *Biokhim. i Fiziol. Drevesnykh i Kustarnykh Yuzhnykh Porod* **21**, no. 2 (1939), 3. Knishevetskaya, *Trudy Gosudarst. Nikitskogo Botan. Sada* **21**, no. 2 (1939), 29.

[122] Mirov, *J. Forestry* **27** (1929), 13; **30** (1932), 93; **44** (1946), 13.

[123] Kurth, "The Extraneous Components of Wood," "Wood Chemistry," edited by L. E. Wise (1944), 385.

[124] Urinson, *Bull. Appl. Bot., Genetics, Plant Breeding U.S.S.R.*, Ser. III, no. 13 (1936), 67.

as are caused by drying, distilling and harvesting procedures. Storage for a few hours even in the shade may in special cases cause a considerable decrease in the oil content. Russian workers found for nutmeg sage that its volatile oil content decreases 33 per cent after storage for 3 hr., and 55 per cent after 6 hr. in the shade, while in the sun it decreases 62 per cent after 6 hr. Their conclusion in this case is that the material should be collected at night and immediately distilled.[125] The losses in volatile components from intact plants are well known and have been measured quantitatively through micro combustion. The number of excreted products is considerable. These results[126] serve as a warning that external circumstances may easily modify quantity and quality of oils, with the result that changes due to other variables cannot be distinguished. On exposure to air, and especially to sunlight during drying of the plant material in the fields, a considerable amount of volatile oil may be lost by oxidation, polymerization and resinification.

For practical purposes, certain compromises have to be made; nevertheless it should be our goal to choose conditions and experimental material so carefully that reproducibility is assured, and the many factors involved can be changed individually. Only in such a way can we expect to unravel the fate of the plant metabolites secreted as essential oils. Such experiments might well throw light on another intriguing problem, i.e., the function of the essential oil in the plant.

A discussion of this subject invites a look at plant metabolism from a more general viewpoint.

III. THE FUNCTION OF ESSENTIAL OILS IN PLANTS

When the plant organism is alive and in process of development, external substances are constantly absorbed and transformed into "building stones." This reshaping of the foreign substances and their incorporation into the plant system, known as *assimilation*, requires energy, which is obtained by a series of reactions, whereby a part of the assimilated products is oxidized. The balance of these two series of reactions appears in the growth of the plant. Therefore, while some of the plant material is in a continuous flux, undergoing degradation and rebuilding, another important part of the reaction products can be expected not to take part in an uninterrupted chain of reactions.

Some of these products—such as cellulose—will be deposited in cell walls, the plant thereby acquiring a more rigid structure. Other substances—such as starch—are stored as energy and organic material sources, to be drawn upon when circumstances arise which cause the re-entrance of these

[125] Gaponenkov and Aleshin, *J. Applied Chem. U.S.S.R.* **8** (1935), 1049.
[126] Haagen-Smit, unpublished results.

substances into the reaction chain. We can thus assign a certain function in the plant to these particular compounds; but we find it much more difficult to do this with a number of other substances, such as alkaloids, anthocyanins and flavones, essential oils, resins and rubber latex.

It is a well-known fact that some plants emanate, besides carbon dioxide, a considerable amount of organic material, chiefly the carriers of the smell of the plant. In some rare cases so much oil is excreted that the oil can be set afire, as in *Ruta graveolens* and *Dictamnus*. At the same time, relatively large amounts of these essential oils are deposited in the plant, and our only evidence for the assumption that such compounds are unimportant sources of energy to the plant is the fact that, prior to leaf abscission, the oils are not transferred to the stem, as is the case with a large part of the carbohydrates.

The question has, therefore, repeatedly been asked: Does the plant derive any specific benefit from these oils?[127] Opinions in this field are based on the observations that some oil-bearing plants are attractive to certain animals, whereas others are repellent. In individual cases, therefore, a contribution is made toward more effective pollination through insect visits.[128] In a number of other cases a degree of protection against the depredations of animal[129] and plant[130] parasites may be afforded by the irritating effect of many oils. Some observers maintain that the oils function as reserve food, as a means of sealing wounds, or as a varnish to prevent excessive evaporation of water (cell fluid). These opinions are not beyond question, and do not appear to be supported by experimental evidence often having their origin merely in a teleological approach to the subject. Most investigators, including Tschirch, the famous resin chemist, hold the view that the functions attributed to those substances are more often of accidental, than of essential, importance to the plant. Those who consider these products a result of phenomena accompanying the growth process have used the term "waste product." This, however, rather underrates the value of these secretion products, which, through their formation, may contribute to syntheses important in the continued existence of the plant. Some carbohydrate precursors may serve as hydrogen acceptors, and in doing so may become unusable for further synthesis. Their function, therefore, may arise in their formation, and not in a later stage. Thus we might compare the oils to "Hobelspäne," the shavings of a plane. As reason for their disposal, the opinion has been expressed that substances such as terpenes, mostly hydrocarbons, are so far remote in their chemical

[127] Czapek, "Biochemie der Pflanzen." III Auflage, G. Fischer, Jena (1925). Detto, *Flora* 92 (1903), 146. Gerhardt, *Naturwiss.* 8 (1920), 41.

[128] v. Frisch, *Verh. Zool. Bot. Ges.* Wien 65 (1915), 1; 68 (1918), 129.

[129] Stahle, "Pflanzen und Schnecken," *Zeitsch. Natur. Medicine* 22, NFXV, Jena. Preyer, *Flora* 103 (1911), 441. Haberlandt, "Physiologische Pflanzenanatomie," 4 Aufl.

[130] Verschaffelt, *Kgl. Ak. Amsterdam* (1910), 536. Gertz, *Jahr. Wis. Bot.* 56 (1915), 123.

and physicochemical conduct from the properties of the living substances that they are excreted as "körperfremde" or alien materials.

There are others who refuse to believe in the waste product origin of the essential oils, and Lutz suspects that these oils have constituents which can be hydrogen donors in oxido-reduction reactions. Although they are thereby transformed into neutral compounds as far as catalysis is concerned, they might re-enter the reaction scheme through a reduction in the presence of light. To prove this theory, experiments were carried out on a fungus belonging to the *Hymenomycetes* on which Lutz determined the antioxidant or hydrogen-donor action of oil constituents. Phenols were found to be excellent donors, as is well known from the investigations of Moureu, but secondary and tertiary alcohols and aldehydes also showed strong activity. Hydrocarbons are inactive in the dark, but become active in the light. On the other hand, primary alcohols, terpene oxides like cineole, and ketones are inactive, and perhaps are from this viewpoint the real waste products. Lutz[131] considers the oil components as moderators in intracellular oxidation to protect against the action of atmospheric agents. He also includes the possibility that some of the components may be used as an energy source during a deficiency state caused by an interruption of the normal assimilation of carbon dioxide

It has also been suggested that plants which emanate a considerable amount of oils are prevented from becoming too warm since heat is absorbed in the vaporization of the oils. In this way the oils would function as a water-sparing mechanism. However, measurements of the relatively large amounts of water and small amounts of oil involved, show clearly that such a contribution would be negligible. In the search for some useful function for the terpenes, Teodoresco[132] was one of the few who carried out experiments on the effects of oils on plants. He showed that the absorption of sun radiation by the essential oil atmosphere around the plant was negligible and certainly did not have any influence on the water evaporation. This oft-debated point, based on a misinterpretation of Tyndall's work, was solved by admitting oil vapor around the plant without, however, making direct contact, and determining the loss of weight through transpiration. Neither direct weighing of the loss of water, as shown in Illustration 2.8, nor transpiration measurements with a potometer demonstrated any heat-screening effect. If, however, the oil was allowed to come in contact with the plant, a considerable reduction in transpiration was evident. Although the damaging effect of prolonged exposure to the oils had been observed before, Teodoresco showed that when the vapor is removed soon enough, recovery follows in a few hours. This action is not confined to the oils

[131] *Bull. soc. chim. biol.* 22 (1940), 497.
[132] Audus and Cheetham, *Ann. Botany* (N.S.) 4 (1940), 465.

obtained from the same plant, but is a more or less nonspecific effect for the volatile oils in general.

It would, therefore, appear that a number of essential oils exercise directly or indirectly a definite action on the transpiration in plants. However, experiments carried out on partial saturation of the atmosphere sur-

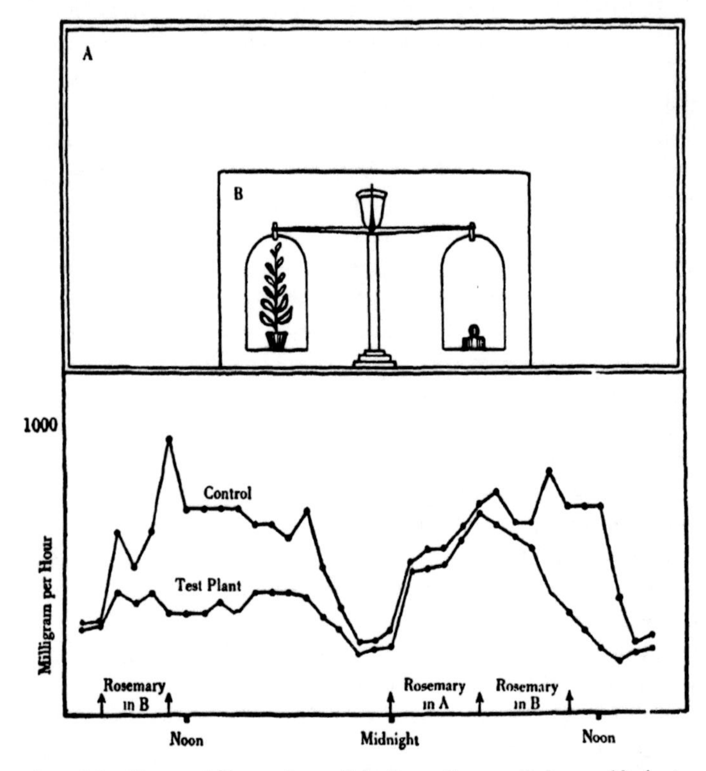

Ill. 2.8. Vapor of *Rosmarinus officinalis* on *Dracocephalum moldavica*.[1]

[1] Teodoresco, *Rev. gen. botan.* **35** (1923), 382. Refutations also by Grijns, *Arch. neerland. physiol.* **3** (1919), 377. Detto, Flora **92** (1903), 147. Nicol, *Compt. rend.* **489** (1929) 289; *Biochem. J.* **26** (1932), 658.

rounding the plant, simulating more closely the outside conditions, showed that the concentration of essential oil vapor would rarely be high enough to cause any significant decrease in transpiration.

The oils inside the plant, although enclosed by special tissues, might have an influence on the transpiration and other important functions of the plant rather than the vapor of the excreted oil, since there is reason to believe that cell walls would not be an insurmountable obstacle for the oil. This effect would result in a general retardation of a number of the plant activities. Teodoresco mentions specifically a decrease in the nyctinastic, seismonastic, phototropic and geotropic movements. The oils inhibit also the formation of chlorophyll in etiolated plants when exposed to light, and cause a decrease in permeability. Continued exposure to the oil vapors causes damage to the living substance, producing a greater permeability, which is in turn followed by death.[133, 134, 135] General toxic action on plants has been observed by Bokorny[136] with oil of turpentine in concentration of 1 : 50,000.

The action of some essential oils is similar in certain respects to that of anaesthetics on animal cells. This problem of anaesthesia is one of the fundamental problems in general physiology, and the results obtained in these studies might well contribute to the understanding of similar effects in plants. The first effect of fat solvents, narcotics and stimulating agents is identical,[137] and it may be assumed that they cause a reversible lowering of the permeability for water and water-soluble substances, in harmony with the findings of Teodoresco and others on the pronounced inhibition of transpiration in plants.

The inhibiting and damaging effect of the oils on many life processes has been turned to our advantage in the use of these compounds as bactericidal and fungicidal agents. However, from the diversity of the compounds in essential oils, it is clear that we have to regard with suspicion any general statement on the bactericidal action of the essential oils. From specific cases which have been studied, it can be concluded that the terpene derivatives, while possessing action bactericidal to certain organisms, are not able to inhibit growth in all of the numerous types of microorganisms. It is, therefore, not astonishing to see aqueous infusions of, for example, lavender, peppermint and juniper drugs fall victim to putrefaction after several days of standing.[138]

On account of their bactericidal action, a number of volatile oils have been employed in the past for the treatment of urogenital infections. The simultaneous irritating effect on animal tissue applied in measured degree may stimulate repairs of tissue, and assist in the removal of mucous from trachea and bronchia and relieve tension of the stomach and colic.

Other toxic effects are reported on cultures of fibroblasts, and further examples of their inhibitory effect on life processes can be seen in the anthelmintic effect of different oils, especially chenopodium oil. The effect

[133] Burgerstein, *Verhandl. Zool. Bot. Ges.*, Wien, 1884.
[134] Heller, *Flora* (1904), 1.
[135] Vandervelde, *Chem. Zentr.* I (1900), 481; II (1901), 440.
[136] *Pflüg. Arch.* **72** (1899), 555.
[137] Heilbron, "An Outline of General Physiology," **37** (1938), 437. Davson and Danielli, "The Permeability of Natural Membranes," (1944).
[138] Kliewe and Hutmacher, *Deut. Apoth. Ztg.* **53** (1938), 952. De Potter, *Compt. rend. soc. biol.* **131** (1939), 158.

of this American wormseed oil on roundworms, hookworms and intestinal amoebas is very similar to that of santonine, a sesquiterpene lactone present in *Semen cinae*. The toxicity of these oils for certain organisms cannot be measured simply by their bactericidal action. Thymol, for example, is much stronger antiseptically but is much less active as an anthelmintic than santonine or ascaridole, the active components of wormseed oil. The toxic effect of some essential oils and oil components is not limited to the organisms which have to be destroyed, and excessive use in higher animals and man causes depression of the higher centers followed by convulsions. A few cases are known where an apparent stimulating effect is observed. This is the case with terpene compounds such as camphor and menthol, which are used as circulatory stimulants in cases of collapse. It is assumed that when the action of the heart muscle is depressed, camphor may improve the cardiac condition and remove arrhythmia.

On the basis of recent investigations, these effects seem also to be due to an inhibitory action on nerve fibers which counteract other fibers belonging to the sympathetic nervous system. Through this effect on the inhibitors, certain muscles are stimulated. A similar explanation might well hold for the acceleration and strengthening of the peristaltic movements of the small intestines of rabbits, according to Haffner,[139] and Sone and Shiro.[140]

In general, we observe a definite toxic effect on the important life processes, and excessive doses, because of depression and paralysis of the central nervous system, are followed by death. The essential oils probably interfere with delicate mechanisms, through their chemical and physical effects, either by entering and disturbing colloidal systems or by taking part in certain reactions. The oils themselves are at the same time exposed to many influences, which may change them in such a way that removal through the kidneys is possible. This so-called detoxication process takes many forms, and may consist of esterification, oxidation, reduction, or conjugation with compounds such as glycuronic acid and amino acids. When borneol is fed to dogs, it appears as glycuronide in the urine; when vanillin is ingested, oxidative processes are responsible for the excretion of vanillic acid. A combination of both processes is evident when camphor is removed in oxidized form. When such a removal is not possible, as, for example, by accumulation of the oils through injection, the organisms react by walling off the foreign material and tumors and sterile abscesses are reported.[141]

This reaction is in principle similar to what happens in a plant lacking the elaborate detoxication and excretion mechanisms present in the higher

[139] *Arch. exptl. Path. Pharmakol.* 186 (1937), 621.
[140] *Tôhoku J. exptl. Med.* 30 (1937), 540.
[141] Saito, *Folia Pharmacol.*, Japan 23 (Breviaria 2) (1936), 6.

animals. A considerable part of the metabolites which are not immediately taken up in further reactions or are not removed by evaporation will have to remain in or near the secretion cells. The interfering action of the oils may then cause lysis of the surrounding cells and changes in the normal metabolism, resulting in the formation of cork and mucilagenous layers, with low permeability for the oil. We may safely conclude that once removed from the continuous chain of reactions, these compounds are a potential danger to all living tissues, and both plant and animal react by walling off the oil from the other tissues. If this is not possible, reactions will take place until the compounds are so transformed that they can be excreted, or until they have become harmless from the point of view of the surrounding tissue.

From a general viewpoint, essential oils, alkaloids, resins, rubber, anthocyanins and many other secreted substances may have in principle a similar history. Their precursors, linked with essential processes in the organism, undergo secondary and further changes when exposed to the medium in which they are left behind.

Due to their commercial importance, our chemical knowledge of these end products exceeds by far our knowledge of the processes from which they have been derived. It is hoped that more fundamental studies are being carried out in this direction which, in turn, will lead to our more rigid control over their formation in the plant.

Suggested Additional Literature

Gildemeister and Hoffman, "Die ätherischen Öle," 3d Ed. Vols. I, II and III, Leipzig (1928–1931).

Simonsen, "The Terpenes," Vols. I and II, Cambridge (1931–1932).

Tschirch, "Die Harze und die Harzbehälter," 2d Ed., Leipzig (1936).

Wallach, "Terpene und Campher," 2d Ed., Leipzig, 1914.

Klein, "Handbuch der Pflanzenanalyse," Wien, 1932.

Semmler, "Die ätherischen Öle nach ihren chemischen Bestandteilen," Leipzig, 1906–1907.

CHAPTER 3

THE PRODUCTION OF ESSENTIAL OILS

METHODS OF DISTILLATION, ENFLEURAGE, MACERATION, AND EXTRACTION WITH VOLATILE SOLVENTS

BY

ERNEST GUENTHER

Note. All temperatures in this book are given in degrees centigrade unless otherwise noted.

CHAPTER 3

THE PRODUCTION OF ESSENTIAL OILS

A. DISTILLATION OF ESSENTIAL OILS

Introduction.—The majority of essential oils have always been obtained by steam distillation or, in the more general sense, by hydrodistillation.[1] The practical problems connected with distillation of aromatic plants are, therefore, of utmost importance to the actual producer of essential oils. Yet our present-day technical literature, especially English literature, is surprisingly meager in regard to data and information which might serve as a really practical and reliable guide. This shortcoming has been felt severely, especially during the years of World War II, when prospective producers in North, Central and South America sought advice concerning the distillation of oils which, due to war conditions, could no longer be imported from Europe and Asia. Encouraged by countless inquiries from almost every part of the Western Hemisphere, the author finally decided to compile a comprehensive paper on this topic which would incorporate not only his own experience of many years in the field, but also the most important phases gathered from the literature published so far. There exist on this subject two really outstanding books, viz., the classical work of Dr. von Rechenberg, who spent a lifetime in the actual distillation of essential oils and on systematic research pertaining to the physical phenomena and laws underlying distillation. These works have never been translated from their German text, are now out of print and, due to the ravages of World War II, not readily available. This author would consider it an irreplaceable loss to our industries if the most important parts of these books, at least those dealing with the practical aspects of essential oil distillation, were not preserved for posterity. Unfortunately, the lucid writings of Professor von Rechenberg have not attained sufficient attention outside of Germany. In more than one way they are so fundamental and exact that they require no modification. This author has, therefore, translated parts of von Rechenberg's treatises, with a view to incorporating some of the most essential features into his own text. These books are recommended:

C. von Rechenberg, "Theorie der Gewinnung und Trennung der ätherischen Öle," Schimmel & Co., Miltitz bei Leipzig, 1910.

This chapter by Ernest Guenther.

[1] The term "hydrodistillation" is used by von Rechenberg as referring to distillation with water vapors (steam).

C. von Rechenberg, "Einfache und Fraktionierte Destillation in Theorie und Praxis," Schimmel & Co., Miltitz bei Leipzig, 1923.

A much smaller book, "Die Fabrikation und Verarbeitung von ätherischen Ölen," by Max Fölsch, Hartleben's Verlag, Wien und Leipzig, 1930, leans on von Rechenberg's text but adds much practical advice.

Those interested particularly in the distillation of colonial oils and in field distillation requiring simple apparatus are referred to Gattefossé's "Distillation des Plantes Aromatiques," Librarie Centrale des Sciences, Paris, 1926.

"Aspects of the Theory of Distillation as Applied to Essential Oils," have been described by Leslie Bloomfield in a series of comprehensive papers which appeared in the *Perfumery and Essential Oil Record*, Vol. 27 (1936), 131, 177, 294, 334, 368, 404, 443, 483; Vol. 28 (1937), 24, 59.

"A Treatise on Distillation," by Thos. H. Durrans, was published also in the *Perfumery and Essential Oil Record*, June, 1920, 154 to 198.

The mathematical and physical principles connected with steam distillation in general are discussed in "Wasserdampf-destillation," by N. Schoorl, which appeared in *Rec. Trav. Chim.* **62** (1943), 341–379.

This chapter will be divided into two parts, the first dealing with the fundamental or theoretical principles underlying all distillation processes, and the second treating more specifically the practical aspects of distillation as applied directly in the essential oil industry.

I. THEORIES OF DISTILLATION

Essential, volatile or ethereal oils are mixtures composed of volatile, liquid and solid compounds which vary widely in regard to their composition and boiling points. Every substance with a determinable boiling point is volatile and possesses a definite vapor pressure, which depends upon the prevailing temperature, and which is very low in the case of very high boiling substances. Hence, the intensity of an odor may be considered, to a certain extent and with many exceptions, as a manifestation of the volatility (boiling point and vapor pressure) of the substance which emits the odor.

Distillation may be defined as "the separation of the components of a mixture of two or more liquids by virtue of the difference in their vapor pressure" (Stephen Miall, "A New Dictionary of Chemistry," London, Longmans Green, 1940). The process of distillation is obviously of considerable importance to the essential oil producer. There are two general types to be considered:

1. Distillation of mixtures of liquids which are not miscible, and hence form two phases. Practically, this applies to the rectification and fractionation of essential oils with steam, and, what is much more important, to the

isolation of volatile oils from aromatic plants with steam. Distillation with steam may also be called hydrodistillation, which general term implies that distillation may be carried out either by boiling the plant material or the essential oil with water, and creating the necessary steam within the still, or by introducing into the retort live steam generated in a separate steam boiler.

2. Distillation of liquids which are completely miscible in each other, and therefore form only one phase. Practically, this applies to the rectification and separation of an essential oil into several fractions (fractionation), without the use of steam.

The difference between the behavior of single-phase mixtures and two-phase mixtures can best be understood by considering what happens when a liquid vaporizes, especially on boiling. Let us consider first the case of a pure liquid in a closed container. At a given, fixed temperature, the average energy of the molecules is fixed. The molecules are in constant and completely random motion. Any molecule in the main body of the liquid can travel only a short distance before it comes under the influence of other molecules at which moment its direction of motion is changed. Any molecule in the surface layer, however, which happens to be moving in a direction away from the main body of the liquid can escape into the space above the liquid, thus becoming a vapor molecule. Now, the vapor molecules, too, are in constant motion, the speed of the molecules of any kind being determined solely by the prevailing temperature. Any vapor molecule hitting the liquid surface has a chance of being captured by the liquid—in other words of being reliquefied (condensed). As the temperature is raised the number of vapor molecules increases. Obviously the chances of a molecule returning into the liquid also increase, so that after a short time the number of molecules vaporizing in a unit of time exactly equals the number condensing (being reliquefied) in the same time. Thus, there arises a condition of dynamic equilibrium, with the total number of molecules in the vapor state remaining constant. If the space filled with saturated vapors is opened, vapor escapes and will be replaced by the same number of molecules, i.e., by the same quantity of vapor newly developed from the liquid mass. This applies not only to liquids but to solids, because, as pointed out above, every substance with a determinable boiling point is volatile.

Let us now suppose that, still at constant temperature, a second liquid, completely miscible with the first one, is added. Since the two liquids form a single phase, the surface of the liquid mixture consists only partially of molecules of the first kind. The number of molecules of the first kind escaping into the vapor space per unit time must certainly depend on the number present in the surface layer, and will, therefore, be smaller now than it was for the pure liquid. However, the molecules being completely miscible, the total number returning from the vapor to the liquid will not immediately be changed. Since the total amount of surface is unchanged and since now more molecules of the first kind are condensing than are being vaporized, temporarily the equilibrium originally established will be disturbed. This process continues until a new equilibrium is established, when these rates again become equal, and this in turn causes a decrease in the number of molecules of the first kind present in the vapor phase at any one time. Exactly the same law applies to the second component of the mixture. In general, the number of molecules of any component of a homogeneous mixture present in the vapor phase will thus be smaller than the number present in the same vapor space if the pure liquid is involved. The fraction of the surface occupied by either liquid is, of course, proportional to its relative amount, and consequently the extent to which the rate of vaporization decreases will depend on the composition of the liquid. The vapor composition of a one phase mixture will, therefore, be determined at any fixed temperature by the composition of the liquid.

Boiling point may be defined as "the temperature at which, under atmospheric or any other specified pressure, a liquid is transformed into a vapor; i.e., the temperature at which the vapor pressure of the liquid equals the pressure of the surrounding gas or vapor" ("Hackh's Chemical Dictionary," Philadelphia, 1944). When distilling at atmospheric pressure, this vapor pressure corresponds to the weight of a mercury column of 760 mm.[2] in height. Any reduction of the pressure above a liquid causes a lowering of the boiling point, any increase of pressure results in a higher boiling point. A liquid consisting of several constituents, completely miscible in one another and possessing different boiling points, in most cases (except the so-called "constant boiling mixtures") does not have a uniform boiling point but a boiling range. As the lower boiling constituents vaporize or distill off, the boiling temperature of the liquid rises and finally approaches that of the highest boiling constituent.

Next, let us consider the effect of adding to a pure liquid in equilibrium with its vapor a second liquid which is completely immiscible with the first one. This brings us to a discussion of the distillation of heterogeneous liquids, as in the case of essential oil distillation with steam or boiling water (hydrodistillation). To facilitate visualization, imagine that the two media are kept well stirred, so that the percentage of each liquid present remains the same in all parts of the mixture, including the surface. Such mixing has little effect on the ultimate result. Again, the rate of vaporization decreases, because the number of molecules of the first liquid in the surface layer is decreased. In this case, however, the liquids are not miscible, and the

[2] Equals 29.922 in.; or a pressure of 14.6974 lb. per sq. in. = 1.0333 kg. per sq. cm.

vapor molecules can only be condensed when they strike a molecule of their own kind, so that the rate of condensation will also be decreased. Now, the rate of vaporization and the rate of condensation both depend upon the percentage of molecules of the first kind present on the surface. These rates will be affected equally, and there will be no change in the number of vapor molecules of the first component present. Applying the same reasoning to the case of the other component leads to the same conclusion. We thus arrive at the important law that *the total number of molecules present in the vapor space above a two-phase liquid mixture at any given temperature is equal to the sum of the numbers of molecules so present if either liquid were dealt with alone.* Furthermore, since the relative amounts of the two liquids present have not in any way entered our reasoning, this conclusion must be true regardless of the relative amounts so long as both liquids are present. In other words, *in the case of a two-phase (heterogeneous) liquid the composition of the mixed vapor, at a given temperature, does not depend upon the composition of the liquid.*

A system of water and essential oil forms a two-phase liquid; therefore, this type of distillation is of primary importance to the essential oil producer. Let us then consider further the results of the above reasoning for our case. The pressure exerted by a vapor, whether it consists of one or several kinds of molecules, is a manifestation of the constant bombardment by the rapidly moving vapor molecules hitting the walls enclosing the vapor. Pressure measures a force acting on a unit area, and this force, in the case of a vapor, results from the vapor molecules striking the wall and rebounding. *The total pressure exerted will be equal to the pressure expended by one molecule multiplied by the number of molecules hitting a unit area of the wall in a unit of time.* The kinetic energy expended by one molecule will depend on the temperature, but the number of collisions with the wall will depend on the number of molecules, of whatever kind, present in the vapor space. In other words, *the pressure will depend on the concentration of the molecules or,* stated differently, *on the concentration of the vapor.*

Now, it has been shown that in the case of a two-phase liquid the total number of molecules present in the vapor phase in equilibrium with it is greater than the number which would be present if either pure liquid were present alone at the same temperature. Hence, the pressure exerted by the vapor mixture will be greater than that exerted by either pure vapor alone. In the distillation of volatile oils with steam or boiling water (hydrodistillation), the pressure in the vapor space is maintained constant, either by connecting the vapor space with the atmosphere or by suitable controls to maintain a reduced or elevated pressure. For definiteness we shall consider an operation at atmospheric pressure. If pure water is heated in a still, it will begin to boil (or in other words, the pressure of its vapor will equal that of the atmosphere), when its temperature has reached 100° C. (212° F.). Let us suppose that an oil insoluble in water is introduced into the still along with the water. If permitted to do so, the pressure in the vapor space would increase as previously shown. But in our case the vapor space is connected to the atmosphere; therefore, the pressure will be reduced to atmospheric pressure, which can be accomplished only by automatic lowering of the temperature. When the temperature of a liquid is lowered, the tendency of the liquid molecules to go into the vapor phase also decreases, thus decreasing the concentration of the molecules in the vapor, and consequently the vapor pressure. Hence, the temperature will be lowered to a value such that the total pressure exerted by the vapor mixture is again equal to the operating pressure (atmospheric pressure in our case). *Thus the boiling temperature for any two-phase liquid will always be lower than the boiling point of either of the pure liquids at the same total pressure.* For example, water (boiling at 100°) and benzene (boiling at 80°) present two such insoluble liquids: when a mixture of the two is brought to a boil at atmospheric pressure (760 mm.), it vaporizes (distills) constantly at 69° so long as both constituents remain present in the liquid mixture. The moment either of the two constituents is completely vaporized (distilled off), the temperature rises to the boiling point of the remaining constituent. Such conditions prevail with all volatile substances, provided they are insoluble in water or only very slightly soluble, and are not chemically reacted upon by water. When brought to boiling together with water, they vaporize at a temperature below that of boiling water and also below those of the boiling points of the pure compounds insoluble in water.

In the preceding discussion we emphasized repeatedly that the vapor in equilibrium with a two-phase liquid consists of two kinds of molecules. The total pressure exerted by such a mixture is due, therefore, to the sum of the pressures of each kind of molecule alone. The pressure exerted by either of the pure vapors at the same temperature would be the vapor pressure of that pure component, while *the total vapor pressure of the mixture is thus equal to the sum of the partial vapor pressures.* By partial pressure we mean the vapor pressure of any one component in a mixed vapor. Obviously for such two-phase liquid systems the partial pressure and vapor pressure of any component are the same. This simple rule of the additivity of partial pressures affords a ready means of estimating the temperature at which any particular steam distillation (hydrodistillation) will occur. The vapor pressures of the two pure components are simply tabulated at a series of temperatures. The operating temperature will then be that temperature at which the sum of the two vapor pressures equals the operating pressure, in the above cited example the atmospheric pressure. In that case, the vapor pressure of water at 69° is 225 mm., the vapor pressure of benzene 535 mm., added together 760 mm. This condition permits the combined vapors of the constituents to overcome the (normal) atmospheric pressure; in other words, the mixture starts to boil at 69° under normal atmospheric pressure. In order to effect the boiling of a volatile compound insoluble in water, it remains immaterial whether the substance in question is brought to a boil with water or whether live steam is injected into the liquid or finely powdered substance. It is the steam (water vapors—whence the term hydrodistillation) that causes the boiling (distillation, in our case) of the compound insoluble in water, at a temperature below the boiling point of the compound itself and below that of water.

The composition of the vapor formed from a two-phase liquid mixture depends on the partial vapor pressures of the pure constituents. Thus, if the vapor pressure of component A is high and that of B low, the mixed vapor will consist very largely of component A. *The ratio between the weights of component A and B will be given by the ratio of their vapor pressures multiplied by the ratio of their molecular weights.* As pointed out, boiling will take place only when the sum of the partial pressures exerted by the components is equal to the pressure maintained in the vapor space; therefore, a heterogeneous (two-phase) liquid boils or distills at a temperature which, at the same total pressure, always lies below the boiling point of the lowest boiling constituent, so long as the latter remains in the mixture. It is for this reason primarily that hydrodistillation has been used for such a long time and so generally in the isolation of essential oils from aromatic plants. By vaporizing (boiling) mixtures of water and essential oils (also from plant material), the temperature will always be maintained lower than the boiling point of water at the same total pressure and, in this way, damage and decomposition of the essential oils by overheating are prevented. The fact that the vapor pressures of most essential oils are low relative to the vapor pressures of water at corresponding temperatures accounts for the fact that the ratio of water to essential oil in the condensate is relatively high. It will make no fundamental difference in the behavior of the mixture whether or not a steam distillation is carried out in the presence of a liquid water phase, but it does influence certain practical aspects of the process, as will be indicated in the second part of this chapter.

In order to isolate an essential oil from an aromatic plant, the material, in actual practice, is packed into a still, a sufficient quantity of water added and brought to a boil, or live steam is injected into the plant charge. Due to the influence of hot water and steam, the essential oil will be freed from the oil glands in the plant tissue. The still, therefore, will contain a mixture of two liquids, viz., hot water and volatile oil which are not mutually soluble, or only very slightly so. Gradually the liquid in the still is brought to a boil, the vapor mixture then consisting of water vapors (steam) and oil vapors. This vapor mixture passes through a connecting tube into a condenser, where it is reliquefied (condensed) by external cooling, usually with cold water. From the condenser the distillate flows into a receiver (separator), where the oil separates automatically from the distillation water. In the course of distillation it is necessary continuously to replace the water evaporating from the still, or to inject a sufficient quantity of live steam to vaporize all the volatile oil contained in the plant material or present in the still. When the last traces of volatile oil have been recovered, only pure water will distill over, and distillation is completed.

As said, the composition of the distillate from a mixture of two insoluble liquids—in other words, the weight quantities of the two substances—depends primarily upon their boiling points, or upon their vapor pressures at the temperature of distillation. If, for example, we distill a water insoluble compound with a boiling point of only 50°, the distillate will consist of a certain volume of water and a larger volume of the water insoluble compound. If, on the other hand, a water insoluble compound with a boiling point of 300° is hydrodistilled, the distillate will contain mostly water and very little of the high boiling substance. Thus, in the distillation of a water insoluble volatile compound, the percentage of the latter in the distillate decreases with rising boiling point of the compound. This decrease, however, is not uniform with all substances. Some substances with similar boiling points will occur in the distillate in different proportions; others with a marked differential in their boiling points may accumulate in the distillate in almost the same proportions. Deviations of this sort are caused primarily by the chemical constitutions and reactivity of the various essential oil components. As explained above, the quantitative composition of the distillate (condensate) can be calculated in advance when hydrodistilling chemically uniform, water insoluble substances. The rule underlying hydrodistillation of essential oils or volatile substances in general may be expressed as follows:

The ratio between the weights of the two vapor components, and therefore of the two liquids in the distillate (condensate), is expressed by the ratio of their partial vapor pressures multiplied by the ratio of their molecular weight.

$$\frac{W_{H_2O}}{W_{oil}} = \frac{P_{H_2O}}{P_{oil}} \times \frac{M_{H_2O}}{M_{oil}}$$

in which W_{H_2O} = weight of water in the condensate;
W_{oil} = weight of oil in the condensate;
P_{H_2O} = vapor pressure of water at still temperature;
P_{oil} = vapor pressure of oil at still temperature;
M_{H_2O} = molecular weight of water (=18);
M_{oil} = molecular weight of oil (assuming that this constant may be determined as an average figure).

Essential oils are not chemically pure substances but consist of several, often many, compounds possessing different chemical and physical properties. The boiling points of the volatile oil components range in most cases from 150° to 300° at 760 mm. pressure. According to the preponderance of lower or higher boiling constituents we speak of a low boiling or of a high boiling oil. Distillation of an essential oil reveals its higher or lower volatility to a very marked degree if the oil is in free, direct contact with the boiling water or with the passing steam: in the early stages of distillation the lower boiling components distill over; the higher boiling ones pass over later.

Let us now study hydrodistillation of a volatile oil with a very simple example: peppermint oil is placed into a glass flask and live steam is introduced into the oil. The external pressure and temperature, in this case, remain immaterial, so long as at least a portion of water remains in steam form. The steam then causes the peppermint oil to form vapors, to vaporize, each steam bubble presenting to the vaporized oil an empty space into which the oil immediately sends vapor molecules. Every volume unit of steam will be filled with an equal volume of oil vapors, rise to the top of the flask and enter the condenser, where steam and oil vapors are condensed. The hydrodistillation of any essential oil is based upon this simple principle which, however, does not fully apply to the oils when they are still enclosed within the plant tissue. There the steam must exert yet another action of considerable influence, i.e., it must transmit heat. Unlike a liquid, the rigid plant matter is not able to conduct the heat from the still walls to all parts of the plant charge. The heat is actually transmitted by water, either as boiling water when distilling immersed plant material or as water vapors when distilling plants by blowing live steam into the charge. Also, the volatile oils occur in special oil glands, sacks or intracellular spaces of the plant tissue; hence the oils must be freed, prior to distillation, by breaking down the plant tissue, and by opening the oil glands as much as possible, so that their volatile content can be readily attacked and vaporized by the passing steam. In unreduced, whole plant material, the oil must be freed during distillation by the force of hydrodiffusion, a very important feature which will be discussed later in more detail.

Let us now return to the more theoretical aspects of hydrodistillation. In steam distillation it is frequently possible to change materially the ratio of water to oil in the condensate by changing the operating pressure. As pointed out earlier, this ratio is determined by the relationship

$$\frac{W_{H_2O}}{W_{oil}} = \frac{P_{H_2O}}{P_{oil}} \times \frac{M_{H_2O}}{M_{oil}}$$

In any hydrodistillation using saturated steam, the sum of P_{H_2O} and P_{oil} will equal the operating pressure and the still temperature will automatically adjust itself until this condition is met. As the operating pressure is lowered below atmospheric pressure, the temperature of the operation will decrease. In general, the vapor pressure of water decreases much more slowly with the temperature than does the vapor pressure of an essential oil, so that the weight ratio of water to oil increases. Conversely, this ratio decreases with increasing temperature. Data for a typical case are given in Table 3.1.

TABLE 3.1. EFFECT OF OPERATING PRESSURE ON WATER TO OIL RATIO IN STEAM DISTILLATION OF CITRONELLAL

Total Pressure mm. Hg	Temp. °C.	Vapor Pressure mm. Hg		Molal Ratio Water/Citronellal	Weight Ratio
		Water	Citronellal		
152.2	60	149.5	2.66	56.2	6.6
238.5	70	233.8	4.70	49.8	5.9
263.7	80	355.5	8.20	43.3	5.1
540.0	90	526.0	14.00	37.6	4.4
782.5	100	760.0	22.50	33.8	3.9
1109.1	110	1075.0	34.10	31.5	3.7

These data demonstrate that operation at reduced pressure results in a lower operating temperature, but also requires the use of more steam per weight unit of citronellal recovered. Operation at elevated pressure (use of high-pressure steam in the still), on the other hand, permits a considerable saving in the amount of steam required per weight unit of oil, but also involves a higher operating temperature. Provided that the higher temperature does not damage the oil, there is evidently some advantage to be gained by the use of high-pressure steam. Details will be discussed in the second part of the chapter on distillation.

Up to this point our discussion has dealt entirely with the use of saturated steam. It is also possible—indeed, in some cases advantageous—to distill essential oils by using superheated steam. Pressure and temperature of superheated steam are no longer mutually dependent. Thus, it is feasible to use superheated steam at a fixed pressure and at any desired temperature above the boiling point at that pressure. The temperature at which such a distillation is carried out can thus be raised without increasing the concentration (partial pressure) of the steam. Since the temperature alone determines the vapor pressure, and consequently the partial pressure of the volatile oil, distillation with superheated steam results in a lower ratio of water to oil, accomplishing a further saving in the amount of steam used. In the above cited case of water and citronellal mixtures the steam would normally be saturated at 90°. If superheated to 100° at a pressure of 526 mm., and then used in the distillation, the molal ratio of water to citronellal is reduced to 23.3 (weight ratio = 1.72), the total operating pressure then being 548.5 mm. By increasing the pressure of the superheated steam any ratio between this and 33.8 (corresponding to the use of saturated steam at 100°) can be obtained.

Two features affecting the use of superheated steam should be pointed out. First, in order to obtain the above cited advantage of superheated steam the still must be completely free of water. When superheated steam comes into contact with water it immediately vaporizes some of the water, being itself cooled in the process and being reconverted into saturated steam. If the quantity of water present is small, it will be vaporized quickly and the process will continue as with superheated steam after the water has been evaporated. Second, the temperature of superheated steam is independent of the pressure; hence the characteristic safeguard against overheating common with saturated steam operation no longer remains operative. The temperature of the charge will reach that of the superheated steam; therefore, the latter temperature must be controlled carefully in order to avoid damage to the essential oil. Also, since there is no water present in the still, the plant charge tends to dry out during distillation with superheated steam, and the forces of hydrodiffusion can no longer play their part. This causes a slowing down in the rate of recovery of essential oil, and in extreme cases may stop it entirely, long before the recovery is complete; in other words, the yield of essential oil will be subnormal. For all these reasons superheated steam distillation may be undertaken only with caution.

It should be mentioned in this connection that for distillation any hot gas (air, flue gas, etc.) could be used in place of steam but, since these gases are not condensable, the size of the cooler required would be so great as to be impracticable.

Let us now again study the behavior of mixtures of liquids which form a single liquid phase. These considerations apply particularly to the fractionation of essential oils after they have been isolated from the plant material. As has already been pointed out, all liquids have a tendency to change to vapors, the extent of this tendency depending on the temperature at which the liquid is maintained. This tendency to vaporize may be gaged by the vapor pressure of the liquid. In general, the components of the liquid mixture will have different vapor pressures at any particular temperature. When such a mixture is vaporized, the component with the greater vapor pressure (the more volatile component) consequently tends to concentrate in the vapor phase, while the less volatile component will be correspondingly concentrated in the liquid phase. This condition holds for all mixtures of liquids which are soluble in one another, and which do not form constant boiling mixtures. Liquid mixtures which form constant boiling mixtures behave somewhat differently and will not be discussed here. The tendency of the more volatile liquid to concentrate in the vapor phase can be observed very readily by reference to the accompanying Diagram 3.1.

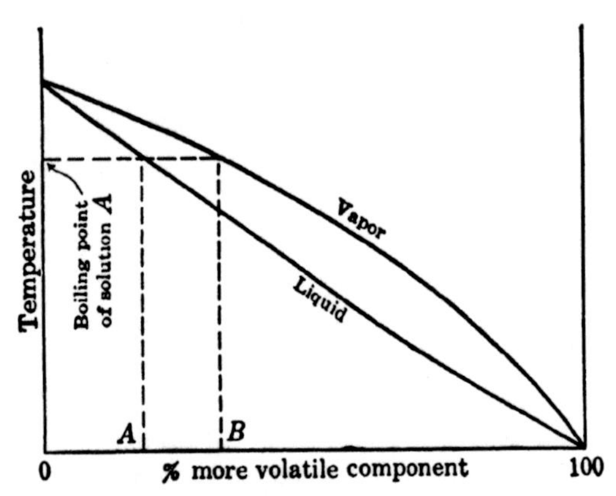

DIAGRAM 3.1. Typical boiling point and vapor-liquid equilibrium diagram for a single-phase binary mixture at constant pressure.

In this diagram the composition of the liquid mixture and its boiling temperature have been plotted. The lower of the two curves represents the relationship between the boiling point of any mixture of these two components and its composition. The upper curve represents the composition of the vapor which is formed from any liquid mixture at its boiling point. Proceeding along a vertical line in the region below the lower curve may be said to correspond to heating a mixture of fixed composition without vaporization. At the temperature corresponding to the point at which this vertical path intersects the lower curve, this particular mixture will begin to vaporize, and the vapors arising first will have a composition represented by the intersection of a horizontal line through the boiling point of this mixture with the upper curve. In the particular case illustrated, a liquid containing *A* per cent of the more volatile constituent would produce an initial vapor containing a percentage of the more volatile constituent represented by point *B*. The vapor produced is thereby enriched with the more volatile constituents. If the distillation is continued without adding liquid to the still, the liquid in the still will become progressively poorer in the more volatile constituents. Furthermore, on condensing and then redistilling the vapor produced, a further enrichment in the more volatile constituents will be achieved. Theoretically, then, it appears possible to obtain a vapor consisting entirely of the more volatile components by a suitable number of redistillations. An effect corresponding to a series of redistillations can be produced in a fractionating column such as that shown in Fig. 3.1.

In this type of system the vapors rising from the still, as always partially enriched with the more volatile component, are essentially condensed and redistilled on the first section above the still. The vapors rising from this section are again condensed and redistilled in the next higher section, this process continuing to the top of the fractionation tower. Such equipment,

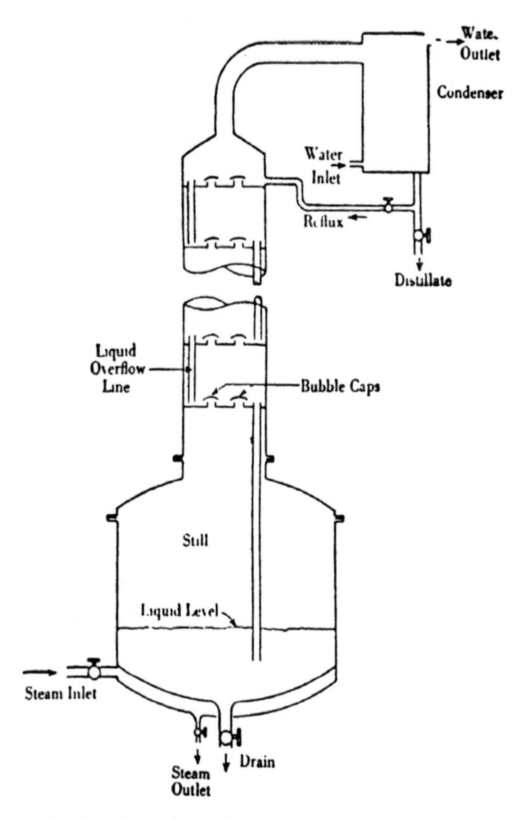

FIG. 3.1. Still with fractionating column. Schematic diagram showing essential parts and typical arrangement.

then, permits obtaining a final distillate which contains a higher percentage of the more volatile components of the mixture than the original material—this, too, in a single piece of equipment. Heat is supplied to such a fractionating system in the still only. On the plates in the tower above the still the heat liberated by condensation of the vapors furnishes in turn the heat necessary to revaporize the material. Of course, the entire system must be insulated thoroughly in order to prevent excessive condensation of vapors due to the heat losses from the tower. In actual operation, such a tower would ordinarily be run by returning part of the condensate at the top to the top plate as reflux. The greater the ratio of reflux to product, the more complete will be the separation of the more volatile from the less volatile components. A system of this kind can be operated at any desired pressure either above or below normal atmospheric pressure. In the final purification of many essential oils (not hydrodistillation), the operation must proceed at very low pressures in order to avoid overheating and consequent destruction of the material. The number of plates required in the fractionation tower is determined largely by two factors:

1. The relative volatility of the components of the mixture.
2. The extent of separation required or desired.

Whenever one component is much more volatile than the other, only a few plates will be necessary to give a high degree of separation, but when the volatilities are more nearly equal, the number of plates must be greatly increased. A rough estimate of the relative volatilities can be drawn from the boiling points at atmospheric pressure of the components of the mixture. There exist quite satisfactory methods for calculating the number of plates required for any particular separation. Details of these methods go beyond the scope of this work and those interested should consult references.[3, 4, 5]

The above considerations show that some separation of the components of a mixture of mutually soluble constituents (such as essential oils) can be achieved simply by vaporizing the mixture and condensing the vapors. Usually, however, this separation will be relatively small, and it will be necessary to resort either to redistillation of the condensate or to the use of fractionating towers as indicated.

In order to consider in more detail the behavior of mixtures of soluble liquids, let us take the case of a mixture of only two constituents. The same principles apply to more complex mixtures, but will be easier to follow in the simpler case. In single-phase mixtures the tendency of either component to vaporize will depend on the temperature of the mixture, and on its composition. In the simplest case, the partial pressure of one constituent will be given by the expression

$$p_1 = P_1 \times N_1 \tag{1}$$

[3] Robinson and Gilleland, "Elements of Fractional Distillation," McGraw-Hill, New York, 1939.

[4] Badger and McCabe, "Elements of Chemical Engineering," McGraw-Hill, New York, 1936.

[5] Walker, Lewis, McAdams and Gilleland, "Principles of Chemical Engineering," McGraw-Hill, New York, 1937.

in which p_1 = partial pressure of constituent 1;
P_1 = vapor pressure of pure constituent 1 at the temperature of the liquid;
N_1 = mol fraction of constituent 1.

$$N_1 = \frac{\frac{w_1}{M_1}}{\frac{w_1}{M_1} + \frac{w_2}{M_2}} \tag{2}$$

w_1 = weight of constituent 1 in mixture;
w_2 = weight of constituent 2 in mixture;
M_1 = molecular weight of constituent 1;
M_2 = molecular weight of constituent 2.

The relationship between the partial pressures of the constituents, total pressure of the mixture (which is equal to the sum of the partial pressures) and the composition of the mixture for a fixed temperature is shown in Diagram 3.2. Systems which follow this rule are ideal systems and are said to obey Raoult's law (Equation (1) above).

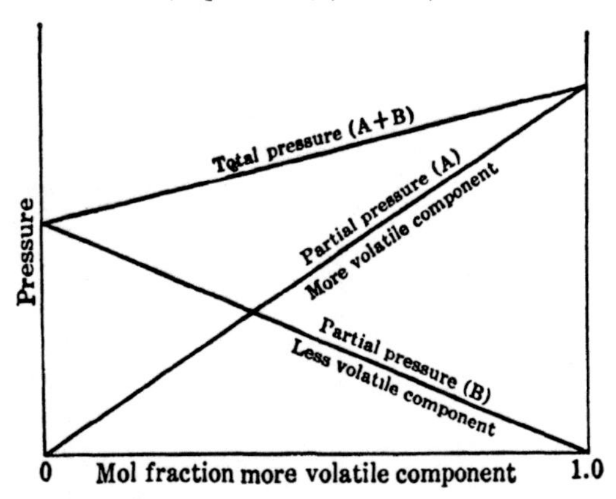

DIAGRAM 3.2. Partial and total pressure curves at constant temperatures for a single-phase binary mixture obeying Raoult's law.

In the more general case, the relationships between these variables are not as simple and can be determined only by experimental methods. A typical case is shown in Diagram 3.3.

Since most distillations are conducted at constant pressure rather than at constant temperature, and since the boiling point of a mixture at a fixed pressure varies with the composition, a somewhat more useful diagram for purposes of analyzing distillation problems is shown in Diagram 3.4.

This diagram represents the composition of the vapor corresponding to the composition of the equilibrium liquid mixture at a constant total pres-

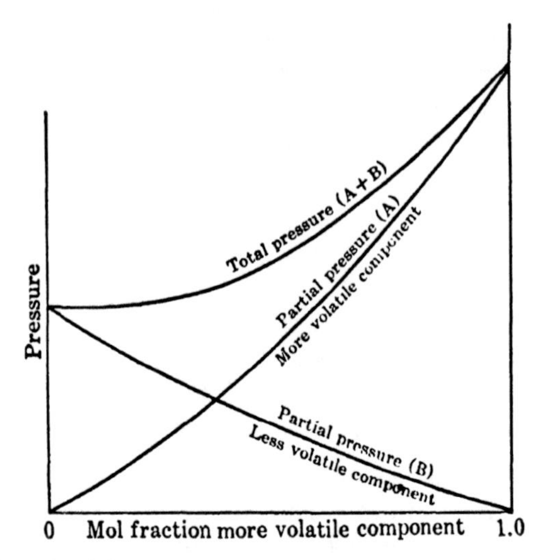

DIAGRAM 3.3. Partial and total pressure curves at constant temperature for a single-phase binary mixture showing one type of deviation from Raoult's law.

sure. Both compositions are expressed in terms of the percentage of the more volatile constituent, and obviously the vapor is always richer in this component than is the liquid from which it originated. Thus, the vapor in equilibrium with a liquid of composition *A* would have the composition *B*. If this vapor were entirely condensed, the resulting liquid would have this same composition *B* and, if redistilled, would give an equilibrium vapor, further enriched and having composition *C*. This is essentially the process which takes place in a fractionation column. The mechanism which accomplishes separation in this type of equipment is evident. The effects of changing reflux ratio and other variables cannot be discussed here in detail.

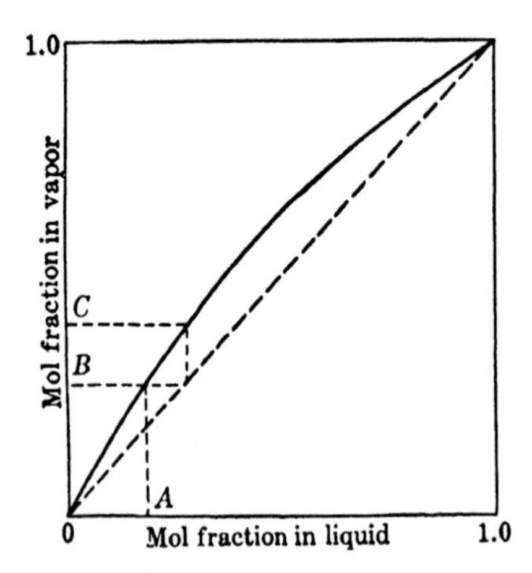

DIAGRAM 3.4. Vapor-liquid equilibrium diagram at constant temperature for a single-phase binary mixture. (All compositions expressed in terms of the more volatile constituent.)

Although a fractionating tower consisting of separate plates as shown in Fig. 3.1 has been used as an example, an equally satisfactory tower for most purposes consists of an open column filled with a suitable packing material. This material can take almost any conceivable shape, but should be characterized by low density (weight per unit volume of the packing), relatively large amount of open space and a large surface area. For example, crushed rock can be used as packing, but because of its high density and low percentage of open space would not be very efficient. Several typical packing materials are shown in Fig. 3.2.

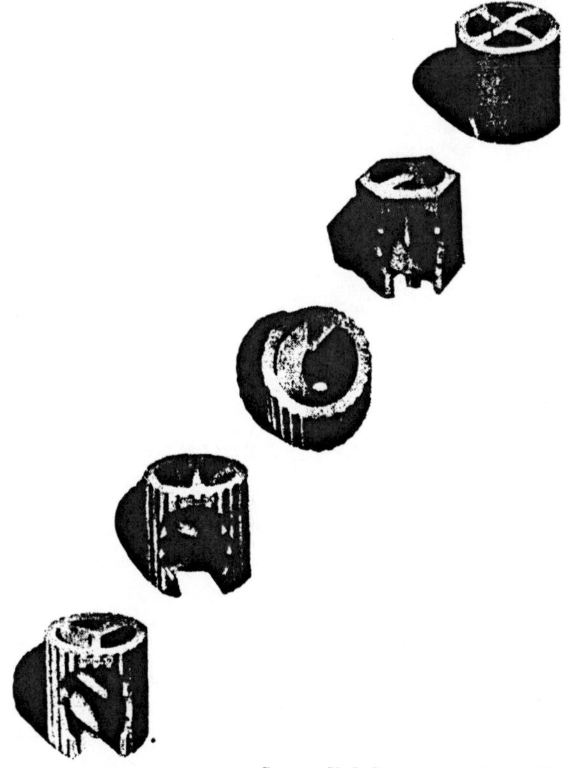

Courtesy U. S. Stoneware Co., Akron, Ohio.

FIG. 3.2. Raschig rings.

In the distillation of single-phase mixtures, it should be kept in mind that changing the pressure in the still has only a minor effect on the overall operation. Since in the distillation of essential oils the principal reason for ever operating at pressures other than atmospheric is to lower the distillation temperature, the pressure will usually vary between atmospheric and some lower pressure, thus limiting the possible variations in pressure. The efficiency of any particular piece of equipment may be changed slightly by operating at different pressures, but the net result will be practically unaffected. This holds true particularly in the case of mixtures such as those encountered in the purification and fractionation of essential oils.

TABLE 3.2. PRESSURE EQUIVALENTS

Lb. per Sq. In.	Kg. per Sq. Cm.	Mm. of Mercury	Atmospheres
1	0.0703	51.7	0.0680
2	0.141	103.4	0.1361
3	0.211	155.1	0.2041
4	0.271	206.8	0.2722
5	0.352	258.5	0.3402
10	0.703	517.0	0.6804
14.7	1.033	760.0	1.000
20	1.407	1034	1.361
25	1.756	1293	1.701
30	2.109	1551	2.041
35	2.461	1810	2.381
40	2.812	2068	2.722
45	3.164	2327	3.062
50	3.515	2585	3.402
60	4.218	3102	4.082
70	4.921	3619	4.763
80	5.623	4136	5.443
90	6.327	4653	6.124
100	7.027	5170	6.804

II. PRACTICE OF DISTILLATION

Having indicated briefly the general principles of distillation of homogeneous and heterogeneous systems, we shall devote the second part of this chapter to a discussion of the practical distillation problems and techniques peculiar to the essential oil industry.

(a) Treatment of the Plant Material.

Comminution of the Plant Material.—The chief application of distillation is in the initial isolation of essential oils from the aromatic plant material. This process involves the handling of predominantly solid products, and the preparation of the material must, therefore, be carried through carefully if the most efficient and complete recovery of the valuable essential oils is to be assured. The essential oils are enclosed in "oil glands," "veins," "oil sacks," or "glandular hairs" of the aromatic plants. If the plant material were left intact, the oils could be removed (vaporized) by the steam only after they had passed through the plant tissues to an exposed surface. This can be accomplished only by hydrodiffusion, a mechanism which will later be shown to play a very important part in plant distillation. Diffusion is always a slow process, and if the plants or parts of plants were left intact, the rate of recovery of oil would be determined entirely by the rate of diffusion. Consequently, before distillation, the plant material must be disintegrated to some extent. This disintegration process, commonly termed comminution, results in exposing directly as many oil glands as is practically possible. It always reduces the thickness of material through which diffusion must occur, greatly increasing the rate or speed of vaporization and distillation of the essential oils. Even in comminuted plant material, only a portion of the oil is freed, the balance remaining enclosed or

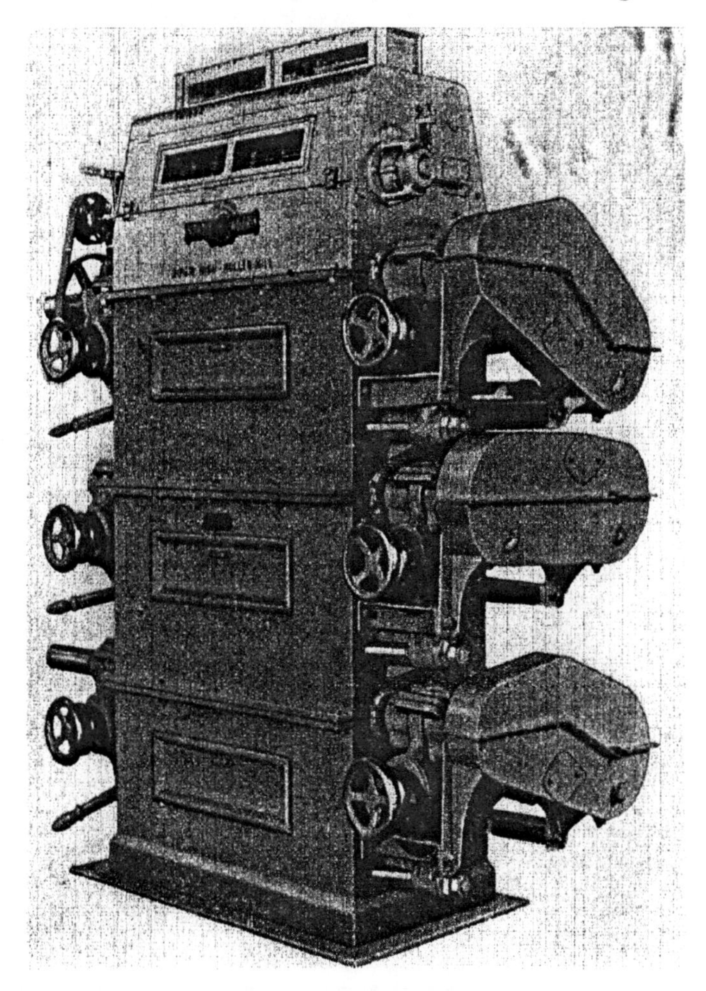

Courtesy of Sprout, Waldron & Co., Muncy, Pa.

FIG. 3.3. 3-Pair high roller mill.

being tightly covered by comminuted plant particles. All actually exposed volatile oil will soon be entrained by passing steam and carried away from the plants.

The extent of comminution required varies with the nature of the plant material. Flowers, leaves and other thin and nonfibrous parts of the plant can be distilled without comminution. The cell walls in these parts are in most cases sufficiently thin and permeable to permit rapid removal of the oil. Seeds (fruits), on the other hand, must be thoroughly crushed in order to rupture as many of the cell walls as possible, to render the oil easily accessible to the passing steam. Roots, stalks and all woody material should be cut into short lengths in order to expose a great number of oil glands.

Seeds can best be crushed by passing them through smooth rolls. These rolls should be arranged so that the distance between them can be varied. The width of this space will determine the extent of crushing obtained. A similar effect can be achieved by regulating the flow of the material upon the distributor above the rolls. If the rolls operate at different speeds, the crushing action is supplemented by a usually advantageous shearing action. Each roll should also be equipped with a scraping device, called a "doctor blade," which serves to keep it free of adhering crushed material. A typical piece of equipment for handling seeds and fruit is shown in Fig. 3.3.

Roots and stalks can best be handled in a hay or ensilage cutter, or similar device. This action simply reduces the long natural parts of the plant to short lengths which are more readily handled in the distillation proper and, above all, assures a more uniform and compact charge in the still. Otherwise the live steam would find ready passages through the wide interspaces of uncut roots or stalk material and escape without coming in close contact with all plant particles. The result, especially in the case of steam distillation, would be a very inferior yield of oil. Woody parts may be sawed into small pieces or chipped mechanically. Typical machines for handling these raw materials are shown in Fig. 3.4.

The principal purpose of comminution being to render the essential oils more readily removable by the passing steam, it is evident that once the plant material has been crushed or reduced in size it must be distilled immediately. Otherwise, the essential oils, being somewhat volatile, will partly evaporate, with two adverse effects: first, the total yield of oil will be reduced by an amount equal to the extent to which evaporation has occurred; second, the composition of the oil will change, thereby affecting its odor. This second effect results from the fact that the essential oils are mixtures of several, often numerous, compounds, the more volatile components evaporating to a greater extent than the higher boiling and less volatile ones. In the case of crushed caraway seed, for example, the evaporation loss consists mainly of limonene, which is lower boiling than carvone; the oil distilled from crushed seed which has been left in contact with open air for some time will, therefore, possess also a somewhat higher specific gravity. The extent of these oil losses by evaporation can be demonstrated easily by crushing a small quantity of caraway seed, weighing it on an analytical balance, airing it for a few minutes and checking the weight. Von

Rechenberg[6] reported a loss of 0.5 per cent which he attributed entirely to evaporation of oil, not of moisture, because air-dried seed was used in the experiment. It is, therefore, imperative that comminution be carried out

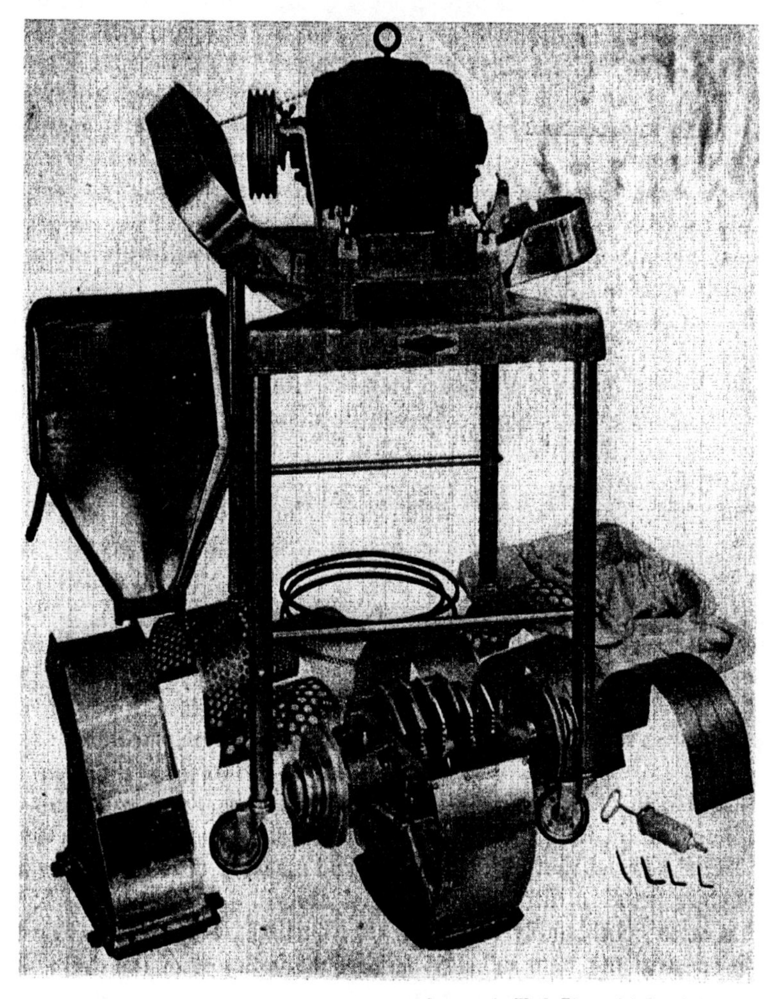

Courtesy the W. J. Fitzpatrick Co., Chicago, Ill.

FIG. 3.4. Stainless, non-corrosive comminuting machine.

immediately before the product is charged into the still if highest yields and best quality oils are to be obtained.

After the plant materials have been properly prepared for distillation, they are packed into the still and distillation can be started. Methods of

[6] "Theorie der Gewinnung und Trennung der ätherischen Öle," Leipzig (1910), 391.

charging and the construction of the still itself will be discussed after the general distillation methods have been presented.

Storage of the Plant Material.—The storage of plant material before comminution also offers some hazard in the way of ultimate loss of volatile oil. The situation here is not quite so serious as in the case of comminuted material and, therefore, if a delay in the distilling of the plant material cannot be avoided, it should be stored in its natural condition. Gradual evaporation results in some loss under these circumstances, the major sources of loss being represented by oxidation and resinification of the essential oils. If the plant material must be stored before processing, it should be kept in a dry atmosphere at a low temperature, and in a room free from air circulation—if possible in an air-conditioned storehouse. All such losses are obviously avoided if the plants are processed immediately.

Loss of Essential Oil in the Plant Material Prior to Distillation.—The volatile oil enclosed in the plant tissue is usually in one way or another affected by the drying of the plant material after the harvest. This effect has been studied and described by von Rechenberg[7] whose findings are so interesting that the author feels justified in quoting a few passages of that work almost verbatim.

Some fresh plants, or parts, with a high water content (e.g., roses, tansy, calamus root) lose much of their essential oil by air drying; others very little. This loss is caused by evaporation, oxidation, resinification and other chemical actions. Contrary to expectation, evaporation here seems to play a subordinate role to oxidation and resinification. Indeed, actual evaporation of the volatile oil through the walls of the plant tissue cannot take place readily because the oil must first be brought to the surface through hydrodiffusion, with water or plant moisture acting as a carrying medium. Thin-walled flowers and leaves present no obstacle to the forces of diffusion, and in most cases evaporation will affect the more water-soluble constituents of a volatile oil rather than the low boiling terpenes. Arrillaga, Colon, Rivera and Jones[8] showed that by field drying and stacking of citronella grass or lemongrass prior to distillation the total acetylizable constituents of the oil decreased considerably with time after cutting. Since losses of acetylizable constituents were sufficient to account for most of the decrease in yield of oil, these authors concluded that the major factor leading to a loss of oil was *oxidation.* Evaporation of whole oil accounted for the additional loss. With both grasses it can be concluded that, for the best results, field drying, with or without subsequent stacking, should not be practiced.

According to von Rechenberg, distillation experiments seldom give reliable data on the loss of volatile oil by evaporation during plant drying.

[7] "Theorie der Gewinnung und Trennung der ätherischen Öle," Leipzig (1910), 279.
[8] *Am. Perfumer* 46 (May 1944), 49.

The reason is simply that distillation of plant material with a high water content usually leaves doubt as to its completeness. Peppermint offers a classical example in this respect. Formerly it was assumed that its oil content increases during the drying of the cut herb, but systematic distillation experiments proved the fallacy of this assumption. Fresh peppermint herb, like most plants or plant parts with a high moisture content, simply cannot be exhausted completely by distillation, or only with great difficulty, and after long hours of distillation. By distilling one portion of peppermint herb in the fresh state right after the harvest, and the other portion in wilted, almost dry ("clover dry") condition, and by calculating the yields upon 100 kg. of fresh herb, it has been shown that the fresh herb contains a little more, possibly much more, oil than the dried herb, but it is very difficult to exhaust the fresh herb completely by distillation.

The loss of oil during the period of wilting and drying of the plant material is much greater than the loss occurring during storage of the plant material after it has been dried. This may be explained by the fact that, during the first stages of wilting and drying, the plant retains a large amount of moisture in the cells, which by diffusion carries the oil to the surface, and aids in its vaporization. Once the moisture has disappeared, and the plant has become air dried, hydrodiffusion can no longer take place. Any loss of oil during storage of the air dried plant material depends upon several factors—condition of the material, method and length of storing, and the chemical composition of the oil. As a rule, but with many exceptions, flowers, leaves and herbs do not endure prolonged storing, whereas seeds, bark, roots and wood, by their very nature, retain their volatile oils much longer. Method of storing (packing tightly in sacks or bales, or spreading on the floor and heaping loosely) plays an important role in this respect. Air currents and extreme variations in moisture content of the atmosphere favor oil evaporation, resinification and, particularly, oxidation. It is possible to keep many types of plant materials for a long period, provided they are stored at sufficiently low temperature and in an air-conditioned room. Under such conditions, caraway seed does not lose volatile oil even over a period of six months. In isolated cases, plant materials—guaiac wood and sandalwood, for example—retain their essential oil for many years, even though exposed to considerable variations of weather.

Von Rechenberg claims that the greatest loss of volatile oil by evaporation and oxidation occurs in comminuting the plant material prior to distillation, especially if this is done in rapidly rotating grinders and mills. The extent of loss depends upon the speed of air circulation in the system, the degree of heat development in the material, and the composition of the volatile oil (its boiling range and resistance to oxidation).

Change in the Physicochemical Properties of Essential Oils During Plant Drying.—Essential oils distilled either from fresh or from dried plant parts show wide variations in physicochemical properties and chemical composition. With many oils it seems advisable, therefore, to state whether they were distilled from fresh, wilted, or air-dried plant material. This is especially true of flowers, leaves, herbs and roots, which in the fresh state contain much moisture.

Peppermint oil, for example, displays marked variations in its properties. Oil from fresh herb, according to von Rechenberg,[9] had a specific gravity of 0.908, that from clover dried herb a gravity of 0.912. (The term "clover dried" means that the stalks are still flexible but the leaves dry.)

A series of interesting distillation experiments were carried out by Schimmel & Company:[10]

Angelica Root Oil

From fresh angelica roots: d^{15}_{15} 0.857 to 0.866
From dried angelica roots: d^{15}_{15} 0.876 to 0.902

The specific gravity of angelica root oil increases in proportion to the length of time the roots have been stored.

Lovage Root Oil

From fresh lovage roots: d^{15}_{15} 1.002 to 1.035
From dried lovage roots: d^{15}_{15} 1.039 to 1.040

Fresh and dried lovage roots exhibit a difference in behavior during distillation. During the distillation of dried lovage root, a yellow, gluey, resinous mass appears together with the oil, especially toward the end of distillation. This mass is largely dissolved in the oil; part of it separates in the condenser pipes, and in the Florentine flask. Fresh lovage roots do not yield this resin, and wilted roots in only a small amount. Oil of lovage from fresh roots, when rectified, is entirely volatile; the oil from dried roots upon rectification leaves in the still considerable quantities of a high boiling residue, which cannot be redistilled with water or steam.

Calamus Root Oil

From fresh calamus roots: d^{15}_{15} 0.962 to 0.968; α_D +20° to +31°
From dried calamus roots: d^{15}_{15} 0.963 to 0.978; α_D +15° to +20°

The oil from the fresh roots is more soluble in 70 per cent alcohol than is the oil from the dried roots. The solubility of the oil decreases with aging (storing) of the root.

[9] "Theorie der Gewinnung und Trennung der ätherischen Öle," Leipzig (1910), 279.
[10] *Ber. Schimmel & Co.*, April (1895), 9.

Estragon Oil

From fresh estragon herb: d_{15}^{15} 0.918 to 0.934; α_D +2° to +4°
From dried estragon herb: d_{15}^{15} 0.890 to 0.923; α_D +5° to +8°

Tschirch[11] reported interesting observations on the resinification of volatile oils in spice plants. Whether the formation of these so-called resins is caused by the polymerization of homogeneous compounds or the addition reactions of heterogeneous compounds, by oxidation, or other forms of conversion of volatile compounds, is not entirely clear.

Natural (not rectified) peppermint oil distilled from fresh herb is more soluble in 70 per cent alcohol than is the oil distilled from dried herb, but the solubility decreases after a few months. If oil from fresh herb is rectified, it resinifies again, whereas oil from dried herb, when rectified, retains its original solubility. Certain constituents of peppermint oil, including possibly menthofurane, seem to resinify during the drying of the herb.

During wilting and drying, the cell membranes gradually break down, and the liquids are free to penetrate from cell to cell, giving rise to the formation of new volatile compounds—e.g., by glycoside splitting. A typical example is found in bitter almond oil, which develops in the course of brief storing of crushed and *moistened* almond or apricot kernels. In the live fruit, the enzyme (emulsin) cannot contact the glucoside (amygdalin) in aqueous solution; but it can readily do so in the crushed and wetted kernels. Analogous reactions and cleavages undoubtedly take place in many other cases. Fresh orris roots, for example, possess a rather disagreeable "green" and "herby" odor; whereas the dried roots, upon aging, develop a faint violet odor. Freshly harvested patchouli leaves are almost odorless; the well-known typical patchouli odor develops only on drying and curing. Vanilla beans constitute another example, the fresh pods resembling to some degree our common garden beans. The odor of grass is very different from that of hay, which develops its typical coumarin note only during the drying process. A phenomenon not yet explained is the disappearance of geraniol in dried roses, while the content of phenylethyl alcohol seems to increase.

(b) General Methods of Distillation.—No investigation has yet been undertaken of the process by which steam actually isolates the essential oil from aromatic plants. It is commonly assumed that the steam penetrates the plant tissue and vaporizes all volatile substances. If this were true, the isolation of oil from plants by hydrodistillation would appear to be a rather simple process, merely requiring a sufficient quantity of steam. However, such is not the case. In fact, hydrodistillation of plants involves several physicochemical processes which will be discussed later.

[11] "Harze und Harzbehälter," 2nd Ed. (1906), Vols. I and II.

There has developed in the essential oil industry a terminology which distinguishes three types of hydrodistillation. These are referred to respectively as:

1. Water distillation;
2. Water *and* steam distillation;
3. Direct steam distillation.

Originally introduced by von Rechenberg, the above terms have become established in the essential oil industry and will, therefore, be retained in our discussion. In order to avoid needless repetition, their significance will be indicated at this point. All three methods are subject to the same

Courtesy of Fritzsche Brothers, Inc., New York.

PLATE 1. A typical old-fashioned lavender still as used years ago by the lavender oi producers in Southern France. Only a few of these stills are being employed today. It is a typical water distillation, the still being heated by a fire beneath.

general theoretical considerations presented in the first part of this chapter which dealt with distillation of two-phase systems. The differences lie mainly in the method of handling the plant material.

Water Distillation.—When this method is employed, the material to be distilled comes in direct contact with boiling water. It may float on the water or be completely immersed, depending upon its specific gravity and the quantity of material handled per charge. The water is boiled by application of heat by any of the usual methods—i.e., direct fire, steam jacket, closed steam coil, or, in a few cases, open or perforated steam coil. The characteristic feature of this method lies in the direct contact it affords between boiling water and plant material. Some plant materials (e.g., powdered almonds, rose petals, and orange blossoms) must be distilled while fully immersed and moving freely in boiling water, because on distillation with injected live steam (direct steam distillation) these materials agglutinate and form large compact lumps, through which the steam cannot penetrate.

Water and Steam Distillation.—When this second common method of distillation is used, the plant material is supported on a perforated grid or screen inserted some distance above the bottom of the still. The lower

Courtesy of Fritzsche Brothers, Inc., New York.

PLATE 2. A field distillery of lavender in Southern France. A typical case of water *and* steam distillation. Many of these stills are in use today in the lavender regions of Southern France. For discharging of the spent plant material the stills can be tilted.

part of the still is filled with water, to a level somewhat below this grid. The water may be heated by any of the methods previously mentioned. Saturated, in this case, wet, steam of low pressure rises through the plant material. The typical features of this method are: first, that the steam is always fully saturated, wet and never superheated; second, that the plant material is in contact with steam only, and not with boiling water.

Steam Distillation.—The third method, known as *steam distillation* or *direct steam distillation*, resembles the preceding one except that no water is kept in the bottom of the still. Live steam, saturated or superheated, and frequently at pressures higher than atmospheric, is introduced through open or perforated steam coils below the charge, and proceeds upward through the charge above the supporting grid.

In so far as the distillation process itself is concerned, and from the purely theoretical point of view, there should be no fundamental difference between these three methods. There exist, however, certain variations in practice, and in the practical results obtained, which in some cases are considerable; they depend on the method employed, because of certain reactions which occur during distillation.

Courtesy of Fritzsche Brothers, Inc., New York.

PLATE 3. Field distillation of lavender flowers in Southern France. The steam is generated in a separate steam boiler.

The principal effects accompanying hydrodistillation are:

1. Diffusion of essential oils and hot water through the plant membranes, whence the term hydrodiffusion;
2. Hydrolysis of certain components of the essential oils;
3. Decomposition occasioned by heat.

These effects will be considered in order.

The Effects of Hydrodiffusion in Plant Distillation.—Even after the plant material has been carefully prepared by proper comminution, only part of the essential oil is present on the surfaces of the material and immediately available for vaporization by steam. The remainder of the oil arrives at the surface only after diffusing through at least a thin layer of plant tissue.

The term diffusion, as used in this connection, implies the mutual penetration of different substances until an equilibrium is established within the system. Such diffusion is caused by the live force of molecules. Where two substances are not separated by a wall (diaphragm), the term "free diffusion" is applied, whereas diffusion through a permeable membrane is called osmosis. The diaphragm may be permeable by only one substance, or by all.

The distillation of plant material is connected with processes of diffusion, and principally of osmosis. In the steam distillation of plant material the steam does not actually penetrate the dry cell membranes. This can easily be proved by distilling plants with superheated (dry) steam. The plant charge, in this case, finally dries out completely, and yields the retained volatile oil only when saturated (moist) steam is applied, after superheated (dry) steam no longer vaporizes the oil. Thus, dry plant material can be exhausted with dry steam only when all of the volatile oil has first been freed from the oil bearing cells by previous very thorough comminution of the plants.

Entirely different conditions obtain if the plant tissue is soaked with water. The exchange of vapors within the tissue of living plants is based primarily upon their permeability while in swollen condition. Microscopic studies have led some to believe that the walls of normal plant cells are almost impermeable for volatile oils. According to von Rechenberg, only limited osmosis of volatile oil can take place at ordinary temperatures. This may easily be proved by soaking uncomminuted dried spices (such as cinnamon or cloves) in cold water for a day or two, then pouring off and distilling the water. The yield of oil, if any, will be negligible, all the oil being retained within the plant tissue. If, on the other hand, the spices (or other plant material) are first sufficiently powdered so that the cell walls are broken and the oil liberated, the water poured off contains considerable quantities of essential oil.

Distillation offers better conditions for the osmosis of oil, because the higher temperature and the movement of water, caused by temperature and pressure fluctuations within the still, accelerate the forces of diffusion to such a point that all the volatile oil contained within the plant tissue can be collected. The effect of a higher temperature may easily be demonstrated by repeating the above described experiments, but by soaking the spices in hot, instead of cold water. The hot water will extract much larger quantities of oil.

Von Rechenberg describes the process of hydrodiffusion, in the case of plant distillation, as follows: At the temperature of boiling water a part of the volatile oil dissolves in the water present within the glands. This oil-in-water solution permeates, by osmosis, through the swollen membranes, and finally reaches the outer surface, where the oil is vaporized by passing steam. Replacing this vaporized oil, additional quantities of oil go into solution and, as such, permeate the cell membranes while water enters. This process continues until all volatile substances are diffused from the oil glands and are vaporized by the passing steam.

The speed of oil vaporization in hydrodistillation of plant material is influenced not so much by the volatility of the oil components (or in other words by the differential in their boiling points), as by their degree of solubility in water. If von Rechenberg's assumption is correct, the higher boiling, but more water-soluble, constituents of an oil enclosed within the plant tissue should distill before the lower boiling, but less water-soluble, constituents. That this actually takes place can be demonstrated by steam distilling comminuted and uncomminuted caraway seed. Uncomminuted (whole) caraway seed will first yield the higher boiling, but more water-soluble, carvone and only later the lower boiling, but less water-soluble, limonene. With crushed seed the opposite is true: the first fraction consists of limonene, the following of carvone. The fact that occasionally the final fraction may contain some limonene only goes to show that, as a result of incomplete comminution, the forces of hydrodiffusion come into play anew. Distillation of uncrushed caraway seed requires almost twice as much time as that of crushed. This well-known fact applies to distillation of all seed material. The explanation is simply that hydrodiffusion acts only slowly, and requires time: in the distillation of uncrushed seeds, all volatile oil enclosed within the plant tissue must first be brought to the surface of the seeds by hydrodiffusion.

It is a well known fact, borne out by experience, that comminution (crushing) of seed material increases the yield of oil. This, however, does not imply that uncomminuted plant material always gives a very low oil yield. Von Rechenberg[12] soaked whole (uncrushed) caraway seed in tepid water until it became swollen, and distilled it with direct, saturated steam at pressure of 5 atmospheres in a well-insulated still. He thus obtained a very slightly lower yield of oil than by distilling crushed caraway seed. This small loss consisted exclusively of carvone, which had been resinified during the longer hours of distillation required for uncrushed, thoroughly wetted seed. Such soaking, steeping, or macerating of plant material was frequently resorted to in the old days of small-scale distillation, when saturated steam of high pressure, generated in a separate steam boiler, was not yet available. In fact, steeping in water as a preliminary process should not be condemned in the case of seed material containing relatively low boiling volatile oils—caraway, fennel, coriander seed, for example. Ob-

[12] "Theorie der Gewinnung und Trennung der ätherischen Öle," Leipzig (1910), 430.

viously this process requires more steam, fuel, time and equipment, but the oil yield will be about normal, provided distillation has been carefully carried through. It should be borne in mind, however, that saturated steam of low pressure, if not properly employed, may easily result in a thorough wetting of the plant charge, and that this factor becomes much more troublesome with a comminuted charge than with an uncomminuted one. Von Rechenberg performed experiments in point with dill, ajowan and fennel seed, as well as with cloves and clove stems. His results again prove that, in the case of uncomminuted material, the oil constituents vaporize according to the degree of their solubility in water, and not in the sequence of their boiling points: carvone distills before limonene in the case of dill seed; thymol before pinene, dipentene and *p*-cymene in the case of ajowan seed; anethole before fenchone in the case of fennel seed; methyl amyl ketone before eugenol and caryophyllene in the case of cloves; eugenol before caryophyllene in the case of clove stems. In von Rechenberg's experiments the distillation of uncomminuted material required twice as many hours as that of comminuted material, and the yield of oil was slightly, and in some cases considerably, lower.

The presence of some water is distinctly beneficial in that it increases the rate of removal of essential oils by distillation, and it would appear, from this fact alone, that water distillation or water *and* steam distillation should be preferred to steam distillation. However, the maximum temperature that can be obtained with water distillation, and water *and* steam distillation, is limited entirely by the operating pressure in the still, which in ordinary operation equals atmospheric pressure. A complete summary of the advantages and disadvantages of the three methods of distillation will be given after the other factors affecting distillation have been discussed. It should be remembered, too, that all essential oils are soluble in hot water to at least a slight degree; therefore, the amount of water present will determine the extent to which the yield of oil will be decreased as a result of the retention (by water in the still) of oil, or certain constituents of the oil. This factor is of special importance in water distillation, since all of the essential oil must first go through the water solution stage, and the water in the still will always be very nearly saturated with oil, especially with the more water-soluble constituents of an oil—with phenylethyl alcohol for example, in the case of rose distillation. The situation is not quite so serious in the case of water *and* steam distillation because a little of the oil dissolves in the still water only as a result of drainage from the still charge which is mechanically separated from the still water. The extent of this drainage will depend upon the amount of condensation taking place within the plant charge, and especially along the still walls, but it can be kept at a minimum by suitable insulation of the still.

The Effect of Hydrolysis in Plant Distillation.—The second effect accompanying distillation of plant material is hydrolysis. Hydrolysis in our case can be defined as a chemical reaction between water and certain constituents of the essential oils. These natural products consist partly, and in some instances largely, of esters, which are compounds of organic acids and alcohols. In the presence of water, and particularly at elevated temperatures, the esters tend to react with the water to form the parent acids and alcohols. Two characteristic features are important in determining the effect of these reactions during distillation. In the first place, the reactions are not complete in either direction. Starting with the ester and hot water, only a part of the ester will react, so that when equilibrium is reached there will be present in the system esters, water, alcohols and acids. Similarly, if only alcohols and acids had been present at the start, all four constituents would be present when equilibrium is established. The relationship between the concentrations of the various constituents at equilibrium may be written as

$$K = \frac{(\text{alcohol}) \times (\text{acid})}{(\text{ester}) \times (\text{water})}$$

in which K = a constant value at any fixed temperature;

(alcohol) = molal concentration of alcohol at equilibrium;
(acid) = molal concentration of acid at equilibrium;
(ester) = molal concentration of ester at equilibrium;
(water) = molal concentration of water at equilibrium.

Consequently, if the amount of water, and hence its concentration, is large, the amounts of alcohol and acid will also be large and hydrolysis will proceed to a considerable extent. As a result, the yield of essential oil will be correspondingly decreased. This result is one of the principle disadvantages of water distillation, since the amount of water present is always large, and hydrolysis relatively extensive. In the case of water *and* steam distillation, the degree of hydrolysis is much less; it is even less with steam distillation, particularly with slightly superheated (dry) steam.

As second important characteristic of hydrolysis reactions in the disdillation of essential oils, it should be noted that hydrolysis proceeds at a measurable rate. The fact that these reactions are not infinitely rapid means that the extent to which they proceed will depend upon the time of contact between oil and water; this holds particularly true for short periods of contact. This is another obvious disadvantage of water distillation, since the oil and water have a maximum time of contact under the conditions there employed.

The Effect of Heat in Plant Distillation.—The third important effect accompanying distillation is the influence of temperature on essential oils.

The pressure of distillation (atmospheric, excess or reduced) can be selected at will, but the temperature of the steam/vapor mixture rising through the charge in the still varies and fluctuates in the course of the operation. It is lowest at the beginning because the lowest boiling constituents of the volatile substances, freed by comminution of the plant material, vaporize first. As the higher boiling constituents begin to predominate in the vapors, and as the quantity of oil vapors *per se* in the steam/vapor mixture decreases, the temperature gradually rises, until it reaches that of saturated steam at the given pressure. Practically all constituents of essential oils are somewhat unstable at high temperatures. In order to obtain the best quality of oil, it is therefore necessary to insure that during distillation the essential oils (or the plant material) are maintained at low temperature or, at worst, that they be kept at a high temperature for as short a time as possible. So far as operating temperature is concerned, there is really little choice between the three commonly used methods of distillation. In the case of water distillation, or water *and* steam distillation, the temperature is determined entirely by the operating pressure. If the still is open to the atmosphere—the usual procedure—the temperature will be at, or slightly below, 100° C (212° F.). If a valve is inserted between the still and condenser, and if the apparatus is sufficiently strong to withstand the pressure, the still can be operated at pressures above atmospheric, and at temperatures correspondingly above 100°. In the case of steam distillation, the operating temperature will be at, slightly below, or above 100°, even at atmospheric pressure, depending on whether low pressure saturated or superheated steam is used. Any of the methods may be operated at temperatures below 100° by use of suitable pressures below atmospheric.

Conclusions.—Although the three processes of diffusion, hydrolysis and thermal decomposition have been considered independently, it must be remembered that in practice all three occur simultaneously, and hence they will frequently affect one another. This holds particularly true of the effect of temperature. The rate of diffusion usually will be increased by higher temperatures. The solubility of the essential oils in water—an important factor, as indicated above—in most cases also increases with higher temperatures. The same holds true of both the rate and extent of hydrolysis. Since the products of hydrolysis are in general more water soluble, they will also affect the diffusion process. Hence, a complete analysis of the various processes incidental to distillation offers a difficult problem. In general, observance of the following principles leads to the best yields, and to a high quality of essential oil: (1) maintenance of as low a temperature as is feasible, not forgetting, however, that the rate of production will be determined by the temperature; (2) in the case of steam distillation, use of as little water as possible in direct contact with plant material, but keeping in mind that some water should be present in order to promote diffusion; (3) thorough comminution of plant material before distillation, and very careful, uniform packing of the still charge, remembering, however, that in all but water distillation excessive comminution will result in channeling of steam through the mass of plant material, thus reducing efficiency because of poor contact between steam and charge.

A brief résumé of the advantages and disadvantages of the three distillation methods in the light of the above discussion will be helpful, and is presented below.

For small-scale installations, particularly in portable units, water distillation or water *and* steam distillation offers the advantage of simplicity of equipment. The latter method is rapidly superseding water distillation (except in a few special cases) because of the better quality and yield of oil, and higher rate of vaporization, i.e., speedier distillation.

For larger and fixed installations, steam distillation unquestionably offers the most advantages. In such plants the necessary control can be readily installed, and under these conditions the quality, yield and rate of oil are superior. Also, as a result of possibility of temperature control, the method is more adaptable. Plant materials containing either low or high boiling oils can be handled in the same equipment with equal ease. Because of the auxiliary equipment required steam distillation cannot be recommended for all distillation. It is especially impracticable for the small producer in the field. Whenever conditions permit the construction of a suitably located, modern plant to process raw material from a large area, such distillery should be equipped to carry on direct steam distillation.

Before closing the general discussion of the three principal distillation methods, it should be mentioned briefly that each method can be modified by changing the pressure in the still. Accordingly, distillation can be carried out:

(*a*) At reduced pressure;
(*b*) At atmospheric pressure;
(*c*) At excess pressure.

The effect of these variations may be observed in the ratio of distillation (condensed) water to volatile oil.

Any type of distillation carried out below the prevailing atmospheric pressure (usually with the aid of a vacuum pump) falls into class (*a*). Characteristic of distillation at reduced pressure is a low distillation temperature which has its limit only in the temperature of the cooling water and the efficiency of the condenser. The outstanding advantage of this form of distillation consists in the absence of the decomposition products resulting from heat. On the other hand, the vaporization capacity of high

	A. Water Distillation	*B*. Water and Steam Distillation	*C*. Steam Distillation
Type of Still	Simple, low priced, portable stills; easily intalled in the producing regions.	Somewhat more complicated and higher priced than *A*. The smaller type is also movable and may be installed in the field.	If well constructed, usually more solid and durable than *A* and *B*. Possibility of large size for large-scale distillation.
Type of Plant Material	Most advantageous for certain materials, especially when finely powdered; also for flowers which easily lump with direct steam. Not well adapted for materials containing saponifiable, water-soluble or high boiling constituents.	Well suited for herb and leaf material.	Suited for any charge except finely powdered material through which the steam forms channels ("rat holes"). Especially well suited for seed, root and wood materials containing high boiling oils.
Mode of Comminution	Best results with finely powdered materials.	Plant material must be uniformly but not too finely comminuted. Granulation gives best results with seeds and roots.	Similar to *B*.
Mode of Charging	Material must be completely covered by water.	Material must be evenly charged into the still.	Similar to *B*. Proper charging is very important; otherwise the steam channels through the plant material and low yield results.
Diffusion Conditions	Good, if material is properly charged and moves freely in the boiling water.	Good.	Good, if steam is slightly wet. Distillation with superheated steam or high pressure steam dries out the plant material, prevents diffusion, and causes a low yield of oil. Such distillation must, therefore, be followed with wet steam.
Steam Pressure Within the Still	Usually about atmospheric.	Usually about atmospheric.	Can be modified (high or low pressure steam), according to the plant material.

	A. Water Distillation	*B.* Water and Steam Distillation	*C.* Steam Distillation
Temperature Within the Still	About 100°. Care must be exercised not to "burn" the plant material by contact with overheated still walls. Vaporized water must be continuously replaced.	About 100°.	Can be modified (saturated or superheated steam), according to the plant material.
Hydrolysis of Oil Constituents	Conditions usually unfavorable. High rate of ester hydrolysis.	Hydrolysis fairly low, provided no excessive wetting of the plant charge by prolonged distillation and steam condensation within the still takes place.	Conditions good, hydrolysis usually slight.
Conditions Within the Plant Charge	Good, if plant material is kept covered with water and moves freely in it.	Good, if material is properly comminuted and charged. Prolonged distillation causes excessive wetting by steam condensation and lumping of the charge. Stills should be well insulated.	Conditions good, if plant material is properly charged. Prolonged distillation with wet steam causes excessive steam condensation within the still and lumping of the charge.
Rate of Distillation	Relatively low.	Fairly good.	High.
Yield of Oil	In most cases relatively low, due to hydrolysis, also because water-soluble and high boiling oil constituents are retained by residual water in the still.	Good, if no excessive wetting and lumping of the plant charge occurs. This would prevent steam from penetrating the charge thoroughly and result in abnormally low oil yield.	Good, if plant material is properly comminuted, evenly charged, and distillation properly conducted. Lumping of the charge or steam channeling might cause an abnormally low yield of oil.
Quality of Oil	Depends upon careful operation; "burning" of plant charge must be avoided, especially when distilling with direct fire.	Usually good.	Good, if operation properly conducted all around.
Distillation Water	Distillation water in some cases must be redistilled, or more conveniently returned into the still during distillation (cohobation). Distillation waters contain products of hydrolysis, chiefly.	If properly separated, the distillation water can be discarded in many cases.	Similar to *B.*

boiling substances, especially of those somewhat soluble in water, is considerably reduced.

By inserting a valve into the gooseneck of the retort and by partly closing this valve during distillation, it is possible to throttle the outflow of the steam/oil vapors and to increase the pressure within the still.[13] Such distillation at excess pressure (0.5 to 1.0 atmospheres excess pressure, or 1.5 to 2.0 atmospheres absolute pressure) is occasionally resorted to in the essential oil industry, but its use remains very limited, because of the resulting decomposition of many oil constituents.

(c) **Equipment for Distillation of Aromatic Plants.**—The equipment required for carrying on distillation of plant materials depends upon the size of the operation and the type of distillation to be used. There are, however, three main parts which, in varying size, form the base for all three types of hydrodistillation. A fourth part is necessary for any method of heating the still other than by direct fire. The three universally employed parts are:

1. The retort, or still proper;
2. The condenser;
3. The receiver for the condensate.

The fourth part consists of a boiler for generating steam. The latter is necessary for the process which, in the preceding discussion, we have called steam distillation, since direct live steam, often slightly superheated, is required, and this can be produced only in a separate steam boiler. In the case of water distillation, or water *and* steam distillation, the still may be heated by direct fire but even here heating is frequently, and indeed preferably, accomplished by steam jacketing the retort, or by means of closed (or occasionally open) steam coils. A separate steam boiler becomes indispensable, also, if any one of the latter heating methods, or a combination of them, is used. These four parts of the distillation equipment will be considered in order.

The Retort.—The retort, or still proper, commonly also called "tank," serves primarily as a container for the plant material, and as a vessel in which the water and/or steam contacts the plant material and vaporizes its essential oil. In its simplest form the retort may consist merely of a cylindrical container or tank, with a diameter equal to or slightly less than its height, and equipped with a removable cover which can be clamped upon the cylindrical section. On or near the top of the cylindrical section a pipe (gooseneck) is attached for leading the vapors to the condenser. For water

[13] For the sake of clarity, it should be mentioned that the injection of high pressure steam *per se* does not, to any marked degree, increase the pressure in a still, which, through its gooseneck and condenser, has a free outlet to the atmosphere. Unless throttled by a valve, or by too narrow a gooseneck, or by the heavy mass of tightly packed plant material, any excess pressure of injected steam is reduced almost at once to the atmospheric pressure.

distillation this simple equipment is sufficient, since water and charge can be introduced, the cover put in place, and a fire simply built under the retort. For water *and* steam distillation a grid or false bottom is inserted sufficiently far above the real bottom of the still so that boiling water and plant material (the latter supported by the grid) do not come in contact. The water is brought to a boil either by a steam jacket or through a closed steam coil, or in simpler apparatus by a fire directly beneath the still. In the case of direct steam distillation the grid may be closer to the real bottom. Here live steam is introduced through a steam line, usually a perforated coil or cross below the false bottom. Such a simple retort, while entirely adequate, would be inconvenient to use because of the difficulty of removing the spent plant material. Fig. 3.5 shows a drawing of this type of retort.

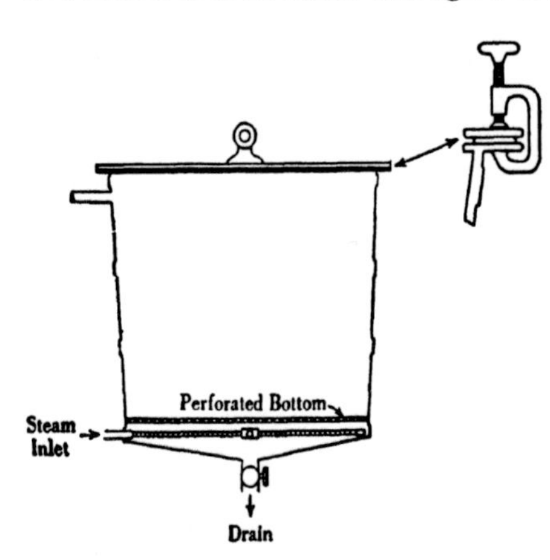

FIG. 3.5. Galvanized iron retort for steam distillation.

The cylindrical section, slightly tapered to facilitate application of the support rings, can be made of 14–20 gage galvanized sheet metal depending on the size of the retort. For larger sizes, heavier metal (smaller gage number) should be used. This is bent to shape, soldered steam tight, and the circular bottom soldered on. Support rings, their number depending on the size of the retort, are fastened outside and around the cylinder at 2 or 2½ ft. intervals, always with one at the top, and one at the bottom, of the section. Except for the top ring, these may be strap metal or angle iron, and in any case about 2 in. wide. The top support ring should be of 3-in. angle iron, to form a suitable contact surface for the cover. Just below the top support ring a 6- to 8-in. length of pipe is soldered to the side of the retort, to serve as a connection to the condenser. Frequently the gooseneck leads from the center of a convex or spherical top cover to the condenser, but the gooseneck should never be high, as it would then act as a sort of reflux condenser. Any unavoidable vertical section of the gooseneck must be well insulated. This connecting pipe should be at least 4 in. in diameter and, if the rate of distillation is to be very rapid, may be even wider. Finally, just below the grid supporting the plant charge a steam inlet line, a 1-in. pipe, enters through the side of the retort. The distance between the bottom of the retort and the steam pipe must be large enough to permit any water condensing within the retort to accumulate at the bottom without contacting the steam pipe. To insure adequate steam distribution, the steam pipe inside of the retort should be arranged in the form of a coil or of a cross, as shown in Fig. 3.6, with small holes, about $\frac{1}{8}$ in. in diameter, drilled in the top of each arm throughout its length. The total surface of these small holes should not be larger than the orifice of the coil or of the arms on the cross, as otherwise the steam will escape from the first holes without reaching the entire length of the coil or cross. In other words, the steam should be injected into the retort in such a way that it will be evenly distributed on the bottom of the retort and in rising will penetrate the plant charge uniformly. Larger stills are equipped with two steam coils, each with a separate steam valve. Prior to injection, the steam is freed from excess water through a water separator. Years ago, it was customary to equip the steam stills, directly above the bottom, with a closed coil, for heating with indirect steam of high pressure. The idea was to keep the injected direct steam as dry as possible, by heating it through this closed (indirect) steam coil. Such precaution, however, is of doubtful value, because dissolved, nonvolatile extractive matter always drips from the plant charge to the still bottom, and is apt to "burn" and decompose in contact with the very hot indirect steam coil. Vapors of disagreeable odor are thus emitted, which might easily affect the odor of the essential oil in the receiver.

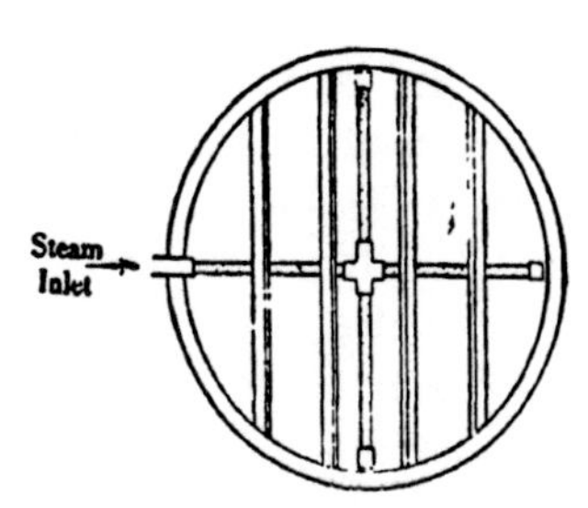

FIG. 3.6. Top view of galvanized iron retort showing crossed tee steam inlet on bottom.

The bottom of the retort is provided with a drain valve sufficiently wide so that any water condensing within the charge and dripping to the bottom can be drawn off in the course of distillation. (This drain valve also serves as an outlet for the wash water, when the still is cleaned.) Otherwise such residual condensed water will accumulate and engulf the steam coil, with the result that the entering live steam will first have to pass through a layer of water, becoming wet in the process. In other words, instead of direct steam distillation we would then have a case of water *and* steam distillation. Wet steam has a tendency to wet the plant charge and agglutinate it. The advantages of direct steam distillation are thus lost. Whether or not this wetting takes place depends also upon the temperature and pressure of the injected live steam. At any rate, it is preferable continuously to remove the condensed residual water from the still bottom. The same result can be achieved by an automatic water separator (steam trap) attached to the still bottom. It is constructed in such a way that only water flows from it, but not steam. This steam trap should be installed in plain view so that its proper functioning is assured at all times. Steam traps are apt to lose their efficiency after a certain time and permit unutilized steam to escape; or they do not separate the condensed water effectively and more and more water accumulates within the still. This means an ever-increasing wetting of the plant charge in the still. Such a condition is recognized by a crackling noise and by trembling of the still. If there is no steam trap, a funnel should be attached beneath the outlet at the still bottom, and the water thus conducted away. Here, too, the water faucet is regulated in such a way that no unused steam escapes, and at the same time, no condensed water accumulates within the still.

This arrangement completes the still proper. Needless to say, all joints must be soldered steam tight, as any steam leak represents loss of essential oil and fuel.

Brief comment should be made about the top of the still and the gooseneck, i.e., the tube connecting the retort with the condenser. The old-fashioned convex or crane-like still heads are becoming obsolete and rare. The top of a modern retort is simply pierced, and a pipe inserted to serve as a gooseneck. The perfect still head is short and well insulated; if convex, it curves gradually and tapers, so that it fits into the gooseneck. Any fancy designs, sudden turns, bends, or too narrow tubing must be avoided, as these would result in a throttling effect and in back pressure within the still.

The gooseneck also is only slightly curved, and, gradually descending, leads from the retort directly into the condenser. It should not ascend, as this would give rise to considerable vapor condensation, the resultant liquid refluxing into the retort. A semicircular gooseneck, such as is sometimes found on old stills, has a purpose only if high boiling and resinous constituents of an essential oil can, by its means, be condensed and returned into the retort. A gooseneck of this type, therefore, may be useful in the rectification of essential oils, but not in the distillation of plant material. In fact, these two operations must never be confused. Ascending and high goosenecks are excusable only if the distillation waters are purposely made to flow automatically back into the retort from the higher placed Florentine flask; but in such a case the gooseneck must be well insulated. It usually is preferable to return the distillation waters into the retort by an injector, which measure makes high goosenecks superfluous. Furthermore, a high gooseneck produces a slight back pressure within the retort; it must, therefore, be amply wide.

The retort cover shown in Figs. 3.6 and 3.7 may be made of sheet metal similar to that used on the retort. It should be strengthened and ringed with strap metal to coincide with the horizontal face of the top angle iron supporting ring on the retort.

Any suitable device for lifting the top may be used. Fig. 3.7 illustrates one such device, which can be attached easily. In order to avoid steam leaks between the retort and cover, they must be held tightly together with a suitable gasket, which may conveniently be a single piece of ½-in. to ⅝-in. soft rope laid all the way around the top angle iron on the retort. Commercial gasket materials (asbestos in rope form) give good service. The top and retort may be held together by external clamps, or by bolts and washers, in which case the top angle iron on the retort and the outer ring of the cover must be suitably drilled with holes to accommodate the bolts. For best results bolts should not be more than one foot apart.

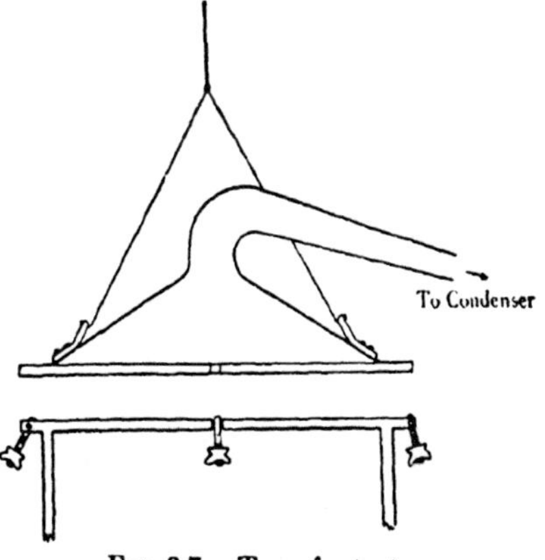

Fig. 3.7. Top of retort.

Another device occasionally used for holding the top to the retort consists of a simple water seal or hydraulic joint (Fig. 3.8). It eliminates all clamps, and saves a good deal of labor, as the still top is easily hoisted in its place after the still has been charged, and can be removed with equal facility. However, the layer of water within the water seal must be sufficiently high to withstand any slight steam pressure developing within the retort; hence a water seal cannot be used when distilling with high-pressure steam. Also, some water evaporates from the seal in the course of the operation, for which reason a water seal is recommended for the distillation of grass or herb material, but not of roots or woods requiring long hours of processing.

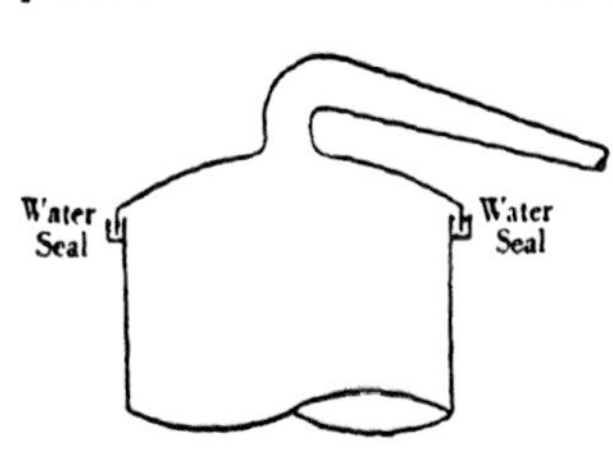

Fig. 3.8. Hydraulic joints or water seals between still top and retort.

The false bottom or grid supporting the plant material may be a circular piece of coarse wire mesh, a tray perforated with many narrow slits, or a wooden platform made in the form of a lattice. In the distillation of seed material—and, especially of crushed seed—it will be necessary to cover the grid with sack cloth, or any other suitable coarse material, to prevent dust and fine particles from falling to the bottom of the retort and clogging the perforations on the steam coil. If the still serves for water *and* steam distillation, the false bottom should be supported about 2 ft. above the bottom of the retort. In the case of direct steam distillation it need be only far enough above the bottom to clear the steam inlet line. Chains or heavy wires attached to three or four equally spaced points around the circumference of the grid may serve as handles so that the plant charge can be easily removed after distillation simply by lifting the grid. If charges in excess of 200 or 300 lb. are to be distilled, it will be convenient to use more than one such section, placing a new one on top of the first layer and continuing the charge above this section. This arrangement prevents excessive packing, assures better steam distribution, and facilitates discharging the spent material, inasmuch as only a fraction of the total charge need be removed at one time. Coarser and specifically lighter material can be packed higher, whereas finer and heavier material should not exceed a certain height.

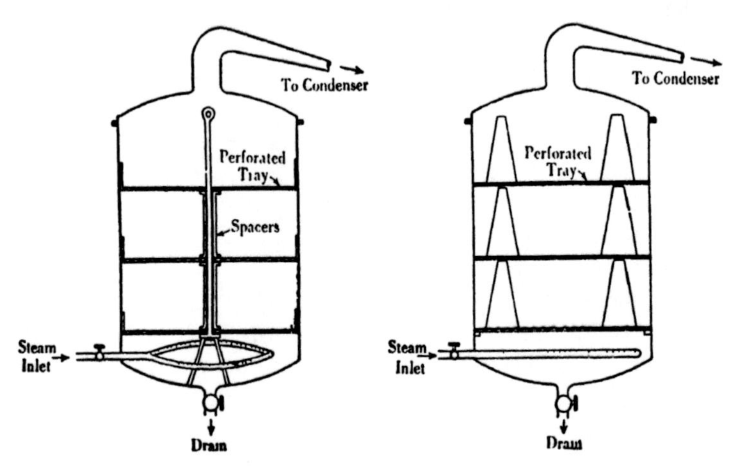

Fig. 3.9. Sketches of two types of multi-tray retorts.

Retorts serving for water distillation should be wider than they are high, so that the plant charge can be kept shallow, avoiding the pressure caused by the weight of a high charge. This will permit the comminuted plant particles to move freely in the boiling water, and assure quicker distillation and a better yield of oil. Retorts serving for water *and* steam distillation may be of approximately equal height and diameter. Retorts for direct steam distillation should be somewhat higher than they are wide so that the rising steam passes as much plant material as possible. As a rule, the diameter should be 6 to 8 ft. at the most; if larger-scale operation requires a larger still capacity it is preferable to increase the height rather than the diameter of the retort. In this case it will be necessary to guard against excessive packing of the charge, which would cause uneven distribution of steam and excessive pressures near the bottom. When calculating the dimensions of a still one should keep in mind not only that some plant materials are very voluminous but also that during distillation the mass often swells and expands by one-third of its original volume. The height of the retort in relation to its width depends upon the porosity of the plant material. A greater height is chosen for voluminous material, and shorter stills are preferred for more compact material. Excessive pressure can be avoided by a construction similar to that shown in Fig. 3.9.

The screen or grid trays may be permanently installed at intervals of 2 to 3, or 3 to 4 ft., according to the size of the retort, and each tray must then be filled or emptied individually through the 2-ft. or 3-ft. manholes. By supporting each section of the charge separately, excessive pressures in any one section are avoided and packing is kept at a minimum. Care must be exercised to fill each tray with only a relatively shallow layer, to insure a uniform distribution of material and, therefore, of the steam. This is particularly true of seed distillation, which requires much more experience and attention than distillation of herbs or leaves.

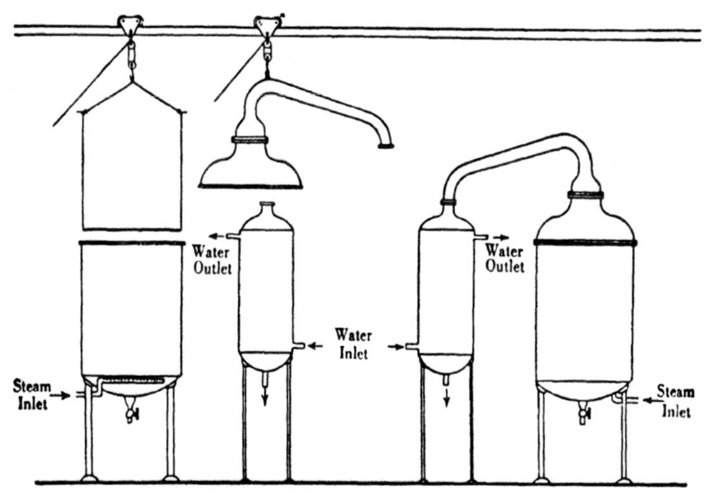

FIG. 3.10. Use of baskets (perforated on bottom) for holding still charge.

As pointed out above, the trays may also be movable, so that they can be lifted from the retort with chains or strong wire. For best results, the trays should not lie directly on top of the charge of the next lower tray but be separated by a space of 2 ft. or more, depending upon the size of the retort. This may be effected in several ways—e.g., by supporting legs, or attaching all of the trays to a central vertical shaft on which the trays may be hoisted from the retort after completion of the operation. The principal precaution is to be sure that the steam actually penetrates the plant charge and does not find an easy passage along the side of the still wall. This may be prevented by coiling ropes around the outer edges of the various trays where they touch the wall of the retort. For the same reason, baskets are not generally to be recommended, particularly those with perforated sides, such as wire baskets. The steam always follows the way of least resistance, and has a tendency to rise along and through the perforations of the wire meshing, between the walls of the basket and the retort.

Baskets with walls of solid (not perforated) sheet metal and perforated bottoms are preferable (Fig. 3.10). However, these should fit quite tightly into the retort proper, leaving only a very small space between the walls of the basket and the walls of the retort. Even this small space, must be completely sealed with rope, so that the steam does not find an easy way from the still bottom to the top by passing outside the basket. If every precaution is taken, such baskets may be useful for the distillation of herb material; while one batch is being distilled, another basket outside of the still can be charged with plants and hoisted into the retort after the first basket has been lifted out. Also, the exhausted contents of such baskets can easily be dumped on a truck and carried away.

In the case of the smaller stills, yet another method of discharging may be found convenient. The entire apparatus may be supported on trunnions located slightly above the middle of the retort. Distillation completed, the retort is disconnected from the condenser and steam line, the top removed, and the entire spent charge dumped out by rotating the retort about the trunnions. Fig. 3.11 shows a typical arrangement.

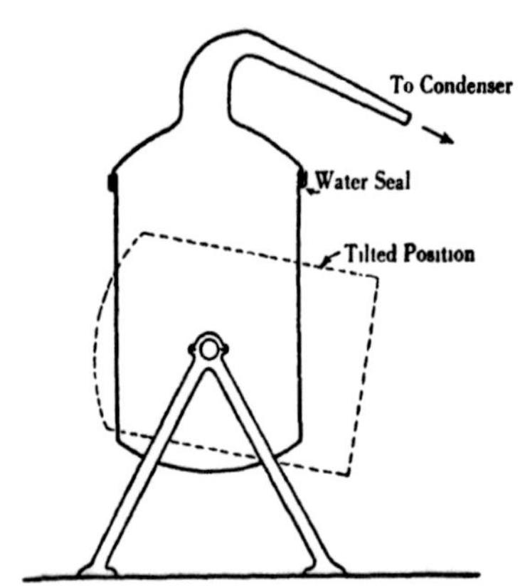

FIG. 3.11. Tilting still on trunions.

In years past most stills serving in our industry were constructed of copper. This metal has the advantage of durability; copper stills retain a certain value even after being dismantled, as the metal can be reworked. The inside of a copper retort, however, should be heavily tinned (lead-free tin!); this is true also (in fact more so) of the gooseneck and condenser. Otherwise the essential oils will contain copper, imparting to the oils a bluish-green color which must be removed before the oil is acceptable to the trade. Sheeted aluminum also can be used for the construction of stills, giving satisfactory results except with phenol-containing oils (phenols attack aluminum). Today most retorts serving for large-scale steam distillation of plant material are made of galvanized sheeted iron, which renders good service for our purposes. Tinned copper is still being favored for equipment used in the water distillation of aromatic plants, and for apparatus employed in the redistillation, rectification or fractionation of essential oils. Some retorts in more primitive countries are made of wood; if solidly constructed they cannot be condemned, but should be used always for distillation of only one type of plant material. Wood has a tendency to absorb a little essential oil, which cannot be removed even by the most thorough washing and boiling with lye. Hence, a certain odor always adheres to wooden retorts which might easily spoil the odor of another type of oil, if the latter were distilled in the same wooden retort.

Insulation of the Retort.—In all cases the retort, including the top, should be well insulated to conserve heat. This holds true particularly of stills exposed to cold air, wind and draft. If insulation is neglected, excessive condensation of steam within the retort will occur as a result of heat losses from its surface. This causes undue wetting of the charge, lumping and agglutinating of the plant particles, excessive steam consumption, prolonged distillation, and, usually, an inferior yield of oil. For small portable units, considerable insulation can be afforded by surrounding the retort with a jacket made of wooden planks and held in place by wire. The interspace may be filled with powdered cork or sawdust. Much better insulators are asbestos and magnesia. Either of these can be applied directly to the retort in the form of a very thick paste in water, which dries to a hard adherent layer. Three to six inches of this material will suffice for most economic operation.

A high grade of insulation of this sort appears particularly important in large installations, where much steam is required. There, all heated sections and steam lines should be well insulated to prevent escape of heat, which represents an unnecessary expense. Probably the most effective insulation material is asbestos, which, in the form of bricks or pipe covering, can be suitably fastened to the still and pipes, or, in the form of powder, can be made into a thick paste with water. This paste may be applied with a trowel to the parts to be insulated. A paste made from ground kieselguhr, water and animal hair, if available, also serves as insulation. In any case, such an insulating layer should be about 2 in. thick. Von Rechenberg[14] suggested the following method of insulating stills and steam pipes:

> "Fifty liters of calcined kieselguhr, ten liters of gritty ground cork waste, and three handfuls of clear pulled pigs' or calves' hair are thoroughly mixed. A thin, hot, stirred soup of rye, wheat, or corn flour is added, to make a viscous, stiff mash. Stones of brick size and strength are then formed and dried on the steam boiler or elsewhere. These bricks serve to cover the stills and steam armatures after they have been covered with a viscous flour soup. If necessary, the bricks are held in place by iron straps. The whole cover is smoothed, and the joints and grooves are filled with a mash of calcined kieselguhr. Finally, cheap, thin fabric is pasted on top and painted over twice with oil paint."

[14] "Theorie der Gewinnung und Trennung der ätherischen Öle," Leipzig (1910), 599.

Very hot steam pipes are more advantageously covered with asbestos fiber. The joints of the steam armatures, which are best made with flanges, should not be insulated.

Charging of the Still.—The problems of charging a retort with plant material, and of discharging it, are more important than is usually realized, and should be attacked by considering the labor involved. Any labor saving device might mean considerable economy in the final calculation. As a rule the plant material should be transported (trucked, hauled, etc.) as near as possible to the still. If the material has to be comminuted, the machines should be located near-by, if possible on a floor or platform above the stills, so that the comminuted material falls or slides by gravity into the retort. The old-fashioned way of charging and discharging with pitchforks and shovels is costly, and, although the initial cost is high, a conveyor belt, or a small crane, will soon pay for itself and in general speed up the operation.

The Condenser.—We shall now proceed to a description of the condenser, the second major part of the distillation equipment. Here again the size and design are variable, and several typical cases will be considered. The condenser serves to convert all of the steam and the accompanying oil vapors into liquid. This requires the removal of an amount of heat equivalent to the heat of vaporization of the vapors plus steam, and a small additional amount of heat to cool the condensed material (condensate) to a convenient temperature below its boiling point. The rate at which heat will be removed from the vapors is expressed by

$$q = UA\Delta t$$

in which: q = heat removed per unit time;
U = a constant depending on operating conditions;
A = the area available for removal of heat;
Δt = the temperature difference between the hot vapors and the cooling medium.

The scope of this work does not permit a full discussion of all factors that affect the value of U. Several of them will be considered in the discussion of condenser operation. Probably the most important ones are the rate of flow of the cooling medium (cold water) past the heating surface, the rate of flow of the vapors, and the material of which the condenser is constructed. The value of U increases as these factors increase, and this fact should always be borne in mind when constructing a condenser. The area available can be made as large or as small as desired, but it is evident from the above relation that the total capacity of a condenser, and therefore of a still, will be directly determined by the area used. The temperature difference can be controlled by the temperature of the cooling medium (hereinafter referred to as water, since water is by far the most commonly used cooling medium) because the temperature of the vapors is fixed within rather narrow limits by the distillation itself. Fig. 3.12 shows the simplest type of condenser, now seldom used, and described here chiefly for its historic interest.

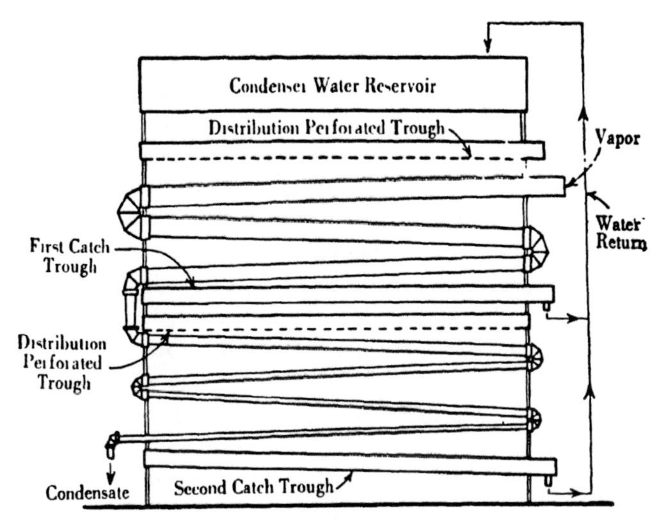

FIG. 3.12. Sketch of an old-fashioned zigzag condenser.

Water is fed to the overhead reservoir from which it flows to a distributor trough which consists simply of a shallow pan with a perforated bottom. This permits the water to trickle over the entire length of the condenser tubes. The water may be caught in an intermediate catch pan, as shown, and a second distributor installed to insure efficient condensation. It will be noted that the condenser tubes are all sloped downward slightly, to insure proper drainage of the condensed oil and steam. Also the size of the condenser tubes becomes smaller as the cold end is approached. In order to avoid excessive back pressures being built up in the still, it is necessary to use fairly large tubes to accommodate the vapors immediately after they leave the retort. Since the volume of the vapors, and, therefore their velocity, decreases rapidly on cooling, as a result of condensation, the size of the condenser pipes can be reduced proportionately. In Fig. 3.12, for example, the first two tubes may be 4-in. pipe, the next two 3-in. and the remainder 2-in. A 4-in. pipe coming from the still will accommodate up to 700 lb. per hr. of condensate (about 85 gal.), in so far as the development of back pressure is concerned. The length and number of tubes to be used will be determined by the amount of vapor to be condensed. An estimate of the pipe area required can be made by using a value of 40 for the factor U in the above equation. The temperature difference will be equal to the average value of the difference between 212° F. (100° C.) (the temperatuer of saturated steam at ordinary pressure) and the temperature of the water in the first and second troughs. For example, if the fresh water in the top trough is 60° F. (15.56° C.) and the water in the first catch trough is 90° F. (32.22° C.), the temperature difference to be used would be the mean of (212 − 60) and (212 − 90) or 137° F. The value of q, the amount of heat to be removed, can be calculated approximately by multiplying the number of pounds of condensate per hour by 1,000. The pipe area required will then be given in square feet. By connecting two or more such zigzag sections in parallel, the same cooling water system can be used for all of them, thus increasing their capacity, conserving height and permitting the use of shorter tubes for a given amount of condensation.

Another very simple and inexpensive type of condenser consists merely of a series of long pipes, usually 2 in. in diameter, laid horizontally in a trough through which water flows. Four 2-in. pipes will have the same vapor capacity as one 4-in pipe as given above, but will offer considerably more cooling surface. Since the value of the factor U in both of these cases is somewhat lower, the length of the pipes must be proportionately greater. Again, the pipes should have a definite slope toward the cool end, to insure adequate drainage of the condensate.

The above described methods of condensing vapors, although cheap and entirely satisfactory, lead to rather awkward and bulky construction.

The most commonly used condenser is that in which coils are inserted into a tank supplied with running cold water, which enters from below and flows against the steam and oil vapors. In order to utilize the cooling water more effectively, it is advisable to insert two adjoining coils into one condenser tank. Fig. 3.13 shows a coil condenser.

In an even more satisfactory condenser arrangement, advantage is taken of the fact that a more rapid flow of cooling water results in more efficient cooling. The condenser tubes are assembled in a single vertical bundle, the number and length depending on the amount of condensation to be accomplished, in such a way that the vapors to be condensed enter the tubes, and cooling water circulates around the tubes. Fig. 3.14 shows a typical construction.

Condensers of this type are available ready built from any equipment supply house, and should be purchased from such a specialist. The construction of a satisfactory leak-proof tubular condenser presents an exceedingly difficult problem for an unskilled workman. The factor U for such a condenser will usually be about 200; thus, for a given amount of condensation and a given cooling water temperature, only one-fifth of the area required in a zigzag condenser will be required. Tubular condensers should be used in a vertical position with vapors entering the top and condensate leaving the bottom. Connection with the retort must again be of adequate size to avoid excessive back pressure in the still. Tubular condensers not only are more efficient and require much less space than spiral condensers, but they also permit easier and more thorough cleaning. If possible they should be fed with soft water to prevent the formation of scale (incrustation), which reduces the exchange of heat, and necessitates frequent cleaning.

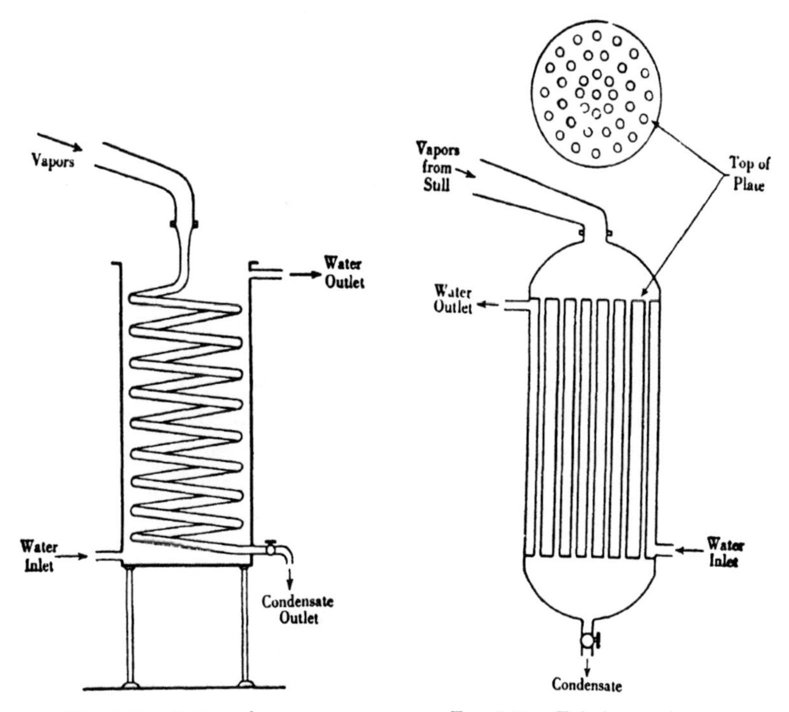

FIG. 3.13. Coil condenser.

FIG. 3.14. Tubular condenser.

It is always better to construct the condenser a little too large rather than too small. Longer tubes or coils require less cooling water, as the contact with the vapors and with the flowing condensate lasts longer and permits the absorption of more heat, so that the temperature of the condensate at the end more closely approaches that of the inflowing cooling water. At any rate, the condenser surface must be large enough to cool the distillate sufficiently, even at a very high rate (speed) of distillation. Slow distillation has many disadvantages, such as hydrolysis of esters, wetting, agglutination and conglomeration of the plant charge, frequently with a concomitantly low yield of oil.

The cooling water in the condenser tank does not need to be cold from top to bottom; such a condition, on the contrary, is rather a disadvantage, because too rapid and excessive cooling of the steam/vapor mixture causes the distillate to run off the condenser unevenly or jerkily. For this reason, the condenser tank should be fed with only as much cold water as is necessary to condense the vapor mixture and to cool the condensate sufficiently—a factor depending also upon the type of oil produced. The maximum efficiency of a condenser is attained when the condensate has been cooled to a sufficiently low temperature by heat transfer to the cooling water, which then flows out at a temperature approaching that of the incoming vapors. This effect, however, is rarely achieved. Usually it suffices if the cooling water flows out at a temperature of 80° C. (about 175° F.) and if the distillate has a temperature of 25° to 30° C. (77° to 86° F.).

If the ratio between condenser surface and heating surface (in the still) is correctly maintained the condenser will permit rapid distillation. But if the condenser surface is too small—and in many of the small field distilleries this is true—the rate of distillation must be adjusted to the efficiency of the condenser. Distillation must then be slow, and this, as pointed out, involves many disadvantages and inadequacies. Otherwise, the vapors blow at high speed through the condenser coils or tubes, which are too short for complete condensation of the vapors or for sufficient cooling of the condensate. Considerable oil may then be lost by evaporation.

The condenser tubes or coils must be made of heavily tinned copper, of pure tin, aluminum, or stainless steel, if discoloration of the oil by iron or copper is to be prevented. Aluminum, however, cannot be used with oils containing phenols.

If distillation is to be carried out at reduced pressure, the tubes or coils must be made strong enough to support a pressure differential of one atmosphere without letting water seep from the condenser tank into the condenser. This is particularly important in the case of oil distillation (rectification, fractionation) *in vacuo*. Condensers serving for distillation at reduced pressures should also be sufficiently wide to permit an unhindered flow of steam and vapors, as any throttling by too small a diameter increases the pressure within the still—in other words, creates back pressure. As a general principle in the construction of distilling equipment it should be kept in mind that the steam and oil vapors should flow easily and smoothly through the system, without encountering any sharp bends or curves in the tubes.

A wire screen, inserted between condenser and gooseneck, prevents plant particles lifted up by live steam from entering the condenser tubes or coils. As the wire screen may become clogged, and would then cause an explosion in the still, the retort should be provided with one or two efficient safety valves.

The Oil Separator.—The third essential part of the distillation equipment consists of the condensate receiver, decanter or oil separator. Its function is to achieve a quick and complete separation of the oil from the condensed water. Since the total volume of water condensed will always be much greater than the quantity of oil, it is necessary to remove the water continuously. The condensate flows from the condenser into the oil separator, where distillation water and volatile oil separate automatically. Many separators are constructed according to the principle of the ancient Florentine flask, hence, are often called Florentine flasks. Volatile oil and water are mutually insoluble; because of the difference in their specific gravities, the two liquids form two separate layers, the usually specifically lighter oil floating on top of the water. Whenever the specific gravity of the oil is greater than 1.0, the oil sinks to the bottom of the separator. The design of the receiver should permit the removal of water whether the oil being distilled is heavier or lighter than water.

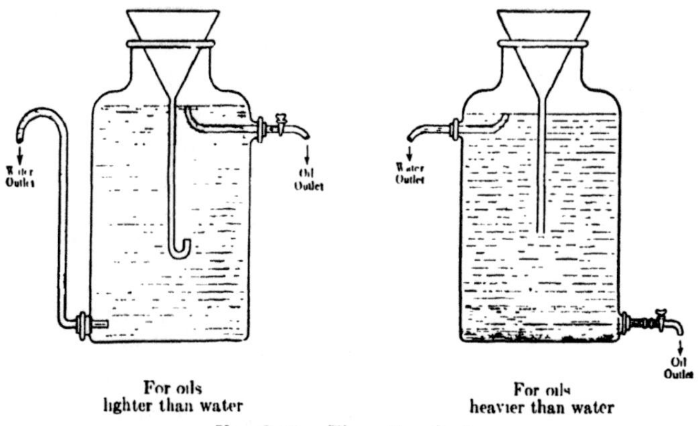

FIG. 3.15. Florentine flasks.

Smaller Florentine flasks are made of glass, larger separators (about 15 liters and more) of metal—usually tin, tinned copper, aluminum or galvanized iron. For all-around use, heavily tinned copper vessels are most practical. Lead must not be employed, as oils containing free fatty acids would form lead salts, which might cause poisoning if the oil were used internally. Rubber tubing or rubber stoppers cannot be used because rubber, being partly soluble in essential oils, gives to them an objectionable odor. Fig. 3.15 shows two oil separators, one for oil lighter than water, and one for oil heavier than water.

Another and quite satisfactory type of receiver operates according to the following principle:

A cylindrical or rectangular vessel is divided into two chambers by a partition which ends a few inches above the bottom of the vessel. The two chambers are connected with one another. The distillate flows into the first chamber, while the distillation water runs off through a tube on the second chamber. Oil lighter than water collects in the upper part of the first chamber and flows out from there, while oil heavier than water sinks to the bottom of the two chambers and is drawn off from there. Fig. 3.16 shows an oil separator of this type.

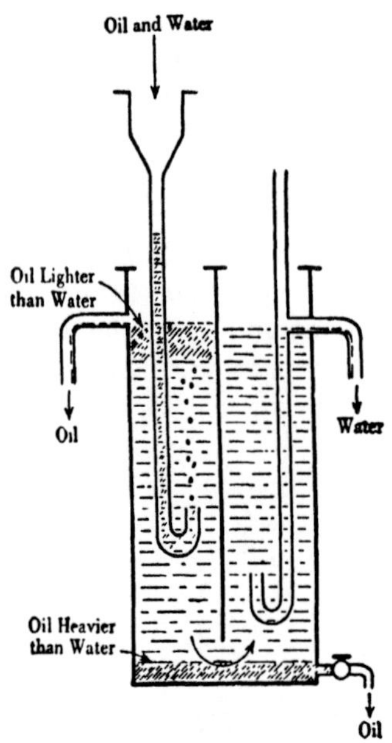

FIG. 3.16. Oil separator for oils lighter and/or heavier than water.

Oil and water often do not separate immediately in the oil separator, especially if the differential between the specific gravities of water and oil is slight. The distillate must not, therefore, flow too rapidly, and any turbulence in the liquids must be avoided; in other words, the separator must be large enough to permit water and oil to separate as completely as possible. Otherwise droplets of oil will be carried away with the outflowing water. A smooth flow of the distillate can be assured by inserting a long-stemmed funnel into the separator, the lower outlet of the funnel being turned upward. The distillate streaming from the condenser thus flows first through the funnel, without disturbing the oil layer, and the oil droplets rise slowly from the orifice of the funnel toward the oil layer, in which they dissolve. If, on the contrary, the distillate is permitted to run from the condenser directly into the oil layer, the distillation water exerts a dispersing effect on the constituents of the oil having a specific gravity close to that of water: a sort of suspension will result. It should be a general rule to separate the oil layer from the water as quickly as possible, and to avoid any agitation of the two media.

If the single oil separator is not large enough, several should be employed, connected serially, usually in the form of step-like cascades, each of the separators being placed a little lower than the one that precedes it, so that oil and water will separate clearly in the last and lowest vessel. Occasionally, distributing bridges are used, in which the distillate flowing from the condenser is distributed into several oil separators.

Some plant materials yield oils which distill over first in fractions lighter than water, and in the later course of distillation, in fractions heavier than water. This is caused by a progressive increase in the specific gravity of the oil fractions. In such cases, two types of oil separators must be employed: the first for separating any oil lighter than water, and the second, connected with the first, for separating any oil heavier than water. The same distillate thus flows through both separators, or the two-chamber separator described previously can be used to advantage.

The oil and water separator shown in Fig. 3.17 offers the advantage of combining several features in a single unit:

The water outlet is placed as far away from the oil layer as is possible.

For oils lighter than water, the oil is drained at A; valves B, C, and D are closed; and water is taken off at E through the automatic drain.

For oils heavier than water, the oil is drained at B; valves A, D, and E are closed; and water is taken off at C through the automatic drain.

For oils separating into two fractions, one heavier and one lighter than water, the lighter oil is taken off at A; the heavier oil is taken off at B; valves C and E are closed; and water is taken off at D through the automatic overflow.

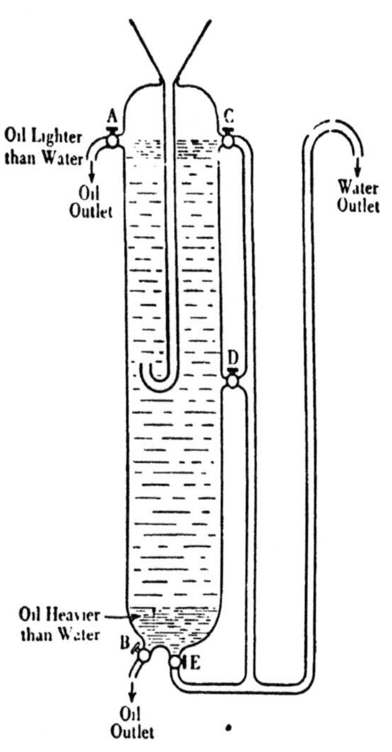

FIG. 3.17. Oil and water separator for oils lighter and/or heavier than water.

Some oils (e.g., rose oil) deposit separate crystals when cooled below a certain temperature. Such oils are liable to clog the condensers if the distillate flows too cold. In this case it is necessary to let the distillate run tepid by reducing the volume of cooling water entering the condenser tank.

Obviously the separation of two insoluble liquids takes place more quickly and more completely the wider the differential between their specific gravities. Therefore, oils or fractions of oils which have a specific gravity only slightly below that of water do not readily separate from the distillation water at room temperature, but form milky suspensions or emulsions. In such cases, the distillate must be forced to run warmer from the condenser into the Florentine flask, because with rising temperature the specific gravity of the oil decreases relatively more than that of the water. The resulting greater differential in the specific gravity between the oil and water at elevated temperature causes the two layers to separate more readily. If, on the other hand, the specific gravity of an oil at room temperature is slightly higher than that of water (oils heavier than water), the distillate should run as cold as possible. Any increase in the temperature, in this case, would further decrease the already small differential between the specific gravity of the oil and that of the water, and separation of the two layers would become even more difficult, if not impossible.

This, however, is the exception. As a general rule, and in the case of most essential oils, the temperature of the condensers should be kept as low as possible in order to prevent evaporation and loss of oil.

The separated oil is finally set aside until suspended water droplets and solid or mucilaginous impurities have separated, when it is filtered clear and stored in well-filled, airtight containers in a cool, dark cellar, or in an air-conditioned room.

It should be remembered that the condensed water will always be saturated with oil. Discarding this water means a loss in yield of oil. In the case of water distillation or water *and* steam distillation this condensed water may be used again as the water supply for the next charge of the same type of plant material, or the distillation water may be returned into the still and redistilled (cohobated) during distillation. For this purpose the oil separator (Florentine flask) must be installed sufficiently high above the still so that the pressure of the flowing distillation water may overcome the slight pressure usually prevailing within the still. In order to avoid excessive height of the gooseneck, the condenser can be set up side by side with the still, the distillation water then being pumped or injected into the still with a steam injector. This procedure prevents loss of oil, since the oil in the water simply means an additional volatile oil charge to the still. It has been suggested that the condensed water be returned to the steam generating equipment (boiler), but this idea cannot be recommended because of the difficulties encountered with the boiler, and also because of the heat in the steam boilers, which would have a deteriorative effect upon the quality of the dissolved oil. In the case of direct steam distillation, the dissolved oil is recovered through redistillation (cohobation) of the distillation water, or through extraction with volatile solvents, both of which will be discussed later in more detail.

Steam Boilers.—Before leaving the subject of equipment, we must make brief mention of the use of auxiliary boilers when water *and* steam distillation, or steam distillation is used. The size of the boiler will depend on the amount of steam required; no generalization can be made. Because of the danger involved in the operation of a steam boiler, it is recommended that such equipment be purchased from an established dealer in power generation equipment. Briefly, besides the usual fire box and tube heater, the system should include gages for determining water level and pressure, safety valves to guard against operation at too high pressure, a pump or injector for circulating the water, and all necessary piping for the particular operation at hand. The supplier should be consulted before ordering any equipment. All reputable suppliers maintain well-trained engineering staffs for the purpose of analyzing customers' requirements, and advantage should be taken of this service.

There are two types of boiler, viz., the so-called low-pressure boiler, developing 40 to 45 lb. of pressure, as measured at the boiler gage, and the high-pressure boiler, which develops a steam pressure of approximately 100 lb. and more. High-pressure steam is used to attain higher temperatures rather than merely to force the steam through the plant material contained in the retort. Theoretically the temperature of saturated steam is a function of the steam pressure. Steam, as developing from boiling water (pressure at the gage = 0), has a temperature of 212° F. (100° C.); at 40 lb. it has a temperature of 287° F. (141.7° C.) and at 100 lb., 338° F. (170° C.). Steam of low pressure and, therefore, of comparatively low temperature, is likely to be recondensed to water in the lower part of the plant charge, whereas steam of higher pressure and temperature penetrates the plant material more effectively and with less condensation in the still. High-pressure boilers are, therefore, more efficient in regard to distillation, shortening its length. On the other hand, it is claimed that low-pressure steam, as a rule, yields more alcohol soluble oils, free of bitter resinous matter.

In actual operation low-pressure boilers produce little pressure but a large volume of steam. They are constructed of appropriate gage sheet metal with cast-iron heads. Even the flues are made of galvanized sheet metal. All of the other boilers are "high pressure." It is true that some distillers use 30 to 100 lb. of pressure, but that depends on the steam requirements. Data collected by experts of Purdue University, Lafayette, Indiana, on retort temperatures in the distillation of peppermint oil, show that there exists little difference between the temperature of the trays at 20 lb. and at 80 lb., but the speed with which the distillation takes place is an important factor economically. The explanation is obvious if one considers that the steam is released into a large retort, not under pressure. There the steam temperature will be reduced to the still temperature immediately without pressure. In some cases, of course, the steam is "pushed" in so fast that a slight back pressure results, but this will seldom cause more than a 10° F. (about 5° C.) rise above 212° F. (100° C.) in the still.

If superheated steam is to be used, a superheater of one form or another must be installed. One method[15] of superheating steam consists of permitting high-pressure, dry saturated steam to expand suddenly to a lower pressure through a well-insulated valve. This will result in a moderate amount of superheating, at least theoretically speaking. A well-designed boiler

[15] For theoretical explanation, see von Rechenberg, "Theorie der Gewinnung und Trennung der atherischen Öle," Leipzig (1910), 400.

should produce very nearly saturated steam and the above method will, therefore, result in slight superheating. If the steam as generated is very wet, it will be necessary to do one of two things in order to accomplish superheating. One method consists of installing in the high-pressure line a water separator, which will remove most of the liquid water from the steam. This dried steam may then be expanded as described above to produce superheated steam. An alternative method is to expose the line carrying wet or saturated steam to a temperature sufficiently above the boiling point of water at the steam pressure to permit the extent of superheating desired. This can be accomplished by running the steam line through a region in which the waste gases from the boiler can transfer part of their heat to the steam. The amount of exposure must be carefully controlled, to avoid excessive superheating. If desired, this heating may also be done in an entirely separate unit, and since the stack gases always contain waste heat, this might just as well be recovered. In the installation of superheating equipment, the boiler supplier can again be of great assistance.

(d) Practical Problems Connected with Essential Oil Distillation.—At this point it is advisable to devote space to a few practical suggestions for the operation of essential oil stills. Many of the points brought out in the following paragraphs have already been mentioned, but it appears desirable to emphasize them, since failure to adhere to them may well represent the difference between successful and unsuccessful operation. The three general methods of conducting steam distillation will be considered in order.

Water Distillation.—Let us first consider the operation of a water distillation system. In every method of plant distillation, whether steam distillation, water *and* steam distillation, or water distillation, only those quantities of the essential oil with which the steam comes in direct contact can be vaporized. Any oil held within the plant tissue must first be extracted from the glands and brought to the surface of the plant by osmosis. But the forces of hydrodiffusion work very slowly whenever the distances to be bridged are relatively long. Water distillation necessitates a thorough comminution of the plant material to the smallest possible size; in other words, the reduction must exceed that required for direct steam distillation or water *and* steam distillation. All interspaces between the plant particles, which in the case of water distillation are filled with water, must be penetrated continuously by rising steam.

The retort is charged with the plant material to be distilled, and sufficient water added to fully cover the entire charge, leaving, however, ample vapor space above the charge to avoid boiling over and carrying over of spray into the condenser. After the cover has been fastened tightly—using a suitable gasket between cover and still to avoid loss of vapors at that point—the retort is connected to the condenser, and the cooling water permitted to flow through the latter. The fire is then started, if a direct fired still is being used, or the steam line opened, if either a steam jacket or steam coils are used for heating. Once the charge has reached its boiling point at the particular pressure used, condensate will begin to issue from the open end of the condenser, and should be run directly into the separator, which was previously filled with water. The rate of distillation can be controlled by the intensity of the fire, by the pressure in the steam jacket, or by the rate of introduction of steam. With direct fire, special care must be taken to avoid overheating the plant material. As the water and oil evaporate, part of the charge will soon cease to be covered with water, and hence no longer will be automatically protected from overheating. It may be advisable to add more water as the distillation proceeds, to prevent any part of the charge from becoming exposed to the full heat of the fire. When a steam jacket or closed steam heating coils are used, there is less danger of overheating unless the water level falls below the top steam coils. Here again the addition of sufficient water will prevent such an undesirable result. With open steam coils this danger is largely avoided, since for every pound of steam injected, a pound of steam condenses as distillation water in the condenser. However, care must be taken to prevent accumulation of condensed water within the retort, or the water level will rise gradually to the top. Therefore, the still should be well insulated and not exposed to draft or cold wind. Furthermore, the water charged into the retort at the beginning of the operation should be hot, as cold water would condense too much of the injected live steam.

The rate of distillation must be adjusted to suit the particular equipment and material being distilled. This rate should, of course, be maintained near the maximum in order to obtain the maximum production of oil. There are other, perhaps less apparent, reasons for maintaining a rapid rate of distillation when using water distillation. Principal among these is the fact that only by rapid distillation can the charge be maintained in a sufficiently loose condition to insure thorough penetration of the plant material by the rising steam. Steam which does not contact the charge, for example steam generated at the water surface as in a slow distillation, cannot carry any essential oil with it, and will be wasted. A lively conduct of distillation prevents to a large extent undesired agglomeration of the plant material, and brings about a more effective contact area between charge and steam. This in turn causes not only an increase in the rate of production, but also a better total yield of oil. It is commonly assumed that during water distillation all parts of the plant charge are kept in motion by boiling water. This, however, is only partly true. Steam bubbles form mainly along the closed steam coils, along the heated bottom and walls of the retort, and rise to the surface by the shortest way, avoiding any obstacles. Provided the distillation material is charged loosely and remains loose in the boiling water, the steam bubbles probably will contact all plant particles quite evenly and vaporize their volatile oil. This is the case especially with woody material, but flowers have a tendency to agglutinate under the influence of steam and form large lumps. True, the volatile oil diffuses quite readily from tender-walled epidermis glands, but when leaves or flowers cling together diffusion is slowed down. Distillation must then be accelerated to a point where all particles of the plant charge are agitated and kept in continuous motion by rising and exploding steam bubbles. The degree of comminution, the weight of the plant charge and the construction of the still should be calculated accordingly. Plant material which contains an essential oil composed of high boiling constituents can be exhausted by water distillation only if comminuted to small particles.

Von Rechenberg pointed out that many years ago distillation was carried out almost exclusively over direct fire, water distillation then being the rule, steam distillation the exception. Experience had been that complete exhaustion of many plant materials could be effected only under great difficulties and after several days of distilling. Extraction of oil of cloves, for instance, seems to have caused a great deal of trouble. Directions dating back to the middle of the last century claim that cloves could be exhausted only by repeated distillation; in other words, the retort had to be opened from time to time, the content stirred, and the evaporated water replaced. In the case of cloves this was repeated from three to eight times. Very probably the plant charge relative to the size of the still was much too large and distillation had to be carried out much too slowly with a small fire; otherwise the cloves in the still would have foamed over into the condenser. Small-scale operators, especially field distillers employing directly fired retorts, still commit the mistake of not putting sufficient water into the retort. Ignorant of the simple rules underlying water distillation, they seem to prefer a slow distillation, or they are handicapped by too small condensers, or by lack of water. Frequently they add to the plant material such a small quantity of water that only the still bottom, which is directly fired, remains covered with water at the end of the operation. This practice is faulty. Plant parts rising above the level of the boiling water in the course of water distillation tend to lump together, to become almost impenetrable for steam, and therefore not to yield their oil completely. For this reason, the retort should be only partly filled with plant material, which should remain fully immersed in water, even when distillation is completed. Only by following this precaution is it possible to exhaust the plant charge by water distillation, as far as this can be done at all.

Water distillation is still quite widely used with portable equipment in primitive countries. There, lack of roads and poor transport facilities prevent hauling of the plant material from outlying growing regions to centrally located distilleries. Therefore, the apparatus must be moved into the growing sections, or in other words, follow the plant material. Small stills, simple, sturdy and low priced, hence retain the favor of many native producers.

Aside from these purely practical conveniences, water distillation possesses one decided advantage. It permits processing of very finely powdered material (root, bark, and wood, etc.) or of plant parts which by contact with direct (live) steam would easily agglutinate and form lumps through which the steam cannot penetrate (e.g., roses or orange blossoms). From such an agglutinated mass, live steam vaporizes the oil only from the outside and not from the inside. Steam distillation, therefore, would remain incomplete. The nascent steam bubbles attack all parts of the plant charge only if the latter moves loosely and freely in boiling water. As a matter of fact, material which readily agglutinates can be processed only by water distillation.

On the other hand, water distillation suffers from several disadvantages. Whether comminuted or not, the plant material cannot always be completely exhausted. Furthermore, certain esters, linalyl acetate, for example, are partly hydrolyzed; other sensitive substances, such as aldehydes, tend to polymerize under the influence of boiling water, etc. Consequently, all other conditions being the same, the quality of product from a rapid distillation will be better, in general, than that of the product from a slow distillation. Water distillation requires a greater number of stills, more space, and more fuel. It demands considerable experience and familiarity with the method and its effect, in fact more experience and care than any other form of plant distillation; otherwise the yield of oil will be affected and fall considerably below that obtained by water *and* steam distillation or by direct steam distillation. Water distillation is the least economical process, water *and* steam distillation giving, in general, better results in the case of field distillation.

Another peculiarity of water distillation lies in the fact that high boiling and somewhat water-soluble oil constituents cannot be completely vaporized from the large quantities of water which must cover the plant charge in the still, or they require so much steam that they can be recovered only partly from the distillation (condensed) water; therefore, the distilled oil will be deficient in regard to these constituents. In other words, distillation remains incomplete. Such compounds are high boiling alcohols (phenylethyl alcohol, cinnamyl alcohol, benzyl alcohol, etc.), phenols (eugenol, etc.), certain nitrogenous substances, and some acids. A typical example is orange blossom oil: the methyl anthranilate present in the flowers cannot be completely recovered by distillation, extraction with volatile solvents giving better results. The case is similar with roses: distilled rose oil lacks the somewhat spicy note of the extracted product (concrete or absolute) and contains much less phenylethyl alcohol, because eugenol and phenylethyl alcohol, as the author[16] proved, remain in the residual still waters. How much of these high boiling, somewhat water-soluble compounds are actually carried over by distillation depends upon their boiling points, their degrees of solubility in water, and the quantity present in the plant material. If the plant charge, despite comminution, contains coarser particles, which during the boiling do not soften and, therefore, are not torn apart, these particles will retain high boiling, water-*in*soluble oil constituents, because diffusion through the greatly swollen tissue layers acts too slowly. These factors explain why essential oils obtained from the same plant material by water distillation or by steam distillation vary considerably in regard to yield, physical properties and chemical composition.

For all these reasons, water distillation is used today in essential oil factories and for large-scale production only in cases where the plant material by its very nature cannot be processed by water *and* steam distillation or, even better, by direct steam distillation.

For most efficient operation, a modern retort serving for water distillation should be flat and wide, thereby offering a large surface of evaporation. The plant material should be filled in evenly, not higher than 4 in. Water is then pumped into the still until it stands about 2 in. above the charge. Steam of at least 3 atmospheres absolute pressure, generated in a separate steam boiler, is injected into the steam jacket beneath the still, so that the water in the still is brought to lively boiling, and each particle of the plant charge thoroughly and continually agitated. The quantity of the plant charge does not necessarily depend upon the size of the still. A somewhat loose charge contains sufficient interspaces to permit an unhindered penetration by the steam bubbles rising from the still bottom; hence the charge can be higher than 4 in. If, in addition, the plant material does not agglutinate or lump while softening under the influence of heat, the charge may be considerably higher. However, complete exhaustion is not always assured; in general, good results in the case of water distillation are obtained only if the charge is sufficiently low to permit the rising steam bubbles to overcome the weight of the plant charge. In other words, the steam should continually agitate the plant particles. In this case it is preferable to work without a perforated grid above the bottom of the retort. If, on the other hand, the charge is high, exercising a marked pressure upon the bottom, the

[16] Guenther and Garnier, "Bulgarian Rose Oil," *Am. Perfumer* 25 (1930), 621.

insertion of a perforated grid is advisable. For certain types of plant materials—e.g., roses, orange blossoms, and ylang ylang flowers—which can be kept floating in the boiling water by lively steam development, much deeper or spherical stills may be employed. Heating coils (instead of a steam jacket) should be avoided in this case because plant particles easily attach themselves to the coils and may give trouble.

Very finely powdered material—such as almond or apricot kernels—has a tendency to "burn" in contact with the hot steam jacket; the water in the still should, therefore, be heated, not by indirect steam, but by direct steam, injected through a steam coil within the still. High, cylindrical stills are better adapted to this purpose than wide, flat ones. In this case, the distillation water is collected separately, and not pumped back into the still during the process, because too much liquid would accumulate in the still by condensation of the injected live steam.

A general rule which applies to all methods of distillation is that each charge should be completed on the same day. The quantity of plant material charged and the rate of distillation must be calculated accordingly. It should be kept in mind that the shorter the distillation, the less the forces causing hydrolysis, decomposition and resinification will come into action. The loss of essential oil arising from these forces may amount to several per cent, as calculated upon the oil. In the case of water distillation, it is not always possible, however, to shorten distillation to a one-day operation.

Fig. 3.18 shows a still for water distillation, with automatic return of the distillation waters.

Water and *Steam Distillation.*—Let us now consider some practical aspects of water *and* steam distillation, a method which in recent years has become quite popular among small producers using portable distillation equipment that can be moved from field to field, following the harvest. The smaller units are heated by direct fire, the larger ones by a steam jacket, a closed steam coil, or in rarer cases by open steam coils. When using direct fire, precaution must be taken to insure that only the bottom of the still, the section containing water below the grid which carries the plant charge, is heated. Otherwise, one of the major advantages of water *and* steam distillation over water distillation, namely, freedom from the danger of overheating the plant material, will be lost. As was stated previously, when this method of distillation is employed the plant charge itself is kept out of contact with boiling water. Hence, if the upper part of the still were exposed to direct fire, the plant material might be dangerously overheated. It is advisable, therefore, to use indirect steam as a source of heat, but not direct fire.

In this type of distillation, observing the precautions mentioned in the last paragraph, steam alone contacts the charge, the steam either being generated from, or passing through, water in the still. Thus overheating or drying of the charge is avoided because the temperature cannot rise above that of saturated steam at the pressure prevailing in the still (at atmospheric pressure never above 100°). Water *and* steam distillation, therefore, represents a typical case of distillation with saturated *low* pressure steam. For

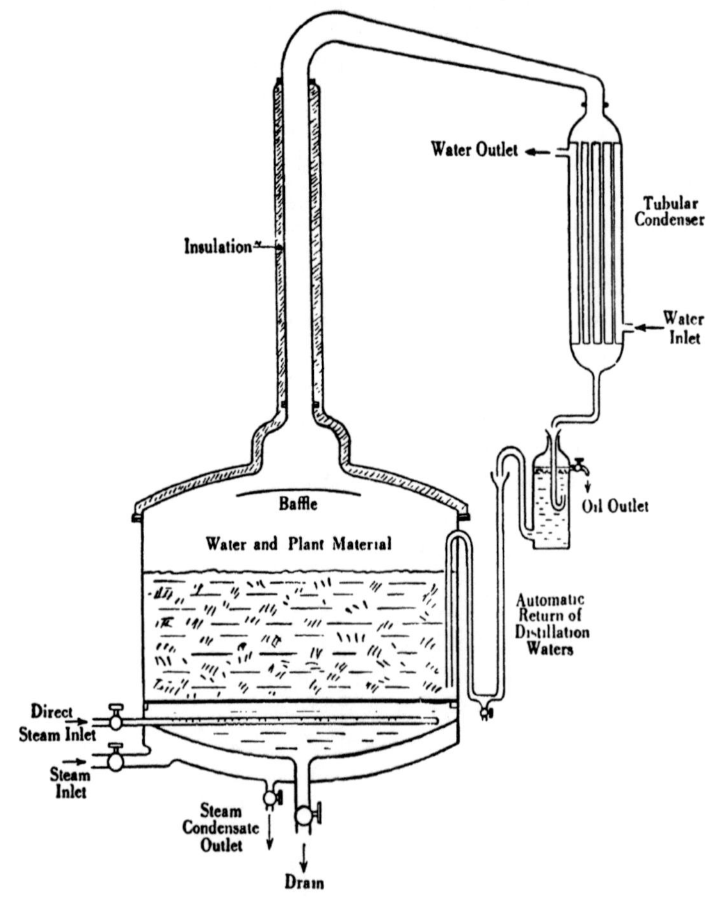

FIG. 3.18. Still for water distillation.

this reason, the condensate contains fewer decomposition products than that obtained by direct steam distillation with live steam, and particularly with high pressure or superheated steam.

Preparation of the plant material is much more important in this method of distillation than in water distillation. Since the steam contacts the material only by rising through it, the plant charge must be so disposed that all parts of it are uniformly contacted, if high yields of oil are to be maintained. This requires that the charge be homogeneous as to size and, furthermore, that the average size of the individual pieces be controlled within rather narrow limits. If, for example, the material is finely ground, it will tend to pack and offer strong resistance to the passage of steam. This in turn may develop steam pressure beneath the charge until such pressure is sufficient to penetrate it. Such penetration, however, will take place at only a few places, releasing the pressure and permitting the steam to escape through only a few passages or channels—sometimes called "rat holes." Obviously, under these circumstances most of the plant material is never contacted by the steam, and the recovery of oil is incomplete. If, on the other hand, a charge consists of, say, whole stalks, leaves and flowers, there obviously will be some fairly large passages through the charge which offer little or no resistance to the passage of steam. Steam will then escape through these and again permit most of the charge to remain unaffected by it. Therefore, in the case of water *and* steam distillation, the plant material should not be too finely ground; nor should it contain excessively long stalks or large roots or pieces of bark. Granulation usually gives the best results. Experience alone can determine the optimum size to which the material should be reduced, and this will vary from plant to plant. At any rate, the preparation of the charge for water *and* steam distillation must always be given most careful attention.

Another problem to be considered in water *and* steam distillation arises from the fact that the charge is cold at the start, and that the first steam to enter it will condense, thus wetting the plant material. This wetting will continue until the entire charge reaches the boiling temperature of water at the operating pressure. With certain types of plant materials—for example, leaves or ground seeds, bark, roots, etc.—excessive wetting may result in lumping or agglomeration of the charge and, therefore, in a subnormal oil yield. Such wetting, again, may cause channeling of the steam. If a charge tends to agglomerate when wet, it is sometimes advisable to add dried twigs or short small pieces of stalk, or any other loose but absolutely neutral material, in limited quantities, so that the charge may be kept porous. To avoid continuation of wetting due to loss of heat by radiation from the walls of the still, the upper part of the retort—in other words, the section housing the charge—should be insulated.

The rate of distillation in the case of water *and* steam distillation is not as important as in the case of water distillation. It affects only the rate of production but not always the quality or yield of oil. A lively pace of distillation recommends itself, however, in order to prevent excessive wetting of the plant charge and in order to increase the production rate. Regarding oil production per hour, water *and* steam distillation is less efficient than steam distillation; it approaches that of water distillation.

Compared with water distillation, water *and* steam distillation has the advantage in that it gives less rise to products of decomposition in the oil (hydrolysis of esters, polymerization, resinification, etc.). As far as portable stills and small stationary posts are concerned, water *and* steam distillation is, in most cases, a better method than water distillation: it requires less fuel, shorter hours, and yields more oil even with a low rate of vaporization. If, however, a plant material—for instance, roses or orange blossoms—forms lumps under the influence of steam, the interspaces disappear and the steam can no longer penetrate the charge and reach every plant particle. In such cases, water distillation must be resorted to.

The great disadvantage of water *and* steam distillation, which limits its adaptation, lies in the fact that, as a result of the low pressure of the rising steam, oils of high boiling range require large quantities of steam for complete vaporization—hence long hours of distillation. In this process much steam condenses in the plant charge, which becomes increasingly wet, agglutinates, and will yield its oil only very slowly.

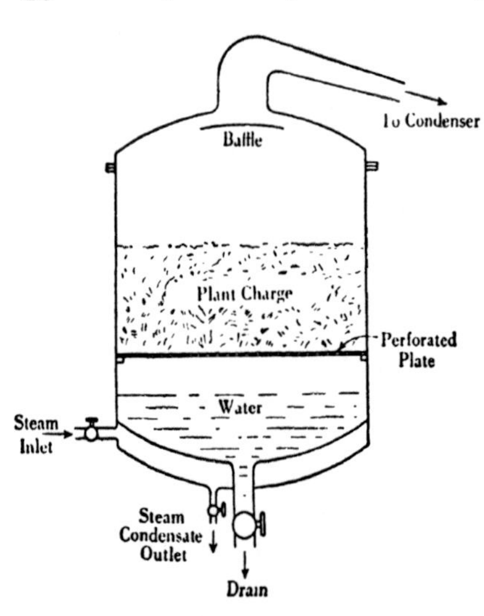

FIG. 3.19. Still for water *and* steam distillation.

As in the case of water distillation, in water *and* steam distillation the condenser can be installed at such a height that the distillation water flows automatically and continuously back into the still. Or the distillation water may be pumped back, or injected into the retort.

After completion of a charge, the water beneath the perforated grid is discarded, and replaced with fresh water. It is not advisable to employ the same water for the next harge because some steam always condenses within the plant charge, and water-soluble extractive matter from the plant charge accumulates in the water beneath the grid. The repeated use and boiling of the same water may cause the extractive plant matter to decompose, and to form volatile products of disagreeable odor, which are liable to impart an objectionable by-note to the volatile oil.

Summarizing, it can be said that water *and* steam distillation must be carried out by observing the following principles: uniform size of the plant particles and sufficiently large interspaces for the rising steam; uniform distribution of the plant material in the retort, so that the charge is penetrated evenly and completely by steam. Although the method of direct steam distillation serves for a variety of plant materials, water *and* steam distillation is suitable only for certain types. It is especially adapted to field distillation in small or medium sized stills.

Water *and* steam distillation can also be carried out under reduced or increased pressure. Indeed, in some cases, reduced pressure gives excellent results.

Fig. 3.19 shows a retort for water *and* steam distillation.

Courtesy of Mr. F. Gutkind, London, England.

PLATE 4. Field distillation of rosemary in Tunis. The stills are directly fired. The plant material is transported by camels to the distillation post.

Steam Distillation.—Live steam, usually of a pressure higher than atmospheric, is generated in a separate steam boiler, and injected into the plant charge within the retort. This type of distillation is referred to as direct steam distillation, or distillation with live steam, or dry steam distillation. Most aromatic plants are distilled today with direct live steam at atmospheric pressure.

The application of steam distillation is subject to exactly the same reservations mentioned in the discussion of water *and* steam distillation, plus one additional factor. When using steam distillation, it is always possible, after the initial period during which the charge in the retort is warming up and condensation taking place, that the steam may be slightly superheated. Indeed, in some cases the steam may be purposely superheated, as already mentioned, in order to improve the oil to water ratio. In expanding from the much higher boiler pressure to the lower pressure prevailing within the retort, the steam tends to become superheated. Two factors then assume importance. First, the temperature of the charge will no longer be maintained at the boiling point of water, under the operating pressure, but will rise to the temperature of the superheated steam. The operator, therefore, must guard carefully against overheating. Second, superheated steam has a tendency to dry out the charge and reduce the rate of recovery of the essential oil. As was pointed out above, a good part of the oil is vaporized only after diffusing, as an aqueous solution, through the cell membranes to the outside of the plant particles. This diffusion, however, becomes possible only by the presence of a certain amount of hot water, and may be stopped altogether, or seriously slowed down, when the charge is completely dried. If, therefore, in the case of direct steam distillation, the flow of oil stops prematurely, it may be necessary to continue distillation with saturated (wet) steam for a time, until hydrodiffusion is re-established. After that slightly superheated steam may again be employed.

In general, it can be said that direct steam distillation excels water distillation, as well as water *and* steam distillation in regard to cost, rate of distillation, and capacity of production. As far as the condition of the plant material and the method of charging are concerned, the same principles apply here as to water *and* steam distillation. Special attention must be paid to the quality of the live steam. The higher the pressure of the steam, the higher is the temperature at which it enters the still; but in this respect the moisture content of the steam plays an important role. Saturated steam usually carries some water in the form of minute droplets, which are condensed by the expanding steam. Hence, the effect of superheating becomes noticeable only if saturated (but dry) steam, of markedly high pressure, is used. The higher the pressure of the steam in the steam boiler, the drier the plant material will remain during distillation. Only the portions of the charge touching the still walls will then become moist through condensation, despite insulation of the still against emanation of heat. In order to limit such loss of heat, and consequent excessive lowering of the temperature, the high-pressure steam, before entering the still, is sent through a water separator, and partly dried. In this connection, it should also be kept in mind that the different systems of steam boilers generate live steam, containing more or less moisture. In cases of prolonged distillation, considerable quantities of steam are condensed in the plant charge, and water accumulates on the bottom of the still. This may give trouble by wetting the lower part of the plant charge. Such condensed water must be drawn off, from time to time, through a stopcock in the still bottom.

Since high-pressure steam causes considerable decomposition, distillation is best started with steam of low pressure, followed by steam of higher pressure toward the end of the operation, when the oil content of the charge has decreased considerably, and when chiefly the high boiling constituents of the essential oil remain in the retort. No general rule can be laid down in this respect, as every type of plant material requires a different and specific method of preparation and also of distillation.

End of Distillation.—As the distillation proceeds, and as the oil content of the charge decreases correspondingly, the ratio of water to oil in the condensate will increase, because the steam can no longer contact the oil in the charge efficiently, regardless of the rate of distillation, and also because the remaining constituents are mostly high boiling. The operator must then decide at what point it is no longer economical to continue the distillation. Several criteria can be applied here. From a knowledge of the size of the charge and the yield to be expected, and from experience or trial distillations in a pilot still, it can quickly be determined whether or not the charge has been nearly exhausted. If yield data on the particular material charged are not available, it usually will suffice to take a small sample of the condensate directly into a test tube or glass cylinder and estimate from this the rate at which oil is being distilled at any particular time. Then, knowing the amount of oil already distilled, and calculating the amount that will be distilled in any additional period of time, it can be decided whether distillation should be continued for that period, or whether it would be more economical to stop and begin a fresh charge. The value of the product also enters into consideration, since a very valuable oil can be run profitably to a much larger water to oil ratio than can a less valuable oil. Certain oils—e.g., vetiver or angelica root oil—contain their most valuable constituents in the last runs (highest boiling fractions), and in these cases distillation must be prolonged for hours even though almost no oil seems to distill over toward the end of the operation. Otherwise valuable, high boiling constituents will be lacking in the oil. This rule, by the way, applies to all types of distillation.

It should also be kept in mind that the oil to water ratio measured at any time during distillation will always be higher than during any succeeding period, since this ratio decreases as the distillation continues. Experience with the distillation of any particular plant material will enable the operator to evaluate these matters properly, so as to obtain a maximum yield, a maximum rate, and a high quality of oil.

Treatment of the Volatile Oil.—The handling of the condensed oil is worthy of brief comment since its quality may deteriorate, particularly if the oil must be stored for some time. Just as the condensed water (distillation water) is always saturated with oil, so the condensed oil will always be saturated with water. There remains also the probability of slow reaction between the oil and water, unless the latter is almost completely removed.

The oil can be brightened (cleared of cloudy appearance) by filtering through kieselguhr or magnesium carbonate on filter paper.. This procedure removes all small droplets of water which cause the cloudiness, but it does not completely dry the oil. Larger quantities of oil may be filtered through mechanical filters, filter presses, or run through high-speed centrifuges. For further details see the section in the Appendix on "The Storage of Essential Oils."

Treatment of the Distillation Water.—The distillation water flowing off the oil separator (Florentine flask) contains some of the volatile oil in solution or suspension, the quantity depending upon the solubility and specific gravity of the various oil constituents. Considering that the distillate presents a mixture of condensed steam and oil vapors, it is evident that the water phase of the distillate actually represents an aqueous solution of oil, completely saturated at the prevailing temperature. Those oil constituents which are somewhat soluble in water will be partly dissolved in the distillation water, and the dissolved portion of this oil will be different in composition from that of the oil separated in the Florentine flask. The latter is usually called main or direct oil, the former water oil. The water-soluble constituents consist mostly of oxygenated compounds, and since these compounds possess a higher specific gravity than nonoxygenated compounds (terpenes, sesquiterpenes, etc.), the water oil usually has a higher specific gravity than the main oil. This difference, however, is not always pronounced, because the distillation water contains not only oil in actual solution, but also in suspended (minute droplets) and emulsified form. A more or less milky appearance of the distillation water thus indicates the presence of oil.

Such distillation water cannot be discarded, but must be submitted to further treatment to prevent loss of oil. In the case of water distillation or water *and* steam distillation, it may be automatically returned into the retort during distillation. For this purpose the Florentine flask must be installed at a sufficient height above the still so that the flow from the flask overcomes the pressure within the still. In the case of steam distillation (with live steam from a separate steam boiler) the distillation water should not be returned into the retort, as too much liquid would condense and accumulate within it and wet the plant charge. The distillation water therefore, is pumped or injected into a separate still for redistillation. The process of recovering the oil from the water by redistillation is commonly called cohobation, the stills serving for this purpose being known as cohobation stills. In its original and stricter sense, the term "cohobation" implies that the distillation water is used over and over for the distillation of a new plant charge (in the case of water distillation or water *and* steam distillation), but today cohobation simply means redistillation of the distillation waters.

The distillation waters are redistilled most efficiently in round stills provided with a steam jacket or a closed steam coil. Indirect heating is preferable, because the injection of live steam into the retort would cause too much water to accumulate within the retort, and hinder the vaporization of the oil from the water. In the case of many distillation waters only 10 to 15 per cent need be distilled off to recover most of the oil dissolved or suspended therein. The residual water may be discarded. Occasionally, however, it is necessary to distill off more than half of the quantity of water; in such case, a considerable portion of the oil distilled over will again be dissolved in the distillation water. To shorten the cohobation and increase the quantity of oil in the condensate, the water in the cohobation still is saturated with common salt (NaCl). This decreases the solubility of the volatile oil in water: the oil distills over more quickly, and with a smaller quantity of water. This procedure is recommended particularly where the distillation water contains slightly water-soluble constituents of high boiling point, which cannot be recovered by mere steam distillation.

The separation of oil and water by cohobation is based upon the simple principle that a mixture of oil vapors and steam possesses a slightly lower boiling point than pure water vapors (steam), and that the vapor mixture arising contains more oil than the liquid phase. By a reduction of the speed of cohobation, the oil content of the distillate may be increased because the rising steam will be more thoroughly saturated with oil vapors.

The following figures cited by Fölsch[17] give an idea of the quantities of volatile oils which can be obtained by the cohobation of various distillation waters:

Plant Material	*Quantity of Water Oil Recovered from 1,000 kg. of Distillation Water (grams)*
Chamomile Flowers	100–120
Coriander Seed	625–650
Dill Seed	360–450
Fennel Seed	175–200
Lavender Flowers	150–200
Peppermint Herb	400–500
Sage Herb	300
Tansy Herb	540

Another method of recovering the oil dissolved or suspended in the distillation water consists in saturating the latter first with salt and then extracting the solution with volatile solvents—e.g., highly purified petroleum ether or benzene. This is usually done twice. The drawn off and united solvent solutions are then concentrated in a still by driving off the solvent,

[17] "Die Fabrikation und Verarbeitung von ätherischen Ölen," Wien und Leipzig (1930), 47.

first at atmospheric pressure, and later *in vacuo*, until every trace of solvent is eliminated from the oil.

Any distillation of aromatic plants, unless conducted at fairly low temperatures, gives rise to products of decomposition in the nonvolatile plant constituents. These products (methyl alcohol, formaldehyde, acetaldehyde, acetone, low fatty acids, nitrogenous compounds, phenols, etc.) are carried into the condensate and present objectionable impurities. Because of their water solubility, they dissolve mainly in the distillation water, since the quantity of water by far exceeds that of oil. Because of the presence of such decomposition products, the crude water oil obtained by cohobation or extraction will in most cases be of dark color, often of disagreeable odor. It should not be combined with the main oil, as it would spoil the odor and flavor of the latter. It is, therefore, advisable to rectify the crude water oil by fractionation in a good vacuum still.

In many cases the great water solubility of the aforementioned decomposition products serves for the purification of volatile oils: when rectifying (redistilling) an oil by hydrodistillation, the distillation water is then simply discarded.

Disposal of the Spent Plant Material.—The disposal of the spent plant material, which represents a rather large bulk, frequently offers an annoying problem. One very economical method of disposal consists in using it as fuel—after air drying, of course, either in the sun or near the still in the case of direct fire stills, or near the boiler when a separate steam generator is used. Since the spent material has a rather low fuel value per unit volume, consideration must be given to the construction of a special fuel box. In many cases the spent material may be used effectively as fertilizer. Certain spent plants make an excellent cattle feed; this is particularly true of seeds which contain a high percentage of protein and fatty oil. The drying is done in dehydrating apparatus or by air drying on shelves. When sweetened with molasses, some spent grasses, such as lemongrass, seem to be relished by cattle.

Trial Distillation.—No discussion of distillation as used in the essential oil industry would be complete without some consideration of the interpretation to be placed upon the results of laboratory distillations, or, as they are frequently called, trial distillations. Since the oil content of plant material to be distilled fluctuates rather widely with such variables as geographical origin, growing conditions, ambient temperature, rainfall, period of harvest, moisture content, etc., it is not usually possible to state any values for the oil content other than by upper and lower limits (which in some cases may be quite widely separated). As already pointed out, handling of the plant material after harvesting, and prior to distillation, also has a marked effect on the oil content. As knowledge of the efficiency with which a large scale distillation is being conducted can be obtained only by comparison of the actual yield with the possible yield, it becomes quite important that the latter value be known with some accuracy. The only means of determining this value is to conduct a laboratory distillation using a sample of the plant material to be distilled in the larger scale operation.

The aim of any commercial distillation is, of course, to recover as large a percentage of the valuable oils in as high a state of purity as possible. Only in the laboratory, on a small scale, and under carefully controlled conditions, can both of these conditions be met. Therefore, the results of such laboratory distillation may be considered as a standard which the large-scale operation should approach as closely as practically possible.

There are two ways of carrying out such trial distillations: (*a*) on a very small scale in a glass flask, and (*b*) on a larger scale in a pilot still.

(*a*) Numerous methods of assaying the contents of essential oil in plant materials have been suggested. The literature offers many modifications of these methods, all of which aim at a quantitative yield of oil. The best and most commonly used method is that of Clevenger, which has found official recognition in "Methods of Analysis," published by the Association of Official Agricultural Chemists, Fifth Edition, 1940. For details of Clevenger's method see below, Chapter 4, on "The Examination and Analysis of Essential Oils, Synthetics and Isolates." This method permits assaying quantitatively the content of essential oil in a small amount (50 to 500 g.) of plant material. Although the amount of oil thus obtained is not sufficient to carry out a complete analysis, conclusions regarding its odor and flavor characteristics can be drawn from the small sample. Occasionally, the oil will have to be set aside for several days, until the slightly "burnt" or "still" odor of the freshly distilled oil has disappeared.

(*b*) A much more satisfactory method consists in distilling a sample of 20 to 50 lb. of aromatic plant material in a regular "pilot" still. Such a still, made of tin-lined copper, should be constructed so as to embody all the characteristics of large stills. It should allow for water distillation, water *and* steam distillation, and direct steam distillation. It will thus be possible to find for each new plant material the most appropriate method of distillation, to study, as well as possible, the rate of distillation and the consumption of steam (by measuring the quantity of distillation water*), and to determine the maximum yield of oil. Interesting observations regarding the effects of hydrodiffusion can be made. In the case of direct steam distillation the use of high-pressure or superheated steam may be studied. The quantity of oil recovered will be sufficiently large to examine the oil analytically, even to fractionate it. The pilot still should be provided with several

* This will be only approximately correct, since heat losses from the distillation system have not been considered.

trays, in order to find out the most opportune way of charging, if seed materials are to be processed. A small crusher and hay cutter will permit trying out the effects of comminuting the plant material according to different sizes. Needless to say, the pilot still should be well insulated—in other words it should resemble large stills in every possible way except size.

For all-around operation the pilot still should also be equipped for automatic return of the distillation waters into the still, in the case of water distillation or water *and* steam distillation, if the return (cohobation) of these waters into the still during operation seems desirable. In the case of direct steam distillation, the distillation water or a small measured part of

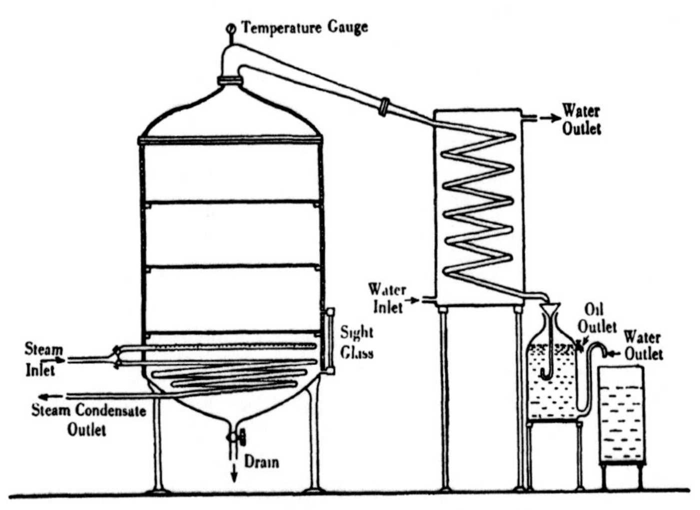

FIG. 3.20. Sketch of an experimental still.

it is saturated with ordinary salt, and three times extracted with low boiling petroleum ether. The drawn off and united petroleum ether extracts are then carefully evaporated on a hot water bath, and the residue dried in a desiccator to constant weight. From this small quantity the oil content of the total distillation waters can be calculated. Obviously, the extraction of only a part of the distillation waters gives an exact result only where the total distillation waters, after completion of the distillation, have been bulked in a tank. The distillation water should always be processed right after distillation of the plant material, because when exposed to the air for some time it loses oil by evaporation. While a small part of the distillation water is extracted experimentally with a solvent, another part should be steam distilled (cohobated). If cohobation yields no oil, the distillation water will have to be extracted with solvents.

Figs. 3.20 and 3.21 show the construction of pilot stills which may have a capacity of about 50 gal.

Steam Consumption in Plant Distillation.—In the distillation of aromatic plants the distillate (condensed water and oil) usually contains much more water than if the isolated oil itself had been hydrodistilled.

The following table indicates the average oil content, by weight percentage, in the distillates of completed operations as established by von Rechenberg,[18] based upon years of experience with industrial distillation of aromatic plants.

Distillation of Plant Material; Average Content of Volatile Oil in the Distillate

Plant Material	*% Oil*
Ajowan Seed	0.77
Angelica Seed	0.19
Angelica Root, Fresh	0.03
Anise Seed	0.81 to 1.16
Arnica Flowers	0.001
Arnica Root, Dry	0.06
Bay Leaves	0.75 to 0.77
Calamus Root, Dry	0.23 to 0.24
Calamus Root, Fresh	0.12
Caraway Seed	2.22 to 3.04
Cedar Wood	0.97 to 1.41
Celery Seed	0.17
Chamomile Flowers, Dry	0.004 to 0.007
Cinnamon Ceylon	0.31 to 0.34
Cloves	0.60 to 0.86
Clove Stems	1.03 to 1.52
Coriander Seed	0.56 to 0.57
Costus Root, Dry	0.01
Cubebs	1.2
Cypress	0.12 to 0.2
Elecampane Root	0.05
Fennel Seed	1.42 to 2.08
Galangal Root	0.05 to 0.08
Ginger Root	0.28
Juniper Berries	0.20
Lovage Root, Dry	0.05
Lovage Root, Fresh	0.02
Lovage Herb, Fresh	0.02
Patchouli Leaves	0.12 to 0.13
Peppermint Herb, Fresh	0.11
Pimenta Berries	0.18
Sandalwood, East Indian	0.05 to 0.16
Sandalwood, West Indian	0.23 to 0.34
Savin	0.25 to 0.31
Vetiver Root	0.015 to 0.02

[18] "Theorie der Gewinnung und Trennung der ätherischen Öle," Leipzig (1910), 362.

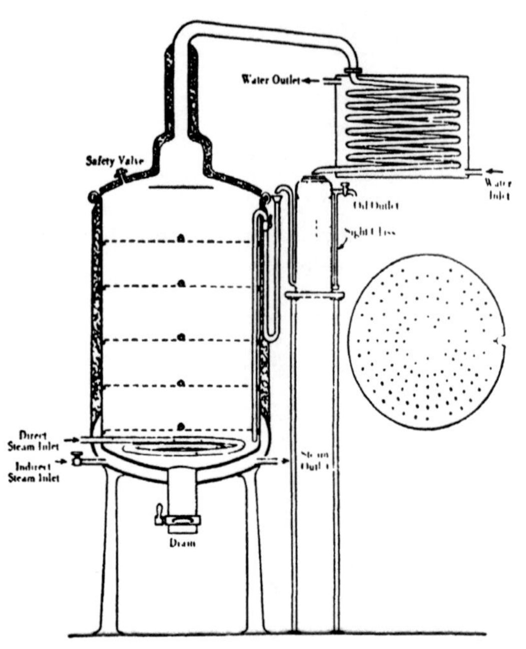

Fig. 3.21. Sketch of an experimental still with automatic cohobation.

Let us now compare these data with those expressing the composition of the condensate resulting from hydrodistillation of some of the pure chemical compounds which occur in volatile oils.

Hydrodistillation of	*Content in the Distillate (%)*	*Occurs in*
Styrene	57.0	Cinnamon oil
p-Cymene	45.7	Ajowan oil
Pinene	55.6	Ajowan oil
Limonene	40.0	Caraway oil
Dipentene	40.0	Ajowan oil
Linalool	18.2	Coriander oil
Menthol	12.0	Peppermint oil
Carvone	9.7	Caraway oil
Anethole	7.1	Anise, fennel and star anise oil
Cinnamaldehyde	3.0	Cinnamon and cassia oil
Eugenol	1.7	Clove, clove stem, pimenta, bay and cinnamon oil
Santalol	0.5	East Indian sandalwood oil

What are the causes of the marked increase in steam consumption during distillation of plant material, as compared with steam consumption during hydrodistillation of essential oils *per se?* Von Rechenberg[19] demonstrated that, in the case of caraway *seed* distillation, the condensate contains only

[19] *Ibid.*, 366.

2.22 to 3.04 per cent oil, whereas in the case of caraway *oil* distillation the condensate contains 8.80 to 10.11 per cent of oil. This obviously implies a much greater consumption of steam, in the first case, for the same quantity of oil, and longer hours of distillation.

The paucity of oil in the distillation of plant material, according to the same author, is caused by several factors:

1. Many aromatic plants contain a quantity of oil insufficient to saturate the relatively large quantities of steam blowing rapidly through the plant charge. On the other hand, it is not advisable to reduce the speed (rate) of distillation below a certain limit. A high steam velocity causes pressure differentials within the still, which prevent the steam from stagnating in the more densely packed parts of the plant filling. For this reason, and in order to increase the efficiency of a still, the operator is always tempted to inject into the still much more steam than is actually required. This results in a large volume of distillation water.

> Example: Let us suppose that a charge of 2,000 kg. of plant material can be exhausted in 11 hours, if we inject 250 kg./hr. steam (250 kg. distillation water in 1 hour). If, instead, we inject twice the amount of steam, i.e., 500 kg./hr., the length of distillation will be shortened at best by one-third, and in most cases only by one-fourth; but not by one-half, as might be expected.

2. In the course of distillation the oil content of the plant charge decreases gradually and the vaporization of oil is not stopped abruptly toward the end of the operation. This does not even take place in hydrodistillation of volatile oils *per se*, and much less with plant material. Plainly, such a prolongation of the distillation greatly increases the steam consumption and depends also upon the individual operator.

3. While retained in the plant material, the volatile oil may be subjected also to forces of adhesion; this seems true even if small quantities of oil are distributed over large surfaces of comminuted plant particles. Experiments to this effect were undertaken by Rodewald.[20]

4. The volatile oil is enclosed within the plant tissue and cut off from direct contact with steam by several layers of membrane, often very tough. For this reason most plant materials must be comminuted prior to distillation. Where steam distillation is practiced, this process of comminuting (grinding, pounding, milling, crushing, rasping) should not be carried too far (certainly not to the point of reducing the material to the size of flour particles), because the interspaces within the plant charge would then become too small. The rising steam must have sufficient space to penetrate all parts of the charge uniformly. Very small interspaces necessitate a slow,

[20] *Z. physik. Chem.* **24** (1897), 193.

ineffective distillation, because any increase in pressure would cause the steam to break channels ("rat holes") through the plant charge, or to hurl parts into the gooseneck and condenser. In other words, too finely powdered material is not penetrated evenly by steam and cannot be completely exhausted by steam distillation.

If, on the other hand, the plant material is not powdered, but granulated, only a portion of the oil is freed, and another portion remains enclosed within the oil glands in the plant tissue. When crushing plant material, such as seed, a portion of the freed volatile oil will be covered again by crushed plant particles. The distillation of excessively crushed seed material, if not properly conducted, may, therefore, require longer hours than that of torn or slightly milled seed, provided the quantity of injected steam is the same.

If the plant material is distilled in uncomminuted condition—as with herbs and leaves, and most flowers—the oil remains enclosed within the plant tissue. Hacking with an axe or machete or cutting in a hay cutter offers an advantage only in that the material can be packed into the still more uniformly; the steam then penetrates the charge more evenly, but very few oil glands will actually be broken up. Since the steam can vaporize only those volatile substances which it touches directly, and will not affect the oil enclosed within the plant tissue, the oil must first be dissolved by hot (liquid) water and carried, by diffusion, through the swollen cell walls toward the outside. Hydrodiffusion, however, requires much more time than vaporization, which takes place almost immediately, because *all* the enclosed volatile oil must be brought to the surface, and that is a slow process. This fact is primarily responsible for the paucity of oil in the condensate, and for the relatively long duration of distillation in the case of uncomminuted leaves and herbs possessing a tough fiber.

5. If the plant material is comminuted prior to distillation, very high boiling or practically nonvolatile substances, such as resins, paraffins, waxes, fatty oils (contained in other cells or glands), mix with and dissolve in the freed volatile oil, thereby substantially lowering its vapor pressure, and reducing its rate of vaporization. This occurs particularly in the case of seeds, most of which contain large quantities of fatty oils. The Agricultural Experiment Station of Möckern, near Leipzig, Germany, reported the following content of fat and fatty oil (ether extract) in seeds from which the volatile oil had first been removed by steam distillation:

Seeds	*Fatty Oil* (%)
Ajowan	33.20
Anise	18.59
Caraway	16.06
Celery	31.32
Coriander	26.40
Fennel	16.71

Assuming that air-dried caraway seed, such as is used for distillation, contains about 15 per cent moisture and 5.5 per cent volatile oil, we arrive at a ratio of 5.5 per cent volatile oil to 12.8 per cent of fatty oil in the seed. For practical distillation, this implies that 5.5 parts of volatile oil must be vaporized from 12.8 parts of fatty, nonvolatile oil. In other words, it is necessary to distill a mixture of fatty oil and volatile oil, which holds 30 per cent of volatile substances in solution. Assuming a content of 5.0 per cent volatile oil in fennel seed, 3.0 per cent of volatile oil in anise seed, 3.5 per cent in ajowan seed, 2.5 per cent in celery seed and 1.0 per cent in coriander seed, we find that we would have to distill:

Ajowan oil mixture containing	11.5%	volatile oil*
Anise oil mixture containing	16.0%	volatile oil
Caraway oil mixture containing	30.0%	volatile oil
Celery oil mixture containing	9.7%	volatile oil
Coriander oil mixture containing	4.2%	volatile oil
Fennel oil mixture containing	27%	volatile oil

Such relatively large quantities of fatty, nonvolatile oils are well capable of reducing the vapor pressure, and thereby the rate of vaporization of the volatile oils dissolved in these fatty oils.

Other reasons aside, it is thus practically impossible, when distilling seed with steam or boiling water, to saturate the steam completely with oil vapors, even when packing the plant charge very high in the still. The oil vapor phase in this mixture will always remain unsaturated; the more the content of fat in the seed exceeds that of volatile oil, the less the steam will be saturated with volatile oil vapors. This theoretical consideration confirms practical experience in the case of seed distillation. In actual practice, therefore, distillation of seed material can seldom if ever be completed, because the fatty oil tends to retain small quantities of volatile oil. It becomes necessary, therefore, to halt distillation, since the small recovery of oil no longer warrants the increasing consumption of steam and labor.

6. The steam consumption is influenced further by the moisture content of the plant material, particularly in the case of herbs, grasses and roots, which are processed either in the fresh succulent, or semidry, or dry, condition. When distilling peppermint herb with live steam, for instance, the following quantities of steam will be consumed, the steam consumption being measured by the quantity of distillation water in the condensate:

Fresh herb requires 250 to 350 kg. of steam per kilogram of oil.
Semidried herb requires 60 to 80 kg. of steam per kilogram of oil.
Air-dried herb requires 30 to 40 kg. of steam per kilogram of oil.

* In all cases a 15% moisture content of the seed is assumed.

These figures, cited by Fölsch,[21] are evidently relative, as actual steam consumption depends upon the type of the still, the quality of the steam, the way of packing, and the experience of the operator.

Because of its high moisture content, fresh peppermint herb, when distilled, has a tendency to lump (agglutinate) and to prevent a uniform penetration by the steam. The volatile oil is, therefore, released from the fresh herb only very slowly. Taking the above figures of steam consumption as a basis, the steam/oil vapor mixture (in other words, the condensate) will contain the following quantities of volatile oil:

Fresh Herb	0.3 to 0.4% peppermint oil
Semidried Herb	1.2 to 1.6% peppermint oil
Air-dried Herb	2.5 to 3.0% peppermint oil

The oil content in the condensate is not uniform from the beginning to the end of distillation, but amounts in the beginning to a multiple of the average oil content. In the case of air-dried peppermint herb, the condensate contains in the beginning about 8 per cent of oil, which decreases gradually toward the end until it amounts to only 0.004 per cent. For practical reasons distillation should then be stopped. As mentioned previously, certain plants contain volatile oils, the most valuable parts of which are very high boiling. When applying saturated steam of atmospheric pressure only, distillation must then be continued for very long periods, although only small quantities of oil are recovered toward the end. If this is not done the high boiling constituents are lacking in the oil, and the oil is of inferior quality. In such cases it will be advantageous to speed up and complete the operation by injecting slightly superheated steam toward the end.

Rate of Distillation.—According to Folsch,[22] the ratio between quantity of distillation (condensed) water and time (in other words, the quantity of water distilled over per hour) may be designated as rate (force or speed) of distillation. It must be regulated according to the diameter of the still, and the size of the interspaces within the plant charge (degree of comminution). If the velocity of the rising steam is too low, the steam will stagnate in the denser portions of the charge, and complete exhaustion by distillation will be impossible. If, on the other hand, the velocity is too high, the steam may break through the charge, form steam channels ("rat holes") and even hurl plant particles into the condenser, partly clogging it. By collecting the distillation water running off the condenser from time to time, and over a period of some minutes, and then weighing it, the rate of distillation can be controlled. For practical purposes the volatile oil may be ignored. The

[21] "Die Fabrikation und Verarbeitung von ätherischen Ölen," Wien und Leipzig (1930), 40.

[22] *Ibid.*, 62.

quantity of distillation water collected during these few minutes is calculated in terms of kilogram/hours per square meter. (See example below.) This figure is then compared with the optimum rate of distillation, as established by trial distillation or by experience with the plant material in question (and taking into consideration its degree of comminution). The steam velocity in the actual operation may be regulated accordingly.

For example, if we obtain in one minute 8 kg. of distillation water, and if the smallest area covered by the charge on the perforated grid in a cylindrical still is 1.2 sq. m., the rate of distillation will be

$$\frac{8 \times 60}{1.2} = 400 \text{ kg./hr. per sq. m.}$$

Once the average oil content of the mixed vapors (steam plus oil vapors) has been established for a certain type of plant material and a certain degree of comminution, and once the most favorable rate of distillation is known, the amount of steam necessary for complete exhaustion of a plant charge can be calculated, and the steam supply adjusted accordingly. By weighing the quantity of distillation water from time to time, by converting the figures to the total length of distillation and by relating this to the quantity of oil expected, the operation may thus be regulated according to optimum conditions. Let us suppose, for example, that 1,000 kg. of coriander seed must be distilled, and that the seed, according to assay, contains 0.8 per cent of oil, in other words that the 1,000 kg. of coriander seed contain 8 kg. of oil. We know from experience or from trial distillations that the average oil content of the vapor mixture (condensate) in the case of coriander seed distillation is 0.5 per cent. Therefore, 1,600 kg. of steam are required to distill over 8 kg. of coriander seed oil. If we work with a distillation rate of 200 kg./hr., i.e., 200 kg. distillation water per hour, the charge should be exhausted in 8 hr. In order to shorten the time of distillation, the rate of distillation must be increased. However, in this case attention must be paid to the fact that, on increasing the speed of distillation, the average oil content of the vapor mixture decreases to a certain extent, because the quicker the steam penetrates the plant charge the less it has occasion to become saturated with oil vapors. In other words, much more steam will be consumed than is calculated theoretically.

Pressure Differential Within the Still.—The velocity of steam flow is caused by differences in pressure. In the case of plant distillation with live steam, which in the boiler is usually at a pressure above atmospheric, the plant charge in the retort prevents the injected steam from expanding immediately and completely. For this reason, the steam pressure cannot fall immediately to the level of the atmospheric pressure. Thus, there arises a certain excess pressure beneath the charge in the retort; but a gradual equalization with the atmospheric pressure takes place toward the top of the still. The degree of this excess pressure is a function of the force (speed) of distillation and of the interspaces within the plant charge. According to the height of the charge or the number of layers, this excess pressure can be increased by 0.3 atmospheres, and, in some cases, even more. But if the pressure exceeds a certain limit (which depends upon the type and height of the plant charge), the steam forms fine, often scarcely visible channels through a powdered charge, whereas coarser masses are torn apart, or even hurled into the gooseneck of the still. An excess pressure ("back pressure") within the retort may be caused also by a gooseneck or condenser pipes too narrow for the volume of steam injected into the still or by sharp bends in the pipes.

Irregular heating of the boiler and variations in the steam consumption (such as are occasioned by the turning on and turning off of neighboring stills) may cause the pressure in a steam generator to undergo continuous fluctuations. High-pressure steam has a tendency to blow into a still somewhat jerkily, giving rise to pressure variations even within the retort. Such fluctuations, however, are by no means harmful; on the contrary, they may exert a beneficial influence, as far as the yield of oil is concerned, by forcing the injected steam to loosen and penetrate the more densely packed portions of the plant charge, where the steam would otherwise stagnate.

Pressure Differential Inside and Outside of the Oil Glands.—As it rises through the plant charge, the steam at first vaporizes all the freed volatile oil which by comminution of the plant material is within reach of the passing steam. Saturated steam (not superheated!) will at the same time condense a certain quantity of water within the retort. Consequently, the temperature of high pressure steam will be reduced to that of saturated steam, in other words, to the boiling point of the water/oil mixture. It must be remembered that this boiling point is slightly lower than that of the saturated steam. As the volatile oil vaporizes from the plant material, the temperature of the steam rises again to that of pure saturated steam, at the pressure prevailing in the charge. If the plant charge is somewhat tightly packed, the temperature of the steam will show a certain range from the bottom to the top of the charge. This differential in temperature depends upon the force of distillation and the drop in the steam pressure from the lower to the upper section of the retort; in other words, the lowest part of the charge will have the highest, and the upper part the lowest, temperature. Gradually the temperature of the steam equalizes itself throughout the charge and, despite poor heat conduction, will prevail, even inside of all plant particles. As has been said, the boiling point of a water/oil mixture is somewhat lower than that of steam alone, the total vapor pressure a little higher. Since the temperature inside and outside of the plant particles has become equalized, a certain excess pressure will develop within those oil glands which still contain volatile oil *and* water enclosed. This pressure differential inside and outside of the oil glands probably has some influence upon the vaporization of the volatile oil through the cell walls. A sufficiently large pressure differential may well cause some cell membranes to burst (provided they are not too thick and strong) or at least to expand the cell walls, to enlarge the pores and to loosen agglomerated particles of the charge, thus opening new passages for the steam. The more the pressure differential is reduced, the more it loses its significance as a loosening agent; but it remains important for the isolation of oil, in so far as it supports the forces of hydrodiffusion. The pressure differential inside and outside of the oil glands is more effective when first heating the retort and toward the end of distillation, provided temperature and pressure fluctuations actually occur inside of the still. A pressure differential, however, can be created only if water is present in liquid form, or by partial condensation of steam when first heating the retort: the water thus formed will penetrate the plant tissue and seep also into the oil glands. (von Rechenberg).

In the hydrodistillation of plant material at reduced pressure, the pressure differential inside and outside of the oil glands exerts itself to a marked degree only with low boiling substances. In the case of distillation above atmospheric pressure, however, the pressure differential assumes considerable importance.

Effect of Moisture and Heat upon the Plant Tissue.—Any plant material serving for distillation contains a certain quantity of moisture, even air-dried material retaining 10 to 20 per cent of water. If saturated steam of atmospheric pressure is injected into the plant charge, condensation of steam will take place until the temperature of the still content has risen to that of the steam.

Heat in conjunction with mòisture soon causes the plant tissues to swell, the cells and pores to enlarge, and the total volume to expand. Completely swollen seed material, for example, may have expanded by one-fourth of its original volume. In actual distillation, this loosening of the plant material may, however, be partly counteracted by the weight and pressure of the softened plant charge.

An actual bursting of the plant membranes by the action of steam takes place probably to a limited extent only. The hot steam undoubtedly exerts a certain preparatory effect important for the vaporization of the enclosed volatile oil, but the action of steam *per se* is not sufficient to liberate that part of the oil which remains protected by resistant cell membranes.

Influence of the Distillation Method on the Quality of the Volatile Oils.—The quality, as well as the physicochemical properties, of a volatile oil are greatly influenced by the condition of the plant material (age, dried or fresh) and by the way distillation is carried out. Many factors enter the picture, viz., the method of distillation (water distillation, water *and* steam distillation, and steam distillation), the degree of comminution of the plant material, the quantity of the plant charge, the length of distillation, the pressure applied, the quality of the steam, the treatment of the distillation waters, whether the oil of cohobation is added to the main oil or not, etc.

The effects of water distillation and steam distillation differ considerably, in that high boiling constituents of the volatile oil are recovered only incompletely in the case of water distillation, if the plant material is insufficiently comminuted. Even leaf material yields volatile compounds of high boiling point only incompletely by water distillation. Von Rechenberg[23] reported that patchouli leaves yielded 3.27 per cent of volatile oil on steam distillation, and only 2.98 per cent on water distillation. The latter oil contained only a small quantity of the high boiling constituents, which incidentally possess also a high specific gravity and a high odor and fixation value. Oil constituents which are slightly soluble in water, phenols and certain alcohols and acids for example, are retained in the water, with the result that water distillation and steam distillation yield different types of oil, if the plant material contains only small quantities of oil.

General Difficulties in Distillation.—Essential oils consist of volatile compounds which are more or less sensitive to the influence of heat. It is doubtful, therefore, that all the volatile constituents present in the living plants can be isolated as such by distillation. In addition, distillation of certain plant materials is connected with difficulties of hydrodiffusion. The oil in part resinifies, and in part remains in the plant tissue. Hence every type of plant material requires a particular method of distillation.

Because of these difficulties, and because of the high cost of distillation in certain cases (through excessive steam and fuel consumption), it has been suggested that such materials be extracted with volatile solvents, and the concentrated extracts steam distilled. The oils obtained usually contain small quantities of resinous and waxy matter; such oils may be soluble in a certain volume of dilute alcohol, but the solutions often become turbid when more of the dilute alcohol is added.

(e) Hydrodistillation of Plant Material at High and at Reduced Pressure, and with Superheated Steam.

Steam Distillation of Plant Material at High Pressure.—Certain plant materials —orris root, sandalwood, cloves, caraway seed, pine needles, for example—are occasionally distilled with steam of a pressure higher than atmospheric, in order to obtain a more favorable ratio of oil to water in the distillate, i.e., to shorten the length of distillation and to increase the total yield of oil. Purely physical considerations, a discussion of which would lead too far, show that a substantial gain

[23] "Theorie der Gewinnung und Trennung der ätherischen Öle," Leipzig, (1910), 441.

can be achieved only with a pressure of several atmospheres within the retort. This, however, usually causes such profound decomposition of the plant material and of the volatile oil that the method cannot generally be applied in practice.

The actual pressure within the retort, when using high pressure steam of 4 atmospheres as measured in the steam boiler, is certainly less than one atmosphere above normal atmospheric pressure. If, notwithstanding, such modest excess pressure leads to favorable results—primarily to a shortening of the distillation process—the explanation must be sought in other, perhaps purely mechanical factors. If the steam were throttled by a valve in the gooseneck and the pressure thus increased, a manometer would indicate continuous pressure fluctuations within the still. These fluctuations prevent the steam from stagnating in the too densely packed portions of the plant charge, and seem to loosen all parts of the charge. This is particularly true of direct steam distillation and, to a certain extent, also of water *and* steam distillation, if the plant material has been packed high, and not sufficiently uniformly or tightly. Water distillation, on the other hand, does not seem to be affected by excess pressure. The effect of high pressure appears to be more pronounced when the plant material has been charged improperly into the still, and when a less efficient distillation, at atmospheric pressure, has been carried out previously.

The use of high-pressure steam for the rectification of volatile oils *per se* is not advisable, nor is it necessary, because for this purpose superheated steam gives better results. Nor should it be made a general practice to distill plant material with high-pressure steam, as this will increase the quantity of decomposition products in the plant material and in the oil, the degree of decomposition being influenced by the height of the pressure applied, the resulting rise of the temperature, and the length of distillation. Ordinary steam distillation, even at atmospheric pressure, affects some of the constituents of the essential oil and of the plant material itself (the latter being even more sensitive to high pressure steam than the oils). The nonvolatile plant matter may thus undergo more or less profound decomposition, with accompanying formation of undesirable volatile substances, which may considerably impair the color and odor of the oil. The distillate may become so much contaminated with foreign matter that even rectification no longer yields a normal oil. (von Rechenberg).

For all these reasons distillation with high pressure steam is not recommended, if the operation aims at obtaining a volatile oil containing delicate constituents. It may, however, be advantageous in some cases, where the distillate is to be further processed—as with oil of camphor and steam distilled pine (stump) oil.

Water Distillation of Plant Material at High Pressure.—It is not advisable to employ this method, because the resulting higher temperature gives rise to decomposition products which impart a disagreeable "burnt" odor to the oil. Neither is there any appreciable gain in the ratio of oil to water in the distillate, except perhaps in cases where previous distillation under atmospheric pressure has been carried out inefficiently.

Steam Distillation of Plant Material at Reduced Pressure.—This method may be subdivided into two types, viz., (*a*) steam distillation at slightly reduced pressure, and (*b*) vacuum steam distillation at such a low pressure that the temperature remains just enough above that of the cooling water to permit sufficient condensation of the steam/oil vapors.

(*a*) It is a known fact that a pressure reduction within the still often shortens the length of distillation. Even a slight reduction may shorten the duration to only one-half the time required for steam distilling at atmospheric pressure. Von Rechenberg[24] demonstrated that this effect is caused by fluctuations in the steam/oil vapor pressure which, as in the case of distillation at high pressure, exert a continuous loosening effect upon the plant charge.

(*b*) The principal advantage of steam distillation of plant material *in vacuo* consists in the pure odor of the volatile oil thereby obtained. It will be free from any off-odor caused by decomposition, which accompanies most oils distilled above 70°.

If the hydrodistillation *in vacuo* is not carried out with steam generated by boiling the water within the still (water distillation, or water *and* steam distillation) but by steam generated in a separate steam boiler, *a distillation with superheated steam at reduced pressure will result.* Even high boiling constituents of the volatile oil will then readily distill over; a previously air-dried plant charge, under these circumstances, may, however, gradually dry out until the volatile oil enclosed within the oil glands can no longer be vaporized, because the forces of hydrodiffusion no longer play their important role. It will then become necessary to apply saturated steam at atmospheric pressure, so that steam condensation within the plant charge again forms (liquid) water, which will permit the forces of hydrodiffusion to act anew.

When hydrodistilling at reduced pressure it is preferable to employ spiral condensers rather than tubular ones, because the former can be tightened better. The surface of condensation should be about five times larger than when distilling at atmospheric pressure. This increase is necessary for several reasons: (1) The differential in the temperature of steam and cooling water is much smaller at reduced pressure. The rate of condensation, therefore, decreases. (2) The volume of a given quantity (weight) of steam is much larger at reduced, than at atmospheric, pressure. For instance the volume of 1 kg. of steam at the following pressures is:

Millimeter Pressure	*Cubic Meters*
760	1.650
380	3.150
150	7.650
76	14.530

The velocity at which the steam enters the condenser will affect the transfer of heat to the cooling surface. Therefore, depending on other variables, an appropriately designed condenser (as to type, length, etc.) will have to be employed. Too long a condenser being impractical, several spiral condensers connected with a T tube may be installed side by side. Since an efficient vacuum pump creates a higher vacuum than is actually required for the distillation of plant material, the pressure within the still should be regulated by a valve permitting enough air to enter the still to sustain the desired pressure. The pressure should be measured by two manometers, one reaching into the receiver and one directly into the retort.

Steam distillation of plant material *in vacuo* is limited in application by the fact that cooling and condensation of the vapors become increasingly difficult as the pressure and temperature of distillation are lowered. The general application of hydrodistillation *in vacuo* to plant material is restricted by another factor. With

[24] "Theorie der Gewinnung und Trennung der ätherischen Öle," Leipzig (1910), 392.

lowered pressure in the still, the partial pressure of the oil vapors decreases relatively more than that of the water vapors (steam); hence, the ratio of the volatile oil in the distillate is smaller than when distilling at atmospheric pressure. In other words, more steam will be consumed when hydrodistilling a certain quantity of oil *in vacuo* than at atmospheric pressure. This lower rate of vaporization of the volatile oil is particularly pronounced in the case of water distillation of plant material containing high boiling and partly water soluble constituents. (See below.) In this case, a multiple volume of steam (as compared with distillation at atmospheric pressure) is often required to attain the same yield of oil. Any increased steam consumption also results in higher working cost, since much more distillation water must be redistilled or extracted.

When processing of plant material by water *and* steam distillation is practiced at reduced pressure, pressure variations in the still may cause loosening of the plant charge, so the rising steam is better saturated with oil vapors. This factor occasionally results in a lower consumption of steam than when working at atmospheric pressure. The most suitable method of distilling plant material at reduced pressure is with water *and* steam, provided the nature of the plant material permits its application. In general, it can be said that steam distillation of aromatic plants, under reduced pressure, remains very limited in practice.

Water Distillation of Plant Material at Reduced Pressure.—According to established thermodynamic principles and to the explanation given in the preceding pages, hydrodistillation at reduced pressure has the effect that, with equal quantities of condensate, the steam volume in the distilling space, and therefore the steam velocity, will increase enormously as the pressure in the still is reduced. For example, a given quantity (weight) of totally saturated steam and benzaldehyde vapor fills a certain volume at atmospheric pressure (760 mm.); at 76 mm. pressure the volume will be approximately ten times larger, at 31 mm. approximately twenty-four times larger than that occupied under atmospheric pressure. The velocity under which the vapor mixture rushes through the condenser increases in the same ratio. Hence water distillation of plants at reduced pressure is connected with certain inconveniences with which the operator should be familiar.

Any increase in the speed of distillation affects the purity of the distillate because minute plant particles are carried over mechanically. As a precaution against this, speed must be moderated as much as possible; flat, wide, rather than tall, stills should be selected for this purpose. It should also be borne in mind that the steam is to some extent throttled in the gooseneck (the narrowest part of the still). This may result in a slight back pressure within the retort, relative to the pressure in the receiver, which differential might easily amount to 10 mm.

It is, therefore, advisable to adjust the speed of distillation to the temperature prevailing within the retort. This will prevent a rise in the distillation temperature above a desired point.

The great advantage of water distillation of plant material at reduced pressure lies in the fact that it can be carried out at relatively low temperatures—e.g., at 50°—which reduces decomposition of the essential oil. It is not advisable to operate at lower temperatures, because the oil vapors can then no longer be sufficiently condensed, and considerable losses of oil might occur. Furthermore, higher boiling, slightly water-soluble compounds are retained partially in the plant material and in the water, and the oil will be deficient in these constitutents. This phenomenon, already discussed under water distillation of plants at atmospheric pressure, is even more pronounced in its effects when reduced pressure is employed. This very factor limits the application of water distillation *in vacuo* to only a few plant materials.

Temperatures of only 30° to 50°, and the presence of water offer favorable conditions for fermentation of the plant material, for which reason distillation of this type should not last longer than a few hours. The oils obtained by this method will never possess a "still" or "burnt" odor, but rather a slight "fermented" one.

Superheated Vapors.—As was pointed out in the theoretical part of this chapter, a vapor is saturated so long as it remains in contact with the liquid from which it originates. Saturated vapors possess characteristic properties by which they differ sharply from vapors separated from their liquid sources. The slightest cooling of a saturated vapor causes partial condensation, the slightest heating results in increased vaporization. So long as it remains in contact with its liquid, a saturated vapor is seldom absolutely dry; usually it contains admixed particles of the liquid in the form of spray. Moderate vaporizing, even evaporating of the liquid phase, carries microscopically small droplets upward into the vapor space. Vigorous boiling ejects larger quantities from the turbulent liquid; these are kept suspended by the flow of the vapors, or they drop back into the boiling liquid, to be replaced by new ones. Very wet vapors are more or less hazy. The transparency of a vapor, however, merely proves that it does not contain larger quantities of the liquid phase; it does not prove that the vapor is absolutely free of liquid, since minute droplets floating in the clear vapors are invisible to the eye. Their actual presence in the vapors is proved by the fact that the condensate of plant materials or of volatile oils is usually contaminated with dust or with nonvolatile colored substances.

Let us assume that we continue to heat and vaporize a liquid at constant external pressure to the point where the last molecule of the liquid phase is transformed into vapor. At this very moment the vapors are still saturated, dry saturated. Further heating no longer induces the formation of vapors, it only increases the temperature of the formed vapors, with a resulting expansion of their volume. The vapors then become superheated. Thus, superheated vapors possess a higher temperature, a larger volume and a lower density than saturated vapors at the same pressure. Superheating of a vapor may also be interpreted as a heating beyond the point of saturation. Saturated vapors, as compared with superheated vapors at the same pressure, therefore, contain a maximum of mass, as well as the highest specific gravity and the lowest specific volume (the specific volume being the reciprocal value of the specific gravity). When comparing the two types of vapors at the same temperature, superheated vapors possess a lower pressure than saturated vapors. A saturated vapor exerts the maximum pressure at the given temperature. Cooling merely lowers the temperature of superheated vapors, without causing condensation (as would be the case with saturated vapors). Only by further cooling, to and below the point of saturation, will a portion of the vapor be condensed. The moment a superheated vapor is brought into contact with the liquid phase from which it originated, vaporizing will take place, until the saturation point is reached once more. The superheated vapor thus passes into a saturated vapor. (von Rechenberg).

Distillation of Plant Material with Superheated Steam.—Relative to its weight, superheated steam can vaporize and entrain more volatile substances than saturated steam. In practice, steam may be superheated by passing it through fire tubes in a boiler—in other words, through a superheater. This superheated steam, mixed with high-pressure and saturated steam, is then injected into the still beneath the grid which supports the plant material. The mixing of superheated steam with saturated steam serves as a precaution against "burning" and decomposition of the plant material. Thus dry, *slightly* superheated steam is obtained. However, there remains the danger of decomposition, at least to a certain degree. The distillation will be shortened, but the oil yield may suffer, because the plant charge easily dries out as the forces of hydrodiffusion no longer play their important role.

Although rectification of certain essential oils with superheated steam at atmospheric and especially at reduced pressure has been found valuable, its use in the distillation of plant material is limited, and often connected with more disadvantages than advantages.

Advantages and Disadvantages of High-Pressure and Superheated Steam in Plant Distillation.—When high-pressure or superheated steam is employed in distillation with direct, live steam (but not in water distillation, or water *and* steam distillation), the condensation of water vapors in the plant charge may be greatly reduced, if not prevented altogether, except in the part of the charge along the walls of the retort, which usually becomes moist despite good insulation. This feature permits a more complete exhaustion of the plant charge. Furthermore, the use of high-pressure steam with its elevated temperature increases the partial pressure of the volatile oil, and the ratio of oil to water in the condensate becomes more favorable. In other words, distillation will be shortened. To exploit this effect of high-pressure steam, any condensed water accumulating beneath the steam coil in the retort must be prevented from rising above the coil, since high-pressure or superheated steam would then be transformed into low-pressure, saturated steam, of 100°—direct steam distillation thus being transformed into a water *and* steam distillation. Therefore, any condensed water must be drawn off from time to time. Such condensed water always contains extractive matter dissolved and dripping down from the plant charge, and this matter has a tendency to undergo decomposition. Some of the resultant products are volatile and of disagreeable odor, and when carried over into the receiver will contaminate the volatile oil. As mentioned previously, not only the volatile oils themselves, but also the plant materials, are very sensitive to the influence of heat and easily decompose. This takes place even at a temperature of 100°, but the effect is much more pronounced at a higher distillation temperature. High-pressure steam or superheated steam gives rise also to resinification and to the formation of insoluble compounds, parts of which vaporize and distill over. Such oils are less soluble in dilute alcohol or, when soluble, cause turbidity on further addition of dilute alcohol. Hence, the use of high-temperature steam in the distillation of aromatic plants cannot be recommended generally.

As was explained in our discussion of hydrodiffusion, the volatile oil enclosed in the cell membranes of aromatic plants must first be dissolved by hot water; and then, by forces of diffusion, be brought to the surface of the plants or plant particles, where the volatile oil may be vaporized and entrained by the passing steam. The exuded water must be replaced, so that the process of hydrodiffusion is not interrupted. The water necessary for this purpose comes partly from the moisture contained originally in the plant material itself, partly from steam condensation (which takes place particularly at the beginning of distillation). When high-pressure, or dry, superheated steam is used, only that part of the volatile oil is vaporized which has been freed by comminution; at the same time the moisture present in the plant material vaporizes, and the plant charge dries out. Then any oil remaining within the plant tissue can no longer reach the outside by hydrodiffusion, as there is no longer any water present or available; distillation will therefore be incomplete, and the yield of oil subnormal, unless saturated steam of low pressure is injected after the application of high-pressure or superheated steam.

There are a few cases in which distillation with superheated steam becomes advantageous—e.g., with plants that contain much moisture (60 to 80 per cent) and are difficult to dry. If such material is distilled with low-pressure saturated steam, the high moisture content of the charge will cause much steam condensation: the plant charge lumps and is difficult to exhaust. This can be prevented by applying superheated steam, a smaller or larger portion of the water within the plant charge then vaporizing while hydrodiffusion still functions.

In general, it can be said that plant material containing low boiling essential oils is preferably distilled with low-pressure steam, whereas high-temperature steam recommends itself for the distillation of high boiling oils.

(f) **Field Distillation of Plant Material.**—In primitive countries, where aromatic plants grow wild, or are cultivated by natives as patch crops, essential oils are obtained by a form of distillation which may most appropriately be termed field distillation. Lack of roads prevents transport of the plant material to centrally located larger distilleries, and the distillation equipment has to follow the plant material into the interior of the growing region. Small portable or movable stills must be used; but they serve only for a certain time of the year, and remain unused for the remainder of the time. They must, therefore, be low priced, sturdy, simple, easy to transport and to install in the fields, and simple to operate. In many cases this type of distillation is old; it has developed along purely empirical lines as an "art" inherited through generations. One should not summarily condemn this industry as antiquated and too primitive, however, because, in many instances and in view of the circumstances, a change to more modern and more expensive equipment is difficult, if not impossible. Indeed, such a change in some cases, might be for the worse, so far as prices of the oils, particularly, are concerned. On the other hand, this method of operation is frequently faulty, although it could be improved readily by only a few slight modifications.

Distillation may be carried out either by heating the still with direct fire or by steam generated in a separate small steam boiler. The former is simply an example of water distillation, or a water *and* steam distillation. Direct steam distillation, in this case, represents a stage in the transition to larger distilleries, because steam distillation is economical only if the steam generator is connected with several stills.

Despite the often primitive apparatus, the quality of oil resulting from water distillation in some instances has been good. However, the yield of oil in field distillation is often far below that obtained by water distillation on a large scale in more modern factories. The principal reason is probably that the small distillers do not always observe the fundamental rules of efficient water distillation, i.e., a small plant charge and a quick distillation. Most small operators are inclined to charge their retorts as high as possible in order to utilize them fully; furthermore, the speed of distillation is usually limited by too small a condenser. Also, in primitive operation the plant material is seldom comminuted, although a thorough comminution in the case of water distillation is often of prime importance for a normal yield of oil.

The following cases of actual distillation in the field will prove to what degree the yield, as well as the quality, of an essential oil depends upon the method of distillation. They also show that in many countries production of essential oils remains utterly primitive, and that the introduction of better methods would result in a considerable improvement in the yield and quality of the oils. The data are cited partly from von Rechenberg's "Theorie der Gewinnung und Trennung der ätherischen Öle," Leipzig, 1910, but have been confirmed by the author during his own investigations in the interior of China, Mexico, France, Spain and many other countries.

Distillation of Lavender in France.—Years ago lavender oil used to be produced in Southern France in numerous small distillation posts, distributed throughout the growing regions of the Départements Basses-Alpes, Drôme, Vaucluse, Alpes-Maritimes and Var. These posts consisted of old-fashioned direct fire stills,

Courtesy of Fritzsche Brothers, Inc., New York.

PLATE 5. An old-fashioned direct fire still as used years ago for the distillation of lavender in Southern France.

holding about 60 kg. of plants and 60 liters of water. An operation was completed after about 15 liters of distillation water had been collected. The action of the boiling water upon linalyl acetate, the main constituent of lavender oil, resulted in considerable hydrolysis of this ester, and the lavender oils obtained by this method were relatively low in esters. The introduction of water *and* steam distillation, in which the plant material is packed on a perforated grid above the boiling water, resulted in a marked increase of the ester content. This effect was even more pronounced when Schimmel & Company showed by systematic experiments in their modern distillery in Barrême (B.A.) that oils containing 50 per cent and more of esters could be obtained by rapid distillation with direct steam.

Distillation of Petitgrain Oil in Paraguay.—Similar conditions prevail in regard to the distillation of petitgrain oil in Paraguay. There, too, the leaf material is charged into primitive field stills and, during distillation, is partly submerged in boiling water. As a result, linalyl acetate, the main constituent of petitgrain oil, is partly hydrolyzed. For this reason, principally, the bulk of Paraguay petitgrain oil has an ester content averaging from 43 to 54 per cent only, whereas experiments with direct steam distillation in modern stills have proved that oils containing up to 80 per cent ester can be produced without too much difficulty.

Distillation of Linaloe Wood in Mexico.—The distillation of linaloe wood in Mexico furnishes proof that the yield and quality (physicochemical properties and chemical composition) of an essential oil depend a great deal upon the method of distillation.

The trunks and branches of the felled trees are reduced to chips with axes and machetes. The chips are then charged into galvanized iron retorts 1.10 m. wide and 2 m. high. In past years water was added to the chips and distillation of each batch carried out for about 18 to 20 hr., the heat being supplied by an open fire beneath the still. Distillation of linaloe wood in past years was thus a typical case of water distillation. The yield of oil then varied from only 0.6 to 1.0 per cent, seldom exceeding 2 per cent. This low yield was undoubtedly the result of insufficient reduction of the wood material, and in general of water distillation which in this case should not be applied.

In order to prove this contention, Schimmel & Company[25] imported Mexican linaloe logs to Europe and submitted the mechanically and properly comminuted material to direct steam distillation by modern methods. Yields ranging from 6.0 to 11.0 per cent of oil were obtained. The resulting oils differed considerably from the Mexican distilled oil. The latter contained more linaloöl, the Schimmel oil more of the high boiling constituents. Evidently in Mexico the wood was not sufficiently comminuted, with the result that little oil was liberated from the oil glands. The old Mexican method of distillation seemed to depend primarily upon the forces of hydrodiffusion, which means that the more water-soluble oil constituents—such as linaloöl—were freed from the wood, while the water-insoluble compounds remained, and partly resinified during the long hours of distillation.

During the years of World War II, the author visited the linaloe oil producing regions in Mexico and observed that the method of distillation has been improved considerably. Today the stills are equipped with a perforated tin plate, 60 cm. from the bottom of the still, the perforated plate supporting the chipped wood material. The section below the plate contains water which does not come in contact with the charge. Thus we have here a typical case of water *and* steam distillation, the water being heated with an open fire beneath the still. As a result of this method of distillation, the yield of oil today ranges from about 2.2 to 2.6 per cent for chips, and from about 3.5 to 4.4 per cent of oil from sawdust. Each operation now requires 8 to 9 hr. of distillation. Each charge consists of 230 kg. of wood material. In the states of Puebla and Guerrero, it is customary to reduce the linaloe wood into chips, while the producers in the state of Colima reduce the wood into coarse saw dust and thereby obtain a considerably higher yield.

Distillation of Cassia Leaves and Twigs in China.—Large quantities of cassia oil are produced yearly in the south Chinese provinces of Kwangsi and Kwangtung. The stills used by the natives are of antiquated Chinese design. Their principal fault lies in the loose connection of the joining parts and in the insufficient condensers. A charge consists of about 60 kg. of cassia leaves and twigs, and approximately 180 liters of water, which is brought to a boil by an open fire beneath

[25] *Ber. Schimmel & Co.*, October (1907), 55.

the retort. Distillation of one batch lasts about 2½ hr. The condensate is collected in a series of pots, arranged in the form of cascades. Cassia oil is heavier than water.

The yield of oil from leaves alone averages 0.10 to 0.13 per cent, and that from a mixture of 70 per cent leaves and 30 per cent twigs 0.15 to 0.17 per cent.

Because of insufficient cooling in the condenser, the distillate usually runs quite warm, if not hot, and therefore a part of the oil remains emulsified or suspended in the water. The milky distillation water is added to the next batch of plant material, a procedure which entails a certain loss of oil by evaporation and particularly by resinification. The principal cause of the subnormal yield of oil lies in the use of water distillation in the case of cassia leaves or, more exactly, in the faulty method of carrying it out. Cassia leaves possess a leathery consistency, remaining tough even in boiling water, and, therefore, if not sufficiently comminuted, cannot be completely exhausted by mere water distillation.

In order to study the problem by practical experiments, Schimmel & Company[26] imported dried cassia leaves and twigs from China, and submitted them to distillation tests in modern direct steam stills. The leaves yielded 0.7 to 0.8 per cent of oil, the twigs 0.2 per cent. These percentages are much higher than those obtained by the native Chinese distillers. True, the plant material arriving in Europe had lost considerable weight from drying; the Chinese producers use fresher leaves and twigs. Assuming the loss of moisture through drying to be about 50 per cent of the original plant weight, the yields of oil, as calculated upon the fresh plant material would, therefore, be as follows:

Fresh leaves, distilled in Europe	0.35 to 0.40%
70% fresh leaves plus 30% fresh twigs, distilled in Europe	0.31 to 0.34%
Fresh leaves, distilled in China	0.10 to 0.13%
70% fresh leaves plus 30% fresh twigs, distilled in China	0.15 to 0.17%

This differential in yield is actually even greater, because as a result of the long transport and desiccation, a part of the cinnamic aldehyde, the main constituent of cassia oil, had been oxidized.

These experiments prove that the native distillation of cassia oil is carried out in such a primitive and faulty way that quantities of oil amounting to about twice the actual production per year are lost in the residual plant material. The use of water distillation is not the only cause of this waste. Another reason for the subnormal oil yield obtained by the Chinese distillers, appears to be this:

Because of insufficient condensation of the steam/oil vapors, distillation must be carried out very slowly. The motion of the plant charge in the boiling water is, therefore, correspondingly slow, and the water between the agglutinating leaves cannot circulate sufficiently. The volatile oil, which diffuses from the leaves into the boiling water partly dissolves therein, and remains, in part, suspended between the agglutinated leaves, without being vaporized by contacting steam bubbles. In other words, presence of the liberated, but not vaporized, oil inhibits further diffusion of oil from the leaves. Evidently, the forces of diffusion can come into play only where there exists a differential in concentration. In other words, the quicker the oil solution is removed from the surface of the leaves, and the quicker the oil is vaporized, the more forcibly diffusion acts. Otherwise an equilibrium in the charge will result, and the distillate will contain very little oil, in spite of the

[26] *Ber. Schimmel & Co.*, October (1896), 11.

fact that considerable quantities of oil are still retained in the leaves. This also explains the relatively short length of distillation in the native cassia stills; the distillers simply stop the operation when they no longer see oil distilling over. It should not be surprising at all that the admixture of twigs to the leaves increases the oil yield, although the actual oil content of twigs amounts to only one-quarter that of the leaves. By the addition of twigs the charge simply becomes looser and the interspaces between the leaves larger. The boiling water, even though moving slowly, can then penetrate the interspaces much better and carry away the oil as it diffuses from the leaves; the oil is thus conducted toward the surface and vaporized.

(g) Rectification and Fractionation of Essential Oils.—Many essential oils, when distilled from the plant material, are contaminated with volatile products arising from the decomposition of complex plant substances, under the influence of hot water or steam. This takes place especially in the case of water distillation in directly fired stills if, through carelessness, the plant charge "burns" on contact with the retort walls touched by the fire. Some of these decomposition products are gaseous—e.g., hydrogen sulfide and ammonia; others—such as methylalcohol, acetaldehyde, acetone, and acetic acid—are very soluble in water. Therefore, they occur mainly in the distillation water, and accumulate in the water oil when cohobating the distillation waters. For this reason the water oil usually possesses a rather disagreeable odor and should not be mixed with the main oil without previous careful purification.

Occasionally the main oil, too, contains as normal constituents substances of somewhat objectionable odor—e.g., certain aldehydes or sulfur compounds. In order to improve the odor of such oils, they must be freed from these undesirable compounds by redistillation. This applies also to crude oils possessing too dark a color, which is often due to the presence of metals, or to fine plant dust carried over by the steam, especially when the live steam enters the still too forcefully or too rapidly. When the steam is injected more slowly, the plant charge becomes somewhat wet by steam condensation, and the dust particles are retained by the plant material.

Redistillation of a volatile oil does not necessarily bring about an improvement in its quality; in fact, in some cases the contrary may be true. This is particularly so with oils possessing easily saponifiable esters, such as bergamot or lavender oil. Linalyl acetate, the main constituent of these oils, is hydrolized by boiling with water, or by rectification with live steam, the freed acetic acid causing further hydrolysis.

For the redistillation of a volatile oil two general methods are employed, viz., rectification and fractionation, both of which will be described in more detail.

Rectification aims at the separation of volatile and nonvolatile compounds if a lighter colored oil is desired; the coloring matter remains as residue in the still. This may be achieved by dry distillation *in vacuo* or by hydrodistillation (with live steam or by boiling with water). Hydrodistillation can also be carried out at reduced pressure.

Fractionation or fractional distillation aims at separating the volatile oil into various fractions, according to their boiling points and odor. In most cases this is achieved by dry distillation *in vacuo.* A volatile oil should never be fractionated at atmospheric pressure, because the high temperatures involved cause decomposition and resinification, the distillate then possessing an odor and physicochemical properties quite different from those of the original oil. The boiling temperature can be considerably lowered by distilling the volatile oil at greatly reduced pressure, a process also referred to as dry distillation *in vacuo.* Decomposition of the oil is thus reduced to a minimum.

Rectification of Essential Oils.—Rectification with water vapors (steam) is the older of the two methods. Retorts employed for this purpose are usually spherical, made of copper, heavily lined with tin, and heated with a steam jacket. To prevent coloring of the oil by contact with metal, the gooseneck and condenser should be made of pure tin or of heavily tinned copper. Condenser and oil separator should be installed at such a height that, if it seems desirable, the distillation water can return automatically into the retort during distillation. Water is poured into the retort to a level of about 4 or 5 in. above the steam jacket and the oil added. Some oils—peppermint and caraway seed oil, etc.—easily assume a disagreeable by-odor when coming in contact with the hot still walls. This by-odor, known as "still odor," may be partly avoided by covering the steam jacket or the steam coils with sufficient water before starting the operation. The water level must be retained throughout the distillation. Flat-bottomed steam jackets are, therefore, preferable for the rectification of volatile oils. A steam coil, provided with many small holes and inserted close to the bottom of the retort, serves for direct heating with live steam (if this modification is preferred) and also for steaming out (cleaning) the still after completion of the operation. Steaming out is usually preceded by a washing with hot water, soap or alkali solution or with volatile solvents.

The speed of rectification is influenced by several factors. If the distillation waters should return automatically into the retort, the speed might be limited by excess pressure developing within the retort; in fact, this might altogether prevent the distillation water from returning automatically into the retort. If the distillate should be absolutely colorless, rectification must be carried out very slowly; otherwise very fine droplets, often invisible in the vapors, are carried into the condenser and oil separator, and color the distillate.

As has been said, some crude volatile oils contain compounds of objectionable odor, which are often more soluble in water than the main constituents. This fact can be taken advantage of by rectifying the volatile oils through hydrodistillation: the distillation water containing most of these objectionable compounds is not returned to the retort, but the water distilling off must be replaced by fresh water; or, instead of heating indirectly, direct live steam may be injected into the oil charge. In the latter case only sufficient water to cover the steam coil need be charged into the retort. However, the danger of oil droplets being carried over mechanically becomes somewhat greater as the live steam entering the retort has a tendency to whirl the oil upward. A short rectification column may be of service in this respect. When rectifying a volatile oil with direct live steam at atmospheric pressure (in other words, with low-pressure steam), some steam will be condensed to water continuously within the retort. The distillation water, in this case, cannot be returned into the retort, but must be cohobated in another apparatus or extracted with volatile solvents. Actually, rectification of a volatile oil with direct steam of low pressure has all the characteristics of a water distillation, because steam continuously separates water as condensate within the retort. If, however, high-pressure live steam (10 atmospheres for instance) is injected into a well-insulated still, condensation of water may be prevented, provided the steam has been carefully dried prior to its entering the still. The distillation then becomes a superheated steam process, because saturated, high-pressure steam, on expansion, changes into superheated steam. In other words, distillation of a volatile oil purely by live steam is not practicable. It turns either into a distillation with superheated steam or into a water distillation, the latter with the modification that there will be only a little water within the retort.

The quantity of oil to be charged into a rectifying still depends upon the final purpose of the rectification. If the oil is only to be decolorized, very little oil need be let into the retort, the vaporizing oil being replaced continuously as new oil is pumped in. This method offers the advantage that the contact of oil and steam is shortened to a minimum, only a small quantity of oil being in the retort at one time. A prolonged contact of volatile oil with boiling water or steam at atmospheric pressure is likely to cause considerable decomposition, resinification, and chemical action, such as hydrolysis of esters, etc.

As has been explained, rectification aims also at freeing the oil from disagreeable by-odors. If these impurities possess a low boiling point—in other words, if they boil below the main portion of the oil—they can be removed in the foreruns of the distillate. Foreruns are then separated so long as they exhibit the objectionable odor. Since a forerun usually amounts to only a small percentage of the total oil, it should be distilled off very slowly. The total amount of oil charged in the still, however, must always be so measured that it can be processed within one day. In order to utilize the capacity of a small still to the fullest, rectification is best carried out with direct live steam; otherwise a part of the retort must be occupied by the water necessary for distillation. After the forerun has distilled over, the speed or rate of distillation may be increased to whatever degree condenser capacity and purity of the distillate will permit.

If the volatile oil to be rectified contains impurities boiling higher than the main part of the oil, the main run should be distilled off slowly, as this will permit better separation and a diminution of the last runs. The speed of distillation may be increased when the last runs containing the impurities start to distill over. To achieve a more complete separation of the foreruns and last runs, a fractionation column may be used and, if necessary, a dephlegmator above the column. Such a dephlegmator causes partial condensation, which affects the higher boiling constituents more than the lower. It thus becomes possible to reduce the quantity of the forerun and to increase the quantity of the main run. As was explained under "Theories of Distillation," rectification columns are equipped with perforated trays, often with Raschig rings or porcelain balls. Columns filled with rings or balls have a practical advantage over columns equipped with bell or sieve plates, in that the former retain less condensed liquid and, therefore, exert less pressure upon the vapors in the still.

The composition of the condensate, i.e., the average oil content of the steam and vapor mixture, depends primarily upon the boiling point or the vapor pressure of the oil constituents. The lower the normal boiling point—in other words, the higher the vapor pressure of the oil constituents at the prevailing temperature of distillation—the greater will be the ratio of oil in the condensate. The average oil content of the steam and vapor mixture in the distillation of oil is much larger than it is in the distillation of plant material. (See section on "Steam Consumption in Plant Distillation.")

Fractionation of Essential Oils.—We shall now proceed to a description of the fractionation, which is carried out at reduced pressure (partial vacuum) and usually by distilling the oil alone, without leading water into the retort or injecting live steam into the oil. This process of dry distillation *in vacuo* is widely applied in the essential oil industry today. By its means pressure can be so far lowered that temperature has no longer any marked influence upon quality. The pressure should not be higher than 5 to 10 mm. Hg as measured in the still above the boiling liquid. How far the temperature of some oil constituents can be reduced is shown by this example: linaloöl, the main constituent of linaloe oil, boils at a temperature of 198°

at atmospheric pressure (760 mm.), and at:

105.4° at 30 mm. pressure
97.2° at 20 mm. pressure
84.4° at 10 mm. pressure
72.8° at 5 mm. pressure.

In practice, any further lowering of the pressure requires that distillation be carried out very slowly; it also necessitates an efficient vacuum still, absolutely airtight joints, and an effective condenser, so that the low boiling constituents of the volatile oil may be recovered, and not lost in the vacuum pump. In the case of almost every vacuum distillation, small quantities of vapors escape into the pumps, especially if the vacuum still is not absolutely tight. The air leaking into the still has a tendency to carry along some volatile oil vapors that may not always be condensed in the condensers. It is advisable, therefore, to insert an absorption vessel, filled with neutral substance which absorbs the vapors, between the oil receiver and the vacuum pump.

In order to distill over the highest boiling oil constituents *in vacuo*, temperatures of 150° to 200° are often necessary. Such temperatures can be obtained by the use of an oil bath which surrounds the lower part of the retort; in a corresponding steam jacket very high pressure or superheated steam would be required. The oil bath offers the advantage that the heat transmission between the two liquids is more gradual than that between superheated steam and volatile oil. Under these circumstances hydrocarbons possessing boiling points up to 300° (at atmospheric pressure) can be distilled off, provided any condensation of oil vapors along the upper walls of the retort is prevented by good insulation.

Between the pressure in the receiver and that within the retort there exists always a differential of a few millimeters. If the pressure in the closed receiver is 1 to 2 mm., the pressure in the vacuum still itself (retort) will be about 5 mm., provided the vapor development remains moderate. The faster the distillation, the lower will be the performance of the vacuum pump; the narrower the condenser tubes, the greater will be this pressure differential.

The stills serving for vacuum distillation of volatile oils are spherical, sufficiently strong to withstand at least atmospheric pressure, made of copper and heavily tinned on the inside, with gooseneck, condenser tubes and oil receivers also tinned. A small and strong glass window permits watching of the boiling liquid within the retort. All joints must be absolutely airtight. A jacket around the lower half of the retort forms an oil bath or a steam bath for heating with high-pressure steam (for at least 75 lb. jacket working pressure). A column directly above the still, equipped with plates, or filled with Raschig rings, or with other packing materials, provides for better fractionation of the boiling liquid (see "Theories of Distillation"). The oil receiver consists of two closed, strong vessels with vertical glass tubes, through which the level of the liquid within each receiver can be gaged.

Courtesy of Fritzsche Brothers, Inc., New York.

PLATE 6. Vacuum stills at Fritzsche Brothers, Inc., Clifton Factory, Clifton, N. J.

These receivers are tightly connected with the condenser outlet through a three-way stopcock, which permits one receiver to remain under *vacuo* and to collect the fraction distilling over at a given temperature, while the other receiver may be opened to draw off the previous fraction. Pressure manometers on the retort and on the oil receivers indicate the pressure within the retort and within the oil receivers. One thermometer held by a nipple reaches within the retort and ends above the boiling liquid, whereas another thermometer registers the temperature on top of and inside of the fractionation column. An airtight suction line connects the receivers with the vacuum pump which should be of high efficiency.

The still shown in Fig. 3.22 serves for the dry (without direct steam) vacuum distillation (fractionation or rectification) of essential oils. Heat-

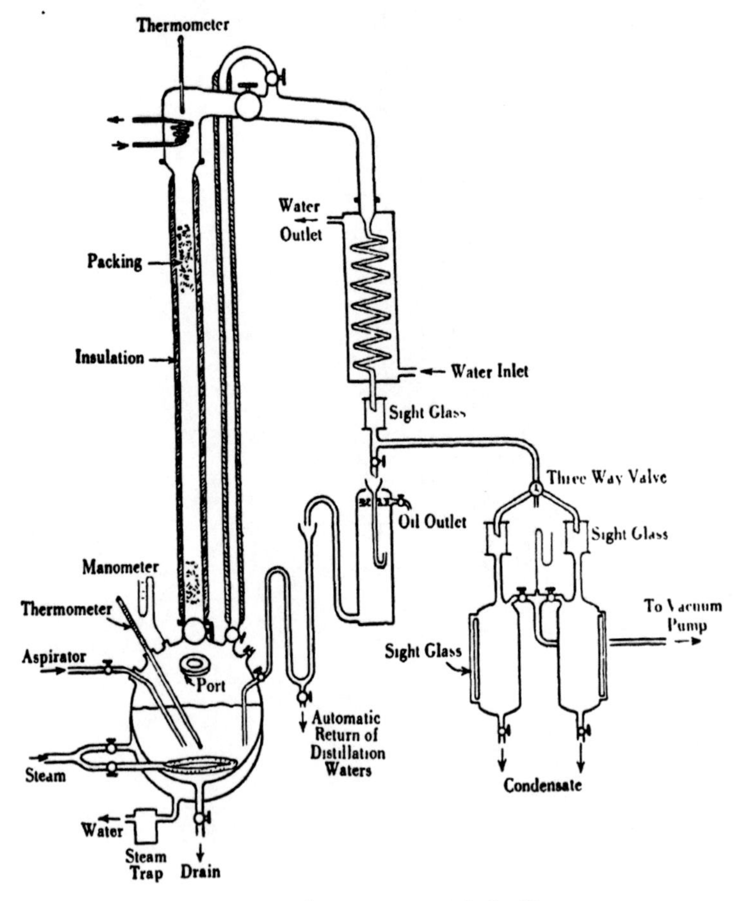

FIG. 3.22. Dual-purpose essential oil still.

ing is achieved by a steam jacket (or oil bath if so desired). The rectification column can be by-passed. Provision is also made for the rectification or fractionation of essential oils by the use of direct steam at atmospheric pressure. In this case, the distillation waters may be automatically returned to the still (cohobated). The same still may be used for the preparation of terpeneless oils.

Inadequacies of Hydrodistillation.—A comparative summary of the advantages and disadvantages associated with hydrodistillation of volatile oils and with dry distillation *in vacuo* would reveal, according to von Rechenberg, an almost general superiority of hydrodistillation over the latter method. Depending upon the nature of the compound to be vaporized, it is possible to adjust the temperature of hydrodistillation to any desired level. The use of dry vacuum distillation remains limited, because high boiling compounds decompose below their boiling points, even *in vacuo*. *Vacuum* distillation *with superheated steam* is more advantageous in this respect. On the other hand, the use of hydrodistillation is restricted for several reasons:

1. As in the case of dry vacuum distillation, the compound to be vaporized should be distilled in liquid form, or should at least melt below the temperature of distillation. However, solid compounds, and even those with very high boiling points, can be vaporized by steam, provided they are reduced to a moderately small size. Comminuted particles should be properly packed on perforated grids within the retort, so that the rising steam penetrates the mass uniformly, just as with plant material.

2. Hydrodistillation cannot be applied to substances which, even at low temperatures, react with water, or are hydrolyzed by water (esters, etc.).

3. Solubility in water, as well as decomposition by water, may, under certain circumstances, present an insurmountable obstacle to the use of hydrodistillation. This is particularly so if the compound to be distilled is high boiling (aside from being water soluble) or, in the case of plant distillation, if the plant material contains only very small quantities of the water-soluble constituent. Solubility in water lowers the vapor pressure of the compound and reduces its capability for vaporization; in other words, relatively much more steam will be required to vaporize the same quantity of oil. Since this lowering of the vapor pressure depends upon the quantity of water present, water soluble and high boiling compounds or corresponding plant matter should be distilled with steam, and not with boiling water. For instance, if it were practically possible to distill rose flowers with steam, the phenyl ethyl alcohol would probably not be retained by the flowers or by the residual still waters.

Solubility in water not only reduces the rate of evaporation, it also impedes the separation of the oil from the distillate. For this very reason the aroma of many flowers cannot be isolated by distillation. Any odoriferous compound is also volatile; any compound which, of itself, dissipates vapors into the air should yield the same, if not a larger quantity to steam, and particularly at a temperature of distillation higher than that of the air. The difficulty is only that the small quantity of volatile substances cannot be isolated from the large volume of distillation waters.

(h) Hydrodistillation of Essential Oils at High and at Reduced Pressure, and With Superheated Steam.

Water Distillation of Essential Oils at Reduced Pressure.—This type of distillation is used to prevent decomposition of the volatile oil, because by its use even easily hydrolyzed esters are retained intact. With certain oils the method gives most favorable results.

On the other hand, it should be kept in mind that the rate of vaporization of water-soluble and high boiling constituents decreases as their boiling point and degree of water solubility increase. Stated differently, in the water distillation of essential oils at reduced pressure, the ratio of oil to water in the distillate is even more unfavorable than when water distillation of the same products at atmospheric pressure is practiced, because any lowering in the external pressure reduces the vapor pressure of all high boiling compounds relatively much more than that of water (steam). Also, the differential between the temperature of distillation and that of the cooling water in this case is slight; therefore, considerable oil losses may be caused by evaporation, particularly when the temperature differential is still further reduced by any excessive and unnecessary lowering of the distillation pressure. The same conditions prevail here as with hydrodistillation of plant materials at reduced pressure.

To achieve a high rate of distillation when hydrodistilling volatile oils at reduced pressure, the empty space above the liquid in the vacuum still should be kept sufficiently large to permit the still content to boil without foaming into the condenser. In addition, the condenser surface must be larger (about five times larger than that required for distillation at atmospheric pressure). In the case of vacuum distillation, the efficiency of the condenser is considerably reduced by the high speed at which the steam and oil vapors rush through the tubes, and also by the fact that with lower temperatures of distillation the capacity of heat absorption by the cooling water diminishes.

In general, it can be stated that hydrodistillation at reduced pressure is especially suitable for the rectification of liquids that possess medium volatility and do not withstand heating, as well as for the purification of high boiling mixtures which are to be freed from lower boiling impurities. The method can also be used for removing traces of a solvent from an extract. Hydrodistillation can be conducted at as low a temperature and pressure as the temperature and the efficiency of the condenser permit.

Water Distillation of Essential Oils at High Pressure.—Pressure within the retort can be increased by inserting a throttling valve into the tube (gooseneck) connecting the retort with the condenser. When operating at a pressure above atmospheric, the unfilled space in the retort above the charge should be sufficient to prevent foaming over of the still content. The use of live steam is preferable, because refilling the still with water during the operation offers some difficulties. When heat is first applied to the retort, no excess pressure must be applied until all air has escaped from the still.

Water distillation of volatile oils at high pressure is useful for certain purposes —for instance, for the hydrolysis of esters, if so desired. This modification, however, by no means represents a general method of rectification. Relative to the steam pressure, the vapor pressure of higher boiling oil constituents increases more as the temperature rises; thus the ratio of oil in the distillate will be more favorable. However, from this angle, and from the practical point of view, water distillation at high pressure is not as effective as distillation with superheated steam, because the latter method vaporizes more oil without necessitating the high pressure of the former method.

Distillation of Essential Oils with Superheated Steam.—This occurs when the steam in the steam/vapor mixture rising from the oil is superheated. As was stated previously, this condition of the water component in the steam/vapor mixture is of great importance for the vaporization of oil. The same unit space occupied by a mixture of oil vapors and steam will contain relatively a much smaller quantity of steam, in a superheated state, than it would contain of saturated steam.

In actual practice, steam can be superheated by two methods:

1. By superheating within the retort.

The volatile oil is poured into the retort (without addition of water) and through a steam jacket or closed steam coil or oil bath, heated above the boiling point of water at the corresponding pressure. If saturated but dry steam is injected into the oil and thoroughly distributed, the steam will be superheated in the hot oil layer.

2. By superheating outside of the retort.

The steam is superheated in a special oven before it enters the retort and, as such, is injected into the oil, which does not have to be specially heated.

A combination of the two methods increases effectiveness of each. The stills serving for distillation of volatile oils with superheated steam should be constructed high with a small diameter; they should be well insulated, and provided with a steam jacket and a many-coiled perforated steam pipe. These precautions permit the injected steam to assume the temperature of the heated oil and to become thoroughly saturated with its vapors. When a distillate of high purity is desired, the force of distillation should be moderate—in other words, the quantity of the injected steam should be reduced. This is especially important in the case of vacuum distillation with superheated steam. A reduction in the rate of the injected steam also permits a more thorough saturation of the steam with oil vapors.

In general, it can be stated that distillation with superheated steam is particularly valuable in the case of those volatile oils or oil constituents which are partly soluble in water, because only a small quantity of water (steam) is required, and this stays in contact with the oil to be vaporized. The vaporizing liquid, therefore, acts like a water-insoluble compound. The method is well adapted to the distillation and purification of benzyl alcohol, cinnamic alcohol, phenyl ethyl alcohol, etc.—in other words, to all high boiling and chemically stable compounds which contain higher boiling impurities.

Distillation of Essential Oils with Superheated Steam at Reduced Pressure.—In the above described process, the steam can be superheated inside or outside of the still. An important modification, however, consists in connecting the retort and the closed oil/water separator (receiver) with a vacuum pump so that the oil vaporizes in the retort at reduced pressure. By this means it is possible to regulate the temperature of the oil vapors at will. According to the chosen temperature, the vapors will be more or less superheated which means a more favorable ratio of oil in the distillate than is the case when the oil is merely steam distilled without superheating. For example, by heating the oil charge in the retort indirectly with steam of 10 atmospheres pressure, by injecting dry live steam of high pressure very slowly into the oil at the same time, and by carefully adjusting the vacuum pump and the direct steam inlet to a distillation pressure of 30 to 40 mm. at a temperature of about 160° within the retort, even high boiling compounds such as glycerin, palmitic and oleic acid will distill over in ample quantities. For the vaporization of high boiling substances, this method therefore exceeds even dry vacuum distillation in efficiency. As for every type of hydrodistillation *in vacuo*, it is necessary to provide for sufficiently large condensers, and to inject the direct steam very slowly into the oil charge, so that no foaming takes place, and the distillate will not be contaminated with impurities mechanically carried over. (von Rechenberg).

B. NATURAL FLOWER OILS

As was stated in the section on "Distillation," most essential oils are today isolated from the respective plants, or parts of plants in which they occur, by the process of distillation. A few essential oils—i.e., those present in the peels of citrus fruit—can be, and in large part are, obtained by cold pressing, which yields products of superior quality.

In our discussion of distillation it was emphasized that the process of distillation suffers from several inadequacies: the relatively long action of steam or boiling water on the plant material affects some of the more delicate constituents of the oil deleteriously; hydrolysis, polymerization and resinification may and do take place; high boiling constituents, especially if somewhat soluble in water, are not carried over by steam, and are therefore lacking in the distilled oil. Other constituents dissolve partly in the distillation water, and cannot readily be recovered. As a result of all these factors, a distilled oil does not always represent the natural oil as it originally occurred in the plant.

A few types of flowers—and this is the case with some very delicate ones—yield no direct oil at all on distillation. The oil is either destroyed by the action of steam, or the minute quantities of oil actually distilling over are "lost" in the large volume of distillation water from which the oil cannot be recovered. This applies to jasmine, tuberose, violet, jonquil, narcissus, mimosa, acacia, gardenia, hyacinth and a few others. When hydrodistilled, these flowers yield either practically no oil, or in such low yield, or of such inferior quality, that for all purposes it is useless. Therefore, flowers of this type must be processed by methods other than distillation. This fact was recognized empirically hundreds of years ago when such flowers were treated by maceration in cold or hot fat, which process yielded fragrant pomades. From this primitive beginning there developed in the Grasse region of Southern France, in the course of many years, a highly specialized industry, employing the processes of maceration and of *enfleurage* and, for the last forty years, the modern process of cold extraction with volatile solvents. Despite similar, but much less important developments in other parts of the world (Bulgaria, Egypt, Algeria, Sicily, Calabria, Madagascar, etc.), Grasse has remained the center of this picturesque and charming industry, which today supplies the perfume manufacturers with a great variety of highly prized so-called "natural flower oils." Representing the authentic scents as exhaled by the flowers, these flower oils are the finest and most delicate ingredients at the disposal of the modern perfumer, enabling him to create masterpieces of his art by skillful application and blending.

The term "natural flower oil," as used today commercially, does not include the distilled essential oils; it applies only to flower oils obtained by the methods of *enfleurage*, maceration and extraction with volatile solvents, which will be described later in detail. A few oils—e.g., those derived from rose petals and from the blossoms of the sour (bitter) orange tree—can be isolated either by distillation or by extraction. The oils are then called essential oils and natural flower oils, respectively, the latter reproducing and representing the original scent of the flowers in a more complete way. It is principally the elaborate apparatus required and the higher cost of manufacturing which prevent a more general adaptation of the process of extraction.

I. EXTRACTION WITH COLD FAT (*ENFLEURAGE*)

In the Grasse region of Southern France, flowers were processed by this method long before the modern method of extraction with volatile solvents was introduced. Generations ago Grasse, an ancient hill town located on the southern slopes of the Alpes-Maritimes and facing the Mediterranean, became the center of extensive flower plantations and, subsequently, of the French perfume industry. Grasse, like few places in the world, is favored by a mild climate, southern exposure and protection against north winds. There the cultivation of flowers for the extraction of their scent became a highly specialized agricultural occupation, passed down from generation to generation.

In the early days of perfumery, flower scents were extracted with fats, and the alcoholic washings of the perfumed fats represented the so-called floral *extraits*. These, blended with certain distilled essential oils and tinctures, constituted the old-style perfumes. In the course of years this simple beginning led to our modern perfume industry with its wealth and variety of raw materials.

Despite the introduction of the modern process of extraction with volatile solvents, the old-fashioned method of *enfleurage* as passed on from father to son, and perfected in the course of generations, still plays an important role. *Enfleurage* on a large scale is today carried out only in the Grasse region, with the possible exception of isolated instances in India where the process has remained primitive.

The principles of *enfleurage* are simple. Certain flowers (e.g., tuberose and jasmine) continue the physiological activities of developing and giving off perfume even after picking. Every jasmine and tuberose flower resembles, so to speak, a tiny factory continually emitting minute quantities of perfume. This phenomenon was first studied by Passy[27] and later by Hesse.[28] Fat possesses a high power of absorption and if brought in contact

[27] *Compt. rend.* **124** (1897), 783. *Bull. soc. chim.* [3], 17 (1897), 519.
[28] *Ber.* **34** (1901), 293, 2928; **36** (1903), 1465; **37** (1904), 1462.

with fragrant flowers readily absorbs the perfume emitted. This principle, methodically applied on a large scale, constitutes *enfleurage*. During the entire period of harvest, which lasts from eight to ten weeks, batches of freshly picked flowers are strewn over the surface of a specially prepared fat base (*corps*), left there (for 24 hr. in the case of jasmine and longer in the case of tuberose), and then replaced by fresh flowers. At the end of the harvest the fat, which is not renewed during the process, has become quite saturated with flower oil. The latter is finally extracted from the fat with alcohol and then isolated.

(a) Preparation of the Fat *Corps*.—The success of *enfleurage* depends to a great extent upon the quality of the fat base employed. Utmost care must be exercised when preparing the *corps*. It must be practically odorless and of proper consistency. If the *corps* is too hard, the blossoms will not have sufficient contact with the fat, curtailing its power of absorption and resulting in a subnormal yield of flower oil. On the other hand, if too soft, the *corps* has a tendency to engulf the flowers so that the exhausted ones are difficult to remove and retain adhering fat, which entails considerable shrinkage and loss of *corps*. The consistency of the *corps* must, therefore, be such that it offers a semihard surface from which the exhausted flowers can easily be removed. Since the whole process of *enfleurage* is carried out in cool cellars, every manufacturer must prepare his *corps* according to the temperature prevailing in his cellars during the months of the flower harvest.

Many years of experience have proved that a mixture of one part of highly purified tallow and two parts of lard are eminently suitable for *enfleurage*. This mixture assures a suitable consistency of the *corps* in conjunction with high power of absorption. The author carried out a series of experiments with various mixtures of vegetable fats, especially hardened vegetable fats which do not easily turn rancid. He also experimented with all kinds of antioxidants and glycoside splitting compounds, incorporating them into the *corps* before *enfleurage*. The result was a variety of interesting qualities and widely different yields of flower oils, but the highest quality of floral oils most true to nature resulted from the old-fashioned mixture of lard and tallow.

Mineral oils, too, have been suggested as bases for *enfleurage* work, and on a limited scale have been practically employed; but they offer no real advantage because their power of absorption is very small as compared with that of animal fats. Furthermore, it is exceedingly difficult to extract and isolate small quantities of absorbed flower oils from the mineral oils with alcohol or by other means.

Many other substances have been suggested as bases for *enfleurage*, and have been patented for this purpose, but none so far has attained any wide commercial application. For instance, according to French Patent

836,172, January 12, 1939 (I. G. Farbenindustrie A.-G.),[29] essential oils and natural flower oils are extracted by treatment of the plant material with esters of polyhydric aliphatic alcohols, containing at the most 6 carbon atoms, with fatty acids of high molecular weight, as obtained by oxidation of paraffin hydrocarbons of high molecular weight. Thus esters of glycol, glycerol, erythritol, mannitol, hexitol or trimethylolpropane may be used.

The fat *corps* is prepared in the Grasse factories during the winter months when they are not busy with the processing of flower crops. The crude pieces of tallow and lard, mostly of French and Italian origin, are purified according to a tedious old-fashioned method. The crude fats are carefully cleaned by hand, all adhering particles of skin and blood vessels removed, then crushed mechanically and finally beaten in a current of cold water. After all impurities have been removed, the fat is melted gently on a steam bath. Small quantities of benzoin (0.6 per cent) and alum (0.15 to 0.30 per cent) are then added. This preservation is very important, as otherwise the *corps* will turn rancid during the hot summer months. While benzoin acts as a preservative, the adding of alum causes impurities to coagulate during the heating; when rising to the surface they can be skimmed off with a spoon. The warm fat is filtered through cloth, then left to cool and stand, so that any water may separate.

(During the past years chemistry has made great progress in regard to antioxidants for fats and oil, several of which could undoubtedly be used for preservation of the *enfleurage corps* employed in the Grasse region.)

The fat *corps* thus prepared is white, of smooth, absolutely uniform consistency, free of water and practically odorless. If well prepared and properly stored, it will resist rancidity for several years.

Some manufacturers also add small quantities of orange flower or rose water when preparing the *corps*. This seems to be done for the sake of convention. Such additions somewhat shade the odor of the finished product by imparting a slight orange blossom or rose note.

(b) *Enfleurage* and *Défleurage*.—Every *enfleurage* building is equipped with thousands of so-called *chassis*, which serve as vehicles for holding the fat *corps* during the process. A *chassis* consists of a rectangular wooden frame 2 in. high, about 20 in. long and about 16 in. wide. The frame holds a glass plate upon both sides of which the fat *corps* is applied with a spatula at the beginning of the *enfleurage* process. When piled one above the other the *chassis* form airtight compartments with a layer of fat on the upper and lower side of each glass plate.

Every morning during the harvest the freshly picked flowers arrive, and having first been cleaned of impurities, such as leaves and stalks, are then

[29] *Chem. Abstracts* **33** (1939), 5132.

Courtesy of Fritzsche Brothers, Inc., New York.

PLATE 7. *Enfleurage* process. (Spreading of jasmine flowers on top of the fat layer on the glass plates of the *chassis*.)

strewn by hand on top of the fat layer of each glass plate. Blossoms wet from dew or rain must never be employed, as any trace of moisture would turn the *corps* rancid. The *chassis* are then piled up and left in the cellars for 24 hr. or longer, depending upon the type of flowers. The latter rest in direct contact with one fat layer (the lower one), which acts as a direct solvent, whereas the other fat layer (beneath the glass plate of the *chassis* above) absorbs only the volatile perfume given off by the flowers.

After 24 hr. the flowers have emitted most of their oil and start to wither, developing an objectionable odor. They must then be removed from the *corps,* which process, despite all efforts to introduce labor-saving devices, is still done by hand. The careful removal of the flowers (*défleurage*) is almost more important than charging the corps on the *chassis* with fresh flowers (*enfleurage*) and, therefore, the women doing this work must be experienced and skilled. Most of the exhausted flowers will fall from the fat layer on the *chassis* glass plate when the *chassis* is struck lightly against the working table, but since it is necessary to remove every single flower and every particle of the flowers, the women use tweezers for this delicate operation. Immediately following *défleurage,* that is, every 24 hr., the *chassis* are recharged with fresh flowers. For this purpose the *chassis* are turned over and the fat layer, which in the previous operation formed the top (ceiling) of the small chamber, is now directly charged with flowers. In the case of jasmine, the entire *enfleurage* process lasts about 70 days; daily the exhausted flowers are removed and the *chassis* recharged with fresh ones.

During the height of the harvest large quantities of flowers arrive every morning, which necessitates certain modifications in the process. Complications result from the fact that at the beginning and at the end of the harvest the quantities of flowers are very limited and, therefore, it is practically impossible to charge the *chassis* each day of the flower harvest with the same amount of flowers.

At the beginning of, and several times during, the harvest, the fat on the *chassis* is scratched over with metal combs and tiny furrows are drawn in order to change and increase the surface of absorption.

At the end of the harvest the fat is relatively saturated with flower oil and possesses their typical fragrance. The perfumed fat must then be removed from the glass plates between the *chassis.* For this purpose it is scraped off with a spatula and then carefully melted and bulked in closed containers. The final product is called *pomade* (*pomade de jasmin, pomade de tubéreuse, pomade de violet,* etc.), the most highly saturated *pomade* being *Pomade No. 36,* because the *corps* on the *chassis* has been treated with fresh flowers 36 times during the whole process of *enfleurage.* At the beginning of the harvest every *chassis* is charged with about 360 g. of fat *corps* on each side of the glass plate, in other words, with 720 g. per *chassis.* Every kilo-

PLATE 8. *Défleurage* process. (Removal of jasmine flowers from the *chassis*.)

gram of fat *corps* should be in contact with about 2.5 kg. (preferably with 3.0 kg.) of jasmine flowers for the entire period of *enfleurage*, which lasts from 8 to 10 weeks. The quantities differ somewhat in the case of other flowers.

At the end of the *enfleurage*, the fat *corps* has lost about 10 per cent of its weight because of various manipulations. In other words, the total yield of the fragrant *Pomade No. 36* is about 10 per cent less than the fat *corps* originally applied to the *chassis*. Most of this loss is caused by fat adhering to the exhausted flowers when they are removed (*défleurage*) every 24 hr.

(c) **Alcoholic *Extraits*.**—In the early days of perfumery, the fragment *pomades* were employed directly; later they were extracted with high proof alcohol, the alcohol dissolving the natural flower oil from the *pomade*. These alcoholic washings are called *Extrait No. 36* when made from *Pomade No. 36*; they reproduce the natural flower perfume to a remarkable degree.

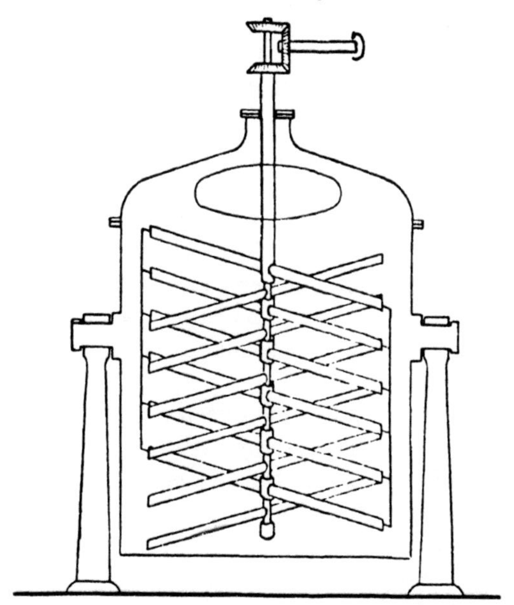

FIG. 3.23. Sketch of a *Batteuse* for the extraction of flower concretes with alcohol. (The agitation is in counter-rotary motion.)

Since no heat is applied during the process of *enfleurage* and during the washing of the *pomades* with alcohol, the *extraits* contain the natural flower oil as emitted by the living flowers. The only disadvantage exists possibly in a slight fatty "by-note" which can be eliminated to a certain extent by freezing and filtering the alcoholic washings. This slight fatty "by-note" is not always objectionable, as it imparts a certain roundness and fixation value to the finished perfumes, especially in conjunction with synthetic aromatics.

In order to prepare the *extraits*, the *pomades* are usually processed during the winter months when the factories are not busy with other work. For this purpose the *pomades* are charged into so-called *batteuses* (Fig. 3.23), closed copper vessels heavily tinned inside and equipped with strong stirrers around a vertical shaft. Several *batteuses* are arranged in batteries, the stirrers of each battery being driven by a powerful motor. The work, which goes on for several months, is carried out in cool cellars in order to prevent loss of alcohol by evaporation. Each batch of *pomade* is stirred for several days, the usual process of methodical extraction being applied. The alcohol employed in the process travels from one batch of *pomade* to the next (constituting in turn the third, second and first washings of successive batches), until it becomes enriched with flower oil and is drawn off as the alcoholic *extrait*. For the last washing, fresh alcohol is used, which also, in its turn, becomes gradually enriched by the continuous process just described. When extended to a fourth and fifth washing, this method extracts the *pomades* so efficiently that the exhausted fat is quite odorless. Being useless for new *enfleurage* it is usually employed for the making of soap.

The fully circulated washing called "*Extrait No. 36*" is run through a refrigerator and cooled to well below freezing temperature, if possible to −15°. Most of the fat dissolved in the strong alcohol separates. The cold alcoholic solution (*Extrait No. 36*) is then filtered, also at low temperatures.

The quantity of alcohol to be employed for the washing of each batch of *pomade* is calculated with a view to obtaining, finally, 1 kg. of *extrait* per kilogram of *pomade*. Obviously some alcohol is lost by evaporation during the process of stirring.

The purified *extraits* reproduce the perfume of the living flowers remarkably well. In fact, during the nineteenth century these *extraits* were widely employed as bases of the classical French perfumes, and several conservative houses still continue this practice. Some of the well-known French perfumes undoubtedly owe their success partly to a high content of *extraits*. The washing of the *pomades* is carried out not only by the factories in Grasse but in some instances also by perfume manufacturers in Paris, London, Berlin and New York who possess the necessary *batteuses* and freezing apparatus.

Since World War I, however, most perfumers have discontinued the cumbersome practice of processing the *pomades* purchased in Grasse; besides, high custom barriers prevented the shipment of alcoholic washings from Grasse into foreign countries. For these reasons the Grasse manufacturers started to offer their *extraits* in a more concentrated and convenient form.

(d) Absolutes of *Enfleurage*.—As mentioned previously, an *extrait* contains not only the natural flower oil, but also a small quantity (about 1 per cent) of alcohol soluble fat, dissolved from the *corps*, which cannot be eliminated, even by cooling the *extrait* far below 0°. When concentrating the *extrait* by distilling off the alcohol, the content of natural flower oil and fat increases correspondingly. Complete concentration in a vacuum still at low temperature results in a concentrated flower oil, free from alcohol, the so-called absolute of *enfleurage*.

The crude absolutes of *enfleurage* are usually of dark color and, because of their fat content, of a semisolid consistency. Lighter colored products of more liquid consistency can be obtained by certain methods of purification whereby more fat is eliminated. Further elimination of fat and purification increases the price of the final absolute. Every manufacturer has his own standards in this respect.

These so-called absolutes of *enfleurage*, absolutes of *pomade*, concentrates of *pomade* or liquid concretes were widely employed before the introduction of the more modern process of extraction with volatile solvents. Even today these absolutes of *enfleurage* find favor with some perfumers because of their lower price. Experts, however, claim that the absolutes of *enfleurage* when redissolved in alcohol are somewhat inferior to the original alcoholic *extraits*. Apparently during the process of concentration certain constituents of the natural flower oil, especially the most volatile and delicate ones, are lost.

A characteristic of absolutes of *enfleurage* is that they have a slight but noticeable "by-note" of vanillin quite alien to the true flower perfume. This note originates from the minute quantities of benzoin incorporated into the fat *corps* for protection against rancidity. Soluble in alcohol, the benzoin dissolves when the *pomades* are extracted with alcohol and upon concentration it accumulates in the absolute.

(e) **Absolutes of *Chassis*.**—When describing the process of *enfleurage* we mentioned that the flowers are removed from the fat *corps* on the *chassis* every 24 hr. These flowers are not thrown away because they still contain that part of the natural perfume which was not absorbed by the fat. It must be borne in mind that the perfume or essential oil of the flowers consists not only of volatile constituents, but also of compounds of higher boiling range which are not so readily released by the flowers. The actual conditions are probably much more complicated and many physiological processes take place, which so far have not yet been fully elucidated.

The part of the natural flower oil which is retained by the flowers after removal from the *chassis* (*défleurage*) can be extracted from these partly exhausted flowers with a volatile solvent—petroleum ether, for instance. Concentration of the solution results in a solid mass. (This product must not be confused with the concretes and absolutes obtained by extracting fresh flowers directly with volatile solvents.) The solid mass thus obtained contains a certain percentage of fat originating from the *corps* with which the flowers were in contact during the process of *enfleurage*; it is purified and made alcohol soluble by eliminating most of the fats at low temperature.

The final so-called absolute of *chassis*, a viscous, alcohol-soluble oil, possesses an odor differing somewhat from that of the absolute of *enfleurage*.

Absolute of *enfleurage* and absolute of *chassis* logically supplement one another because each represents only part of the total natural flower oil present in the living flowers. Yet, they are usually marketed separately, perhaps because the absolute of *chassis* is lower priced than the absolute of *enfleurage*.

Absolutes of *chassis* give excellent results in perfume blends, especially in conjunction with synthetic aromatics, the harsh notes of which are thereby softened and blended.

II. EXTRACTION WITH HOT FAT (MACERATION)

As explained, certain flowers—e.g., jasmine and tuberose—give their greatest yield of flower oil upon extraction with cold fat (*enfleurage*) because their physiological activities continue for 24 hr. and longer after harvesting. During this period, the fat on the *chassis* absorbs the perfume emitted by these flowers.

However, the physiological activities of other flowers—roses, orange blossoms, acacia, and mimosa, for instance—are stopped by picking. When extracted or distilled, they yield, therefore, only as much oil as is contained in the flowers at that moment. Since no further oil develops in these flowers, the long and rather complicated method of *enfleurage* would prove ineffective. Hence, other methods must be resorted to, whereby a medium actually penetrates the plant tissues and dissolves all flower oil present in the oil glands.

Hesse and Zeitschel[30] studied methods of distillation, cold *enfleurage*, maceration with hot fat, and extraction with volatile solvents as applied to various flowers, and the effect upon the yield of flower oils. Applying *enfleurage* to orange blossoms, for instance, Hesse found that this method yields only one-fifteenth of the amount of volatile oil obtained by steam distillation. Hesse thereby confirmed what had been known empirically in Grasse for generations.

Generations before the modern process of extraction with volatile solvents had been introduced (probably even in classical times), the perfumes of roses, orange blossoms, violets, acacia, mimosa and others had been obtained by treating the flowers with hot fat. The principle is simple:

The flowers are extracted by immersion in hot fat. In other words, the same batch of hot fat is systematically treated with several batches of fresh flowers until the fat becomes quite saturated with flower perfume. The exhausted flowers are removed and the fragrant fat, called *Pomade d'Orange*, *Pomade de Rose*, etc., is sold as such, or the *pomade* may be treated further

[30] *J. prakt. Chem.* [2], 64 (1901), 250, 258; 66 (1902), 506, 513.

by washing it with strong alcohol, exactly as with jasmine or tuberose *pomades*, obtained by cold *enfleurage*. The alcoholic *extraits* (*Extrait d'Orange, Extrait de Rose*, etc.) may be marketed as such, or they are concentrated *in vacuo*, giving thereby the corresponding absolutes of *pomade*.

The process of maceration, therefore, is somewhat analogous to that of *enfleurage*, with the fundamental differences that, in the case of maceration, *hot* fat is employed, and that the actual macerating of the flowers in the hot fat is done quickly.

Maceration was an important process before the introduction of the more modern method of extraction with volatile solvents. Fifty years ago, orange blossoms, if not distilled, were treated by maceration; acacia blossoms, which do not lend themselves to steam distillation, had to be processed exclusively by maceration. Similarly, roses were macerated in Southern France because French roses, unlike Bulgarian roses, give only a very low yield of oil upon distillation. However, today, the process of maceration with hot fat is employed very little. Its products, especially those from orange blossoms, find application only in a few old-fashioned perfume formulas. Otherwise the concretes and absolutes made by volatile solvent extraction have almost completely replaced the former *extraits* and absolutes of maceration.

For completeness, however, we shall give a brief description of the way this old-fashioned process is carried out:

As solvent a highly purified fat base is employed. It should be prepared most carefully and in the same way as described under *enfleurage*.

A batch of 80 kg. of *corps* is heated to about 80° and at that temperature macerated with charges of 20 kg. of fresh flowers each time. This is repeated until 1 kg. of *corps* has been treated with about 2 to $2\frac{1}{2}$ kg. of flowers. Every extraction lasts about one-half hour, at 80°, when the mass is left standing for about an hour during which it cools but continues macerating the flowers. The mass is then reheated, melted and strained through metal sieves and filter bags, whereby the exhausted flowers are eliminated. Since they retain some adhering fat, they are, while in the sieves, treated with scalding water, which liquefies the fat. The water easily separates from the fat layer. In order to remove all adhering fat, the flowers are finally packed between filter cloth, placed in a hydraulic press and submitted to pressure ranging up to about 3,750 lb. per sq. in.. Scalding water is thrown on the filter bags during the process so that any fat still retained by the flowers is melted and expressed. Expressed fat and water again separate easily. Instead of hydraulic presses, some manufacturers employ centrifuges for removing the exhausted flowers from the fat *corps*.

The method of maceration is rather cumbersome but it served its purpose in the old days when no better process was available. Its products (*extraits* and absolutes of maceration) often show a fatty "by-note" which originates from the fat *corps* and modifies the character of the original flower perfume. A further disadvantage consists in the fact that, on account of this fat content, absolutes of maceration easily turn rancid, thereby developing a sharp, disagreeable note. Because of their high alcohol content, the *extraits* are better protected against rancidity and spoilage in general.

III. EXTRACTION WITH VOLATILE SOLVENTS

This method was first applied to flowers in 1835 by Robiquet.[31] Somewhat later Buchner,[32] and Favrot[33] experimenting independently, processed flowers with diethyl ether. Around 1856, Millon[34] in Algeria extracted flowers with various solvents; Hirzel[35] in 1874 suggested petroleum ether as the most suitable solvent and obtained patents for his apparatus in several countries of Europe. Gradually the new method attracted the attention of the manufacturers in Southern France (Grasse and Cannes) and large-scale experiments were conducted independently by several industrial workers such as Piver, Vincent, Roure, Naudin, Massignon, Chiris, Charabot, and Garnier.[36] The latter obtained a patent for a novel type of rotatory extractor and extended his activities from Southern France to Bulgaria, Syria, Egypt and Réunion Island. Finally all the flower oil manufacturers in Grasse were forced to adopt the volatile solvent process, and constructed special extraction plants in addition to their existing *enfleurage* buildings.

The principle of extraction with volatile solvents is simple: fresh flowers are charged into specially constructed extractors and extracted systematically at room temperature, with a carefully purified solvent, usually petroleum ether. The solvent penetrates the flowers and dissolves the natural flower perfume together with some waxes and albuminous and coloring matter. The solution is subsequently pumped into an evaporator and concentrated at a low temperature. After the solvent is completely driven off *in vacuo*, the concentrated flower oil is obtained. Thus the temperature applied during the entire process is kept at a minimum; live steam, as in the case of distillation, does not exert its action upon the delicate constituents of the flower oils. Compared with distilled oils the extracted flower oils, therefore, more truly represent the natural perfume as originally present in the flowers.

[31] *J. Pharm.* 21 (1835), 335. *Buchner's Repert. f. d. Pharm.* 54 (1835), 249. *Pharm. Zentralbl.* (1835), 553.

[32] *Buchner's Repert. f. d. Pharm.* 56 (1836), 382.

[33] *J. Chem. méd.* (1838), 221. *Pharm. Zentralbl.* (1838), 442.

[34] *J. Pharm. chim.* [3] 30 (1856), 407. *Compt. rend.* 43 (1856), 197.

[35] "Toiletten-Chemie," 3d Ed., Leipzig (1874), 77.

[36] For details see *Perfumery Essential Oil Record* 12 (1921), 197–222.

Despite this obvious advantage the volatile solvent process cannot entirely replace steam distillation, which remains the principal method of isolating essential oils. Steam distillation, in most cases, is a simpler process: by employing portable direct fire stills, distillation can be carried out even in remote and primitive parts of the world, whereas solvent extraction necessitates complicated and expensive apparatus, and a crew of well-trained workers. Running expenses are comparatively high; a mistake in operation can be costly; the unavoidable loss of solvent, of which large quantities are employed during the process, is an important factor in the price calculation of natural flower oils. Extraction with solvents can, therefore, be applied advantageously only to the higher priced materials, particularly the flowers. A loss of 10 liters of solvent per 100 kg. of flower charge remains rather insignificant in the calculation of absolute of jasmine which is normally valued at several hundred dollars per pound;[37] but with low-priced oils such as rosemary or eucalyptus, ranging normally below $1.00 per pound, the loss of a few liters of solvent would make extraction prohibitive.

All extracted flower oils are of more or less dark color because they contain much of the natural plant pigments which are not volatile. Steam distilled oils, on the other hand, are in most cases of light color. Furthermore, they usually are soluble even in dilute alcohol, while extracted oils require 95 per cent alcohol for complete solution.

Despite these drawbacks, the products of extraction possess one supreme advantage, i.e., their true-to-nature odor. In addition, certain types of flowers—e.g., jasmine, tuberose, jonquil, hyacinth, acacia, mimosa and violet—do not yield their volatile oil on steam distillation, and must, therefore, be extracted with solvents.

(a) Selection of the Solvent.—The most important factor for the success of the extraction process is the quality of the solvent employed. The ideal solvent should possess several properties:

1. It should completely and quickly dissolve all the odoriferous principles of the flower, yet as little as possible of such inert matter as waxes, pigments, albuminous compounds, etc. In other words, the solvent should be selective.

2. It should possess a sufficiently low boiling point to permit its being easily removed (distilled off), without resorting to higher temperatures; yet, the boiling point should not be too low, as this would involve considerable solvent loss by evaporation in the warm climate of Southern France.

3. The solvent must not dissolve water since the water present in the flowers would dissolve and accumulate in the solvent.

[37] Up to $2,000.00 per lb. in 1946.

4. The solvent must be chemically inert, i.e., not react with the constituents of the flower oil.

5. The solvent should have a uniform boiling point; when evaporated it must not leave any residue. The slightest traces of high boiling compounds, upon evaporation of the solvent, would accumulate and remain in the flower oil and completely spoil its odor. It should be borne in mind that the yield of flower oil is generally very small, and that large quantities of solvent are required to cover the flower material in the extractors. In the case of petroleum ether, for instance, even traces of high boiling impurities are apt to impart to the concretes and absolutes an objectionable off-odor of kerosene, which cannot be eliminated without doing considerable harm to the delicate flower oil.

6. The solvent should be low-priced and, if possible, nonflammable.

The ideal solvent which would fulfill all these requirements does not exist. Considering every feature, highly purified petroleum ether appears to be the most suitable one, with benzene (benzol) ranking next.

Mixed solvents form a fascinating problem which so far has been little touched, but which promises quite interesting results. As compared with straight solvents, mixed solvents can either reduce or increase their dissolving power. Much experimental work along these lines has still to be undertaken.

Petroleum Ether.—Crude petroleum on fractional distillation yields a number of hydrocarbon fractions of different boiling ranges which find certain industrial applications. The fractions, boiling range 30°–70°, commercially called petroleum ether, consist of saturated paraffins, viz., mainly pentane and hexane. Because of their chemical inertness and complete volatility, these fractions are particularly suited for flower extraction. A further advantage lies in their selective power of dissolving: they yield products which contain relatively little wax, albuminous and coloring matter, but correspondingly more of the odoriferous compounds. Certain American petroleums are best suited for our purpose, because they consist mostly of inert, saturated paraffins, whereas Galician, Rumanian, Russian or "cracked petroleum" contain derivatives of benzene and naphthene, as well as unsaturated olefinic compounds, the latter being chemically active and liable to polymerization. They may thus form high boiling compounds of objectionable kerosene odor, especially on prolonged use of the solvent.

The petroleum ether must be free from sulfur and nitrogenous compounds. It is purified by washing in turn with strong sulfuric acid, water, hot dilute sodium hydroxide solution, water, and then drying. Castille and Henri[38] recommended repeated washing with sulfuric acid mono-

[38] *Bull. soc. chim. biol.* 6 (1924), 299.

hydrate, followed by washing with alkaline potassium permanganate solution and drying. They found the hexane fraction of petroleum ether, boiling range 65°–70°, of great advantage in extraction work because solutes remain in their normal molecular state; unstable compounds stay unchanged and no addition compounds are formed.

When testing petroleum ether for use in extraction work, special attention must be paid to the presence (absence!) of a nonvolatile residue. For this purpose a sample of 50 cc. should be evaporated in a glass or porcelain dish at a temperature not exceeding 40°. After complete evaporation the glass dish should show no residual odor whatsoever, but especially no odor indicating the presence of kerosene or sulfur compounds. A similar test can be carried out by permitting the solvent to evaporate on a clean filter paper at room temperature. More detailed methods of testing petroleum ether for purity are described in the "United States Pharmacopoeia," Thirteenth Revision.

In the extraction plants of Southern France the petroleum ether is usually prepared by submitting petroleum fractions to slow and repeated rectification in special stills provided with high fractionation columns and dephlegmators. As a rule, a small quantity (about 5 per cent) of odorless paraffin or fat is added to the gasoline in the still so that higher boiling compounds are retained and prevented from distilling over. According to the quality of the gasoline employed, 20 to 40 per cent remains as residue in the still, while 60 to 80 per cent represents the final *cœur* (heart) of petroleum ether suitable for extraction. Its boiling point should not be higher than 75°.

Although petroleum ether is the best solvent found so far for flower extraction, it possesses some inherent disadvantages—for example, relatively high solvent losses in the course of the extraction process. These losses are due primarily to evaporation of the low boiling, almost gaseous, fractions. Furthermore, petroleum ether is readily inflammable and dangerous to work with.

Benzene (*Benzol*).—Benzene ranks next to petroleum ether as a solvent for the extraction of flowers. It is a coal-tar product made by treating and purifying coal-tar naphtha with sulfuric acid and subsequently with sodium hydroxide. The fractions below 130° contain the lower benzene hydrocarbons which are composed mainly of benzene (C_6H_6), toluene, and other homologues. The industrial "benzol" often contains pyridine, carbon disulfide and thiophene which must be removed by treatment with concentrated sulfuric acid, water and caustic soda solutions. Further fractionation eliminates most of the higher boiling homologues but complete purification is obtained only by repeated crystallization. Thus, pure benzene melting point 5.5°, is obtained, the higher homologues remaining liquid and being separated by vacuum filtration, or other methods.

Crystallizable benzene is of such purity that 95 to 98 per cent of it distills within 1 per cent of the theoretical boiling point 80.1°. This uniform boiling point is of great advantage in extraction work, also because solvent losses are reduced. Yet, 80.1° is a relatively high boiling point, which makes it rather difficult to remove the last traces of solvent from the concentrated flower oil.

A further drawback of benzene in flower extraction work lies in its high dissolving power. It dissolves not only the odoriferous principles but also much wax, albuminous, and coloring matter, so that the final flower oils extracted with benzene are dark, highly viscous, often almost solid masses, which can be purified only under considerable difficulties and by special processes.

Compared with petroleum ether, benzene usually gives much higher yields of concretes, due to the higher amount of inert wax, albuminous, and coloring matter present. As far as the actual odoriferous principles are concerned, the yields obtained by benzene or petroleum ether are usually quite similar.

Summarizing, it can be stated that petroleum ether is preferred for extracting the more expensive flowers, while benzene serves in the case of lower priced plant material such as oak moss and labdanum, where the presence of coloring matter is not considered of too great a disadvantage.

Alcohol.—Alcohol cannot be used for the extraction of fresh flowers because it dissolves the water contained in them and becomes increasingly more dilute. With some flowers (tuberoses, for example) alcohol develops a most disagreeable odor; from others (jasmine, for example) it extracts dark, solid masses which possess an odor similar to molasses.

High-proof alcohol—in some instances dilute alcohol—is widely employed, however, for the extraction of dried plant materials, leaves, barks, roots, and especially gums, from which alcoholic tinctures are obtained. These tinctures find wide application in pharmacy and perfumery.

Concentration of these tinctures, usually by driving off the alcohol in a vacuum still, results in the so-called oleoresins and resinoids. These products are usually viscous, often almost solid, masses of dark color, representing the concentrated odoriferous principles, plus the alcohol soluble resins, coloring matter, etc., contained in the original plant material.

Resinoids of olibanum, myrrh, opopanax, benzoin, etc., are widely employed in perfumery, oleoresins of vanilla, ginger root, capsicum, celery seed, etc., in the flavoring of all kinds of food products and beverages.

(b) Apparatus of Extraction.

General Arrangement.—The extraction buildings of the Grasse region are usually of light masonry, one-storied, painted in light colors. The flat roof serves as a shallow water tank, insulating the building against sunheat and providing the condensers in the building below with water. The steam boilers are housed separately, at a safe distance from the main building, in order to exclude any danger of fire. The employment of volatile, highly inflammable petroleum ether or benzene also necessitates that all electric motors and switches be explosion proof, or be housed outside of the extraction building. The reserve solvent which is not in circulation must be stored in fireproof cellars, separate from the buildings.

The extraction building is equipped with one or two stills or columns for fractionating the solvent, a few batteries for extracting the flowers, and stills for concentrating the flower oil solutions. The batteries are of different size, so that they can be used according to the quantity of available flower material, which varies according to weather conditions and with the progressing harvest.

Construction of Apparatus.—Until some years ago the extractors and stills were constructed of copper, because this metal retains its value after scrapping the apparatus, and the pliable copper can be hammered and repairs are easily made. Lately, however, the extractors are constructed more often of heavily tinned sheet iron which is much cheaper and, therefore, offers the advantage of lower investment and quicker amortization.

The apparatus must be of solid construction to stand wear and tear; all extractor pipes and valves should be within easy reach to save time and exclude mistakes on the part of the operators. Pipes and valves must be sufficiently wide to prevent formation of pressure by air and petroleum ether vapors, one of the principal causes for loss of the solvent in vapor phase. Large diameters also permit quicker flow of solvent and solutions to and from the apparatus, and considerably speed up operation. Pumping of solvents and solutes is done by air pressure created by air compressors.

The extractors are mounted on elevated metal platforms along the inside walls of the building. The platform runs even with the ground outside of the building, so that the arriving flower material may be charged directly into the extractors and the exhausted flowers discharged with equal ease.

Loss of solvent during the operation presents one of the most serious problems. These losses are usually caused by incomplete distillation of the solvent from the exhausted flowers, by insufficient condensation in the condensers, or by too narrow pipes and valves creating pressure and blowing off a mixture of air and solvent vapor. To avoid this as much as possible the whole system of extractors, evaporators and solvent tanks is arranged as a closed circle, with only one outlet where escaping solvent vapors can be condensed. The Société Carbonisation et Charbons Actifs some years ago developed an efficient, rather small sized apparatus in which solvent vapors are absorbed by activated carbon and recovered by blowing live steam through the saturated carbon. A current of hot air then reactivates the carbon. The apparatus is simple and permits considerable economy in factories where large amounts of solvent are in circulation.

Description of Extraction Batteries.—A battery usually consists of three or four extractors, four or five metal tanks holding solvent and solutions, and an evaporator for concentrating the flower oil solutions. There exist two types of extractors—viz., the stationary and the rotatory types. Some factories employ both, but only one type can be used in the same battery.

The stationary extractors usually have a capacity of 1,200 liters, holding about 135 kg. of jasmine flowers, or 180 kg. of rose flowers. From 400 to 450 liters of solvent are required for extracting 100 kg. of flowers. Losses of petroleum ether may range from 12 to 14 liters for 100 kg. of flowers treated, but the loss can be considerably reduced with the solvent recovery apparatus described above.

Courtesy Les Parfumeries de Seillans, (Var), France, through Fritzsche Brothers, Inc., New York.

PLATE 9. The extraction of jasmine flowers with volatile solvents.

The stationary extractors are cylindrical, standing vertical. In the interior they should be provided with several perforated metal grids arranged horizontally around a vertical central support shaft. The flowers are charged upon these grids, thus spreading loosely over a larger surface and preventing lumping. The solvent can thus penetrate the mass freely and uniformly. After the flowers have been charged into the extractor, the metal cover on top is closed tightly with clamps.

Extraction is carried out methodically by successive washings whereby each batch of flowers is treated three times with solvent. A third washing is used as second washing for the next flower batch, then as first washing, and is finally pumped into the evaporator for concentration. The solvent distilled off is used as fresh solvent for a third washing which serves again as second of the next flower batch, etc. A fourth extraction, in most cases, yields at best only small quantities of waxes and other inert substances. The actual flower oil is contained in the first and second washings, while the third one serves merely to wash down parts of the second washing still adhering to the flower material. There exist, however, cases of emergency,

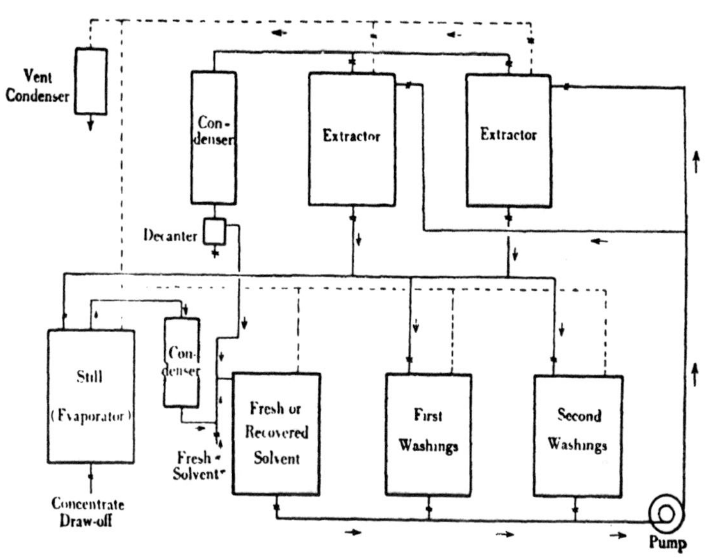

FIG. 3.24. Schematic diagram of an extraction system. (Extraction with volatile solvents.)

especially during the height of the harvest, when great quantities of flowers arrive and must be worked up quickly in order to avoid fermentation. In such cases the third washing is often eliminated altogether, in order to save time. Just how to proceed requires experience and good judgment on the part of the factory manager.

After the third washing the flowers are practically exhausted, and can be discharged. However, they still contain a considerable amount of adhering solvent which, before discharging the flowers from the extractor, must be recovered by steam distillation, i.e., by simply blowing live steam through the mass. Water and solvent, after condensation, separate automatically in a specially constructed Florentine flask.

The first washing requires about 45 min., the second 35 min., the third

25 min. For drawing off the solutions and pumping in the next washing, 5 to 10 min. must be allowed for each operation. Including 90 min. of steam distillation for recovering the solvent still adhering to the exhausted flowers, complete extraction of one batch of flowers requires about 4½ to 5 hr. However, no strict rules can be laid down, as every flower type requires a different method of working, and every manufacturer follows his own ideas.

Fig. 3.24 shows a schematic diagram of a system for extraction of flower material with volatile solvents.

Rotatory Extractors.—Years ago, Charles Garnier invented a rotatory extraction apparatus which was adopted by many factories. In modified and improved form[39] the apparatus represents a simple, solid and moderately priced piece of machinery.

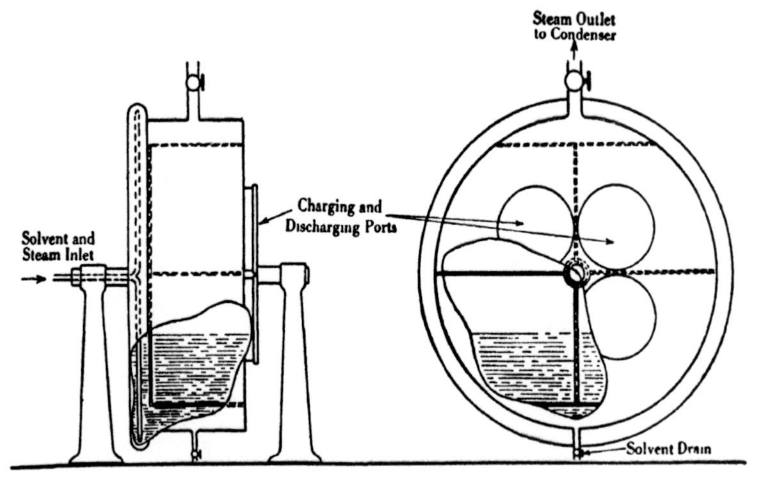

Fig. 3.25. Rotary extractor, Garnier type.

The latest model (Fig. 3.25) consists of a heavily tinned iron drum rotating around a horizontal axle. Four perforated metal partitions, rectangularly and horizontally arranged around the central axle, divide the interior into four compartments into which the flowers are charged through four manholes. While the whole system rotates slowly, the flower material moves, tumbles and dips into the solvent lying on the bottom of the extractor. The liquid seeps through the perforations and drips back to the bottom when one compartment is lifted out of the solvent in the continuous movement of the whole drum. Thus, the solvent does not fill the extractor but only the lowest part, the flowers dipping slowly and continuously into the solvent and rising again in the rotatory movement.

[39] French Patent No. 585199, October 30 (1923). See also Apparatus of Hugues, French Patent No. 508085, December 12 (1919).

Three successive washings are usually made, carried out similarly to the systematic extraction method described under stationary apparatus. The first, i.e., the most saturated washing, is then pumped into the evaporator and concentrated by distilling off the solvent. After the third washing has been drawn off the exhausted flower material, live steam is blown into the extractor to distill over and recover the petroleum ether still adhering to the extracted flowers.

The advantages of the rotatory extractors as compared with stationary apparatus are evident. Through the movement of the flower material in the solvent, its action becomes more penetrating and more effective, resulting in a somewhat higher yield. Figured as concrete, it is about 8 per cent higher in the case of the rotatory apparatus. Since the solvent covers only the bottom of the rotating drum and not the whole flower material, as in the stationary apparatus, much less solvent is in circulation and, therefore, evaporation losses are reduced. Only 160 to 170 liters are required in the rotatory apparatus to extract 100 kg. of flowers. The loss of solvent for 100 kg. of flowers is less than in stationary extractors; it may be 8 to 12 liters but can be reduced by using the previously described solvent recovery apparatus.

One rotatory extractor does the work of three or four stationary extractors arranged in one battery. Although superior in many ways the rotatory extractors suffer from several disadvantages and cannot altogether replace the stationary type; for example, the latter is better adapted to voluminous plant material, such as lavender, which cannot be so easily charged and discharged through the manholes of the rotating drum.

Concentration of Solutions.—The first, i.e., the most concentrated, washing is filtered through a fine screen and pumped into the so-called evaporator in which the greater part of the solvent is driven (distilled) off. These evaporators are of varying construction, representing basically a modified water or steam bath. In other words, the heating is done by indirect steam blown into a steam jacket beneath the still. The solvent, however, should not be completely driven off in this operation. Most manufacturers stop operation when the temperature in the still (evaporator) reaches about 60°, because any higher temperature at atmospheric pressure would be harmful to the delicate flower perfume. The first washing contains, of course, only a small percentage of flower oil. Therefore, concentrating of this washing in the evaporator means driving off 90–95 per cent of the solvent. (The recovered solvent serves as fresh solvent for a third washing.) The concentrated solution remaining in the evaporators is permitted to cool, is filtered, then transferred to a special, smaller vacuum still, and there completely concentrated *in vacuo*.

Final Concentration.—Vacuum stills (Fig. 3.26) of small capacity (50 to 100 liters) serve for this purpose. The final concentration represents a most delicate operation and requires much experience and constant attention on the part of the operator. The concentrating has to be done at as low a temperature as possible, yet any trace of solvent must be eliminated. Every manufacturer has his own, often secret, methods of purification. The completely concentrated and purified products represent the so-called floral

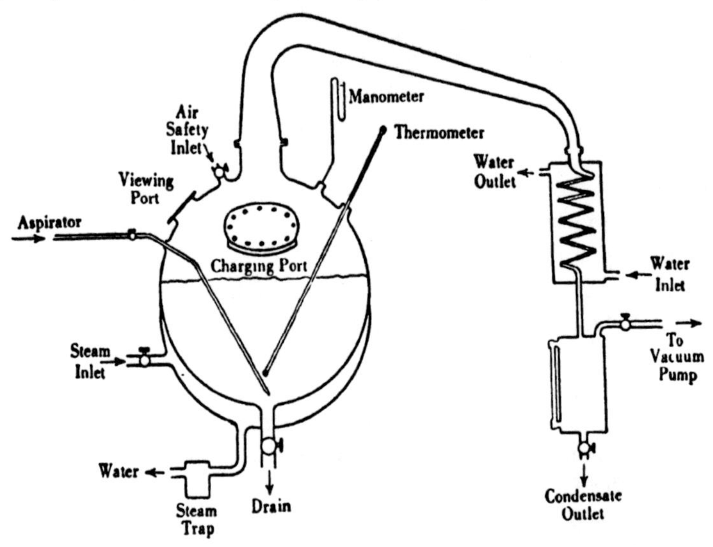

FIG. 3.26. Vacuum still for the final concentration of natural flower oils (removal of last traces of solvent).

concretes, which contain the odoriferous principles of the natural flower perfume, plus a considerable amount of plant waxes, albuminous material and color pigments. The concretes are, therefore, usually of solid consistency and only partly soluble in 95 per cent alcohol.

Concrete Flower Oils. Although these insoluble concretes are more difficult to work with, some perfumers prefer them to the alcohol soluble absolutes, which are obtained by precipitating and eliminating the insoluble waxes with strong alcohol, and concentrating the filtered alcoholic solutions. Distilling off the alcohol from the solutions when making these absolutes undoubtedly entails the loss of some of the most volatile and delicate constituents of the natural flower oil. It is often claimed that an alcoholic washing of a concrete is superior and more true to nature than a simple alcoholic solution of the corresponding absolute. On the other hand, the processing of concretes requires special equipment and considerable time; therefore, the absolutes represent a more convenient form of flower oils than the concretes.

Courtesy Les Parfumeries de Seillans, (Var), France, through Fritzsche Brothers, Inc., New York.

PLATE 10. Vacuum still for the concentration of the alcoholic washings in the preparation of flower oil absolutes.

Conversion of Concretes into Absolutes.—The alcohol soluble absolutes are prepared from the concretes by the following general method:

The concrete is either thoroughly rubbed down in a large mortar with a quantity of high-proof alcohol or, as some manufacturers prefer, melted and dissolved in warm alcohol. Subsequently eight to ten times the amount of alcohol is added and the mass stirred for a prolonged period in *batteuses*, as described under *enfleurage*. Usually five to six washings of the concrete are made in a systematic way, i.e., a third washing serves as a second one for a following batch of concrete; the second is used as first for a batch of new concrete, the first washing consequently representing the most concentrated solution. After standing and drawing off the clear solution from the alcohol insoluble waxes, the first washing is then thoroughly cooled in a refrigerator or in a special room, at temperatures ranging from −20° to −25°, when more wax precipitates and is filtered off in the cold. The resulting clear solution can be used as such in alcoholic perfumes.

Most perfume houses have neither the time nor the facilities to carry out their own washing of concretes, and prefer using alcohol soluble floral oils. For those, the manufacturers in Grasse offer the so-called liquid absolutes as the most concentrated and convenient form of floral oils. These liquid absolutes of extraction are obtained by carefully concentrating the first alcoholic washing of the corresponding concrete at low temperature in a good vacuum still. This process of concentrating involves a loss of several liters of alcohol per kilogram of absolute.

The absolutes are usually viscous oils with a more or less pronounced color, according to the degree of final purification (for which each manufacturer employs his own process). The absolutes are soluble in high-proof alcohol, and represent the most concentrated form of natural flower oils used in practical perfume work. However, they must not be confused with the actual volatile flower oil in the scientific sense. The absolutes usually contain from 50 to 80 per cent of alcohol soluble waxes, and only 20 to 25 per cent volatile oil, which can be isolated from the absolute by steam distillation. However, these volatile oils from the absolutes are not offered on the market because of their excessively high price, and because they completely lack the high fixation value of an absolute, which is due to the presence of alcohol soluble natural waxes, etc., in the absolute.

CONCLUSIONS

It might be worthwhile to review briefly the advantages and disadvantages of the various methods of manufacturing natural flower oils.

1. *Steam Distillation* of flowers yields volatile oils—for example oil of neroli bigarade, rose, ylang ylang. Not all types of flowers, however, can be processed by hydrodistillation, because boiling water and steam have a deteriorating influence upon the rather delicate odoriferous constituents. The flowers of certain plants yield no oil at all when distilled, and hence must be processed by methods other than distillation.

2. *Enfleurage* (extraction with cold fat). This method is carried out only in France, where it is still practiced, but on a much smaller scale than in former years. The method is restricted to those flowers (jasmine, tuberose, and a few others) which, after picking, continue their plant physiological activities in forming and emitting perfume. *Enfleurage*, in these cases, gives a much greater yield of flower oil than other methods. Despite this advantage, *enfleurage* has lately been replaced by extraction with volatile solvents because *enfleurage* is a very delicate and lengthy process, requiring much experience and labor.

3. *Maceration* (extraction with hot fat). This process used to be applied to those flowers which gave a very small yield by distillation or by *enfleurage*. Maceration, however, has lately been almost entirely superseded by the modern process of extraction with volatile solvents.

4. *Volatile Solvent Process.* Of general application, this process is today applied to many types of flowers, and carried out in several countries. It is technically the most advanced process, yielding concretes and alcohol soluble absolutes, the odor of which truly represents the natural flower oil as it occurs in the living flowers, or in the plants.

SUGGESTED ADDITIONAL LITERATURE

Y. R. Naves, "Extraktion von Duftstoffen mittels flüchtiger Lösungsmittel," *Riechstoff Ind.* (1936), 135, 151, 176, 212; (1937), 23, 50, 137.

Y. R. Naves and G. Mazuyer, "Parfums Naturels," Paris, 1939. (English translation, "Natural Perfume Materials," N. Y. Reinhold, 1947. Translated by E. Sagarin).

(c) The Evaluation of Natural Flower Oils and Resinoids.—The assay of distilled volatile oils has made remarkable progress during the last fifty years, probably because such oils are employed in much larger quantities than natural flower oils obtained by extraction with volatile solvents or by *enfleurage*. Moreover, the pharmaceutical profession, which uses many volatile oils, has always endeavored to assay carefully any products employed as medicine. Yet, it seems strange that so little attention should have been paid to the assay of such highly priced products as extracted flower oils, especially since it is common knowledge that sophistication of concretes and absolutes has become quite frequent, causing considerable loss to the often too credulous buyers. The reason for this neglect may be sought in the unfamiliarity of many users with these highly priced yet somewhat ambiguous products, the quality of which may depend upon many factors—methods of manufacturing, solvents used, degree of concentration, care in purification, etc. No wonder then that definite norms of quality do not yet exist and that the manufacturer may offer various explanations for deviations in his products. Indeed, some manufacturers market their natural flower oils in several grades, according to different degrees of dilution, in order to suit the usage and the purse of the users. Too, they claim that such "standardization" will guarantee a uniformity every year which nature alone does not achieve. "Les arts perfectionnent la nature," to quote an inscription on an old fountain in Grasse.

Natural flower oils, therefore, have remained strictly articles of confidence, and the examination of them is usually carried out by simple olfactory tests. Even such tests, however, require an intimate knowledge of the subject, a well-trained sense of smell, and familiarity with manufacturing methods and possible variations in quality, which very few buyers or even perfumers possess. Furthermore, any olfactory test should be based upon the comparison of an offered sample with standard samples of unquestioned purity and, if possible, of the same age. (Such samples, unfortunately, are seldom on hand.) It is surprising how the odor character of a natural flower oil may change during the first six months or year after its manufacture. Some odors improve for a certain period, and then slowly deteriorate, assuming a somewhat sour or rancid note. Something of a parallel may be drawn with wines of young and older vintage. In view of these facts, it seems highly desirable and timely to establish definite and universal standard methods for the physicochemical assay of natural flower oils and resinoids from gums, balsams, and similar plant material.

The adaptation of routine methods as applied to distilled volatile oils cannot *per se* be extended to extracted flower oils or resinoids, as these products contain large proportions of natural substances which, although olfactorily inert, possess a variety of chemical functions which would make the interpretation of the analysis most difficult, if not outright impossible. Logically, any physicochemical assay should, therefore, be applied mainly to the odoriferous portions of the extracted flower oils, which are usually identical with the volatile fraction. In other words, the extracted floral oil is steam distilled and the separated volatile portion examined by the usual tests for specific gravity, optical rotation, refractive index, acid number, ester number, ester number after acetylation, content of aldehydes, ketones, phenols, etc.

Separation of the two portions by dry distillation at reduced pressure is inadvisable, because of the tendency toward pronounced and often destructive pyrolysis of the higher boiling constituents. Consequently, the method of distillation with steam suggests itself for the separation of volatile and nonvolatile portions.

The first attempts toward establishing such a standard method were made by Walbaum and Rosenthal[40] who described an apparatus for the

[40] *Ber. Schimmel & Co.*, Jubiläums Ausgabe (1929), 189.

determination of the content in products distillable with steam from concrete and absolute flower oils. However, the separation of the volatile constituents from the waxes in this apparatus remains incomplete, even after 5 hr. of distillation, and gives much trouble, such as frothing, etc. Furthermore, live steam at atmospheric pressure causes hydrolysis and other chemical reactions. The method, therefore, can at best give only comparative results as far as the yield of volatile constituents is concerned, and only if carried out under absolutely unvarying and most carefully controlled conditions.

Several years later, Naves[41] suggested a more reliable, accurate and practical method, using distillation with superheated steam under reduced pressure. When superheated, dry steam behaves like a gas, follows the gas laws, and in the condensate yields a higher ratio of volatile aromatic constituents to carrier steam. Acting solely by its volumetric effect, dry, superheated steam is chemically less active than wet steam. Thus, with superheated steam it becomes possible to distill delicate esters and other compounds which would undergo hydrolysis with wet steam at the temperature of boiling water. For details of the method and a description of a cleverly constructed apparatus, the reader is referred to an interesting paper by Naves, Sabetay and Palfray,[42] who also examined a number of flower oils for their content of volatile constituents, and recorded the physicochemical properties of the distillable portions. The data given, however, are not yet complete enough to establish reliable standards universally adaptable by the trade. Much work along these lines must yet be done, and many more samples of unquestionable purity will still have to be examined before the essential oil industry can agree on definite norms.

Further progress in the perfection of the assay of natural flower oils was achieved by Sabetay,[43] who suggested that concretes and absolutes should be examined for their content of volatile constituents by codistilling these floral products or resinoids, etc., with ethylene glycol in a partial vacuum. (The same author[44] also suggested applying this method to the determination of volatile oils in drugs and spices.) Ethylene glycol is a more efficient carrier than steam, and the waxes or residues remaining in the distilling flask will be practically devoid of any odoriferous compounds. The application of a partial vacuum reduces the distillation temperature to a degree not harmful to the delicate constituents of the floral oils. Sabetay's method possesses the added advantage of simplicity:

[41] *Documentation scientifique No. 50,* December (1936), 303. *Chem. Abstracts* **31** (1937), 4772.

[42] *Perfumery Essential Oil Record* **28** (1937), 331.

[43] *Ann. chim. anal. chim. appl.* **21** (1939), 173. *Chem. Abstracts* **34** (1940), 3018.

[44] *Ibid.* **22** (1940), 217. *Chem. Abstracts* **35** (1941), 4547.

If a concrete or absolute is mixed with glycol and distilled under 8–15 mm. pressure at a temperature of 90°–100°, all of the volatile oil contained in the concrete or absolute can be driven over, separated and measured. For instance, weigh 1–10 g. of the concrete or absolute, add 25 cc. of ethylene glycol and distill at 90°–100°, from a 50–100 cc. Claisen flask with Vigreux points, fitted with a thermometer, capillary tube and receiver, and with a metallic or oil bath as a source of heat. As a rule, the residue in the flask will have little odor, but, if necessary, 20 cc. more of ethylene glycol can be added, and the distillation repeated a second and, possibly, a third time, until the distillate no longer becomes turbid upon addition of water. The combined distillate is diluted with water (or brine), treated with sodium chloride (if brine is not used) and extracted with three 20 cc. portions of a mixture of equal parts of pentane and ether. Dry the combined ether-pentane extracts over anhydrous sodium sulfate, remove most of the solvent by distillation, rinse the residue with pentane into a small Claisen flask with Vigreux points fitted with a capillary tube, and heat gently under 50–100 mm. pressure to constant weight. The volatile oil thus obtained may then be subjected to the usual physicochemical tests.

By comparing the figures (yield of volatile oil from the absolute or concrete or resinoid, specific gravity, optical rotation, refractive index, acid and saponification number of the volatile oil) thus obtained with those of absolutely genuine products, conclusions can be drawn as to the purity of the flower oil sample investigated.

The method of Sabetay may have to be modified in certain respects, and will have to be applied to numerous lots of unquestionably pure natural flower oils before definite standards can be agreed upon by the trade. Naves[45] prefers the use of superheated steam for the isolation of the volatile constituents from natural flower oils rather than codistillation with glycol, as certain constituents are relatively soluble in glycol-water solution.

Before concluding this chapter it might be well to discuss briefly the interpretation of analytical results, as well as the deterioration and possible adulteration of natural flower oils.

If absolutely pure, and manufactured according to unvarying methods, and from flowers grown in the same geographical location, natural flower oils should be of similar character and show little variation, especially in regard to their content of volatile (distillable) portions, and to the physicochemical properties of the volatile constituents. This, however, is not always the case, particularly with *enfleurage* products. The care exercised in the manufacturing process, and especially in the final purification of the

[45] *Helv. Chim. Acta* **27** (1944), 1103, 1108. *Soap, Perfumery, Cosmetics* **29**, No. 1 (1946), **38**.

product, exerts considerable influence upon its quality. The latter depends primarily upon the ratio between the weight of flowers treated during the entire *enfleurage* season and the weight of fatty vehicle (*corps*) employed. Thus a jasmine *pomade* will contain more volatile constituents and possess a much stronger odor if 1 kg. of natural *corps* has been treated during the flowering season with 2.5 or 3.0 kg. of jasmine flowers rather than only 1.5 kg. of flowers. Concretes obtained by extracting the flowers three times with solvent, instead of only twice, will contain more waxes. An absolute obtained by extracting the concrete four or five times with alcohol, instead of only three times, will contain more alcohol soluble waxes and other inert material, and correspondingly less volatile, odoriferous material.

Concretes and absolutes usually acquire a reddish color upon aging. This color alteration, noticeable particularly in jasmine and orange flower extracts, may be attributed mainly to the presence of indole. The odor improves, usually, for a few months, and assumes a harmonious fullness and depth lacking in the freshly extracted product. After a year or two of stability, depending of course upon proper storage, the product deteriorates gradually, finally acquiring a somewhat acid, rancid note, which is caused by the formation of acetic acid and ethyl acetate. This holds true especially if the product originally contained a small percentage of ethyl alcohol which was not removed during the final concentration. Hence, it is advisable to examine the aqueous phase of the analytical distillate, after extraction with ether, for its acid and ester number.

Pomades and absolutes of *enfleurage* are particularly susceptible to rancidity and development of acidity. In fact, even the freshly prepared absolutes of *enfleurage* show a relatively high acid number (which should not exceed 80), but this is caused by the presence of alcohol soluble free fatty acids extracted from the fat *corps*.

Adulteration of natural flower oils can be carried out in different ways, viz., by substitution with natural flower oils from lower priced geographical sources, by addition of volatile oils, or fractions therefrom, aromatic isolates or synthetic aromatics, or by dilution with inert materials. Thus, an absolute of jasmine marketed under the label of the Grasse region may contain the Egyptian product, a misrepresentation which, at present, can be detected only by olfactory tests, as we do not yet possess sufficient analytical data to differentiate between the products from these two geographical sources. The addition, to concretes, of exhausted natural flower waxes, obtained as alcohol insoluble residues in the preparation of alcohol soluble absolutes, results in a correspondingly lowered content of distillable volatile portions of the concrete. This can be proved by the above described distillation tests. Determination of the congealing point of the concrete may also give valuable hints in this respect.

A dangerous form of adulteration consists in the addition, to concretes or absolutes, of both odorless matters such as exhausted waxes, fats or fatty oils, *and* volatile, odoriferous compounds which occur also in the genuine flower oil, but which can be obtained synthetically or by isolation from lower priced essential oils. Thus benzyl acetate, benzyl alcohol, indole, etc., may be added to jasmine absolute; phenylethyl alcohol, rhodinol, etc., to rose absolute; linaloöl, linaloöl acetate, methyl anthranilate, etc., to orange flower absolute. If cleverly carried out by properly balancing the ratio between odorless nondistillable and odoriferous distillable compounds, such sophistication may give considerable trouble to the analyst, who will have to depend upon olfactory tests—and that, as pointed out, requires a highly trained and experienced sense of smell. Occasionally, natural flower oils are adulterated with odorless solvents such as diethyl phthalate, specific tests for which will be found in the chapter on "Examination and Analysis of Essential Oils, Synthetics, and Isolates."

When evaluating any natural flower oil, it is always advisable to test first for solvents and for alcohol. Traces of alcohol should not be objectionable as they are difficult to remove in the final purification during the manufacturing process without impairing the quality of the product. However, a flower oil should never contain any solvents like petroleum ether, benzene, or, particularly, kerosene, because their presence indicates incomplete purification; they impart to the product an off-note most detrimental to the delicate odor of natural flower oils. Special tests for alcohol and petroleum ether are described in Chapter 4 on "Examination and Analysis of Essential Oils, Synthetics, and Isolates."

SUGGESTED ADDITIONAL LITERATURE

The indexes of refraction of some perfumery concretes:

Chas. L. Palfray, *Ann. chim. anal.* **28** (1946), 94. *Chem. Abstracts* **40** (1946), 4850.

C. CONCENTRATED, TERPENELESS AND SESQUITERPENELESS ESSENTIAL OILS

Most essential oils consist of mixtures of hydrocarbons (terpenes, sesquiterpenes, etc.), oxygenated compounds (alcohols, esters, ethers, aldehydes, ketones, lactones, phenols, phenol ethers, etc.), and a small percentage of viscid or solid nonvolatile residues (paraffins, waxes, etc.). Of these the oxygenated compounds are the principal odor carriers, although the terpenes and sesquiterpenes, too, contribute in some degree to the total odor and flavor value of the oil. The oxygenated substances possess the added advantage of better solubility in dilute alcohol and, with the exception of some aldehydes, of greater stability against oxidizing and resinifying influences. Due to their unsaturated character, the terpenes and sesquiterpenes oxidize and resinify easily under the influence of air and light or under improper storing conditions which means spoilage of odor and flavor, and lowering of the solubility in alcohol.

For many years, therefore, it has been the endeavor of the essential oil industry to supply the users with concentrated, terpeneless and sesquiterpeneless oils. Such oils consist mainly of oxygenated compounds; they are more soluble, more stable, and much stronger in odor, yet retain most of the odor and flavor characteristics of the original oil.

The degree of concentration is automatically limited by the amount of oxygenated compounds present in the natural oil. For example, an orange oil containing only 2 per cent of oxygenated constituents and 98 per cent of terpenes, sesquiterpenes and waxes can, theoretically, be concentrated fifty times at the most, whereas a bergamot oil containing 50 per cent esters, alcohols, lactones, etc., and 50 per cent hydrocarbons can be concentrated only to double strength.

Before discussing in more detail the methods of manufacturing these concentrated, terpeneless and sesquiterpeneless oils, we should point out for clarity's sake that they must not be confused with the so-called isolates or aromatic isolates, or commonly but incorrectly called "synthetics" which are isolated from certain essential oils. For instance, citral can be isolated by fractionation or by chemical means from lemongrass oil, eugenol from clove oil, safrol from sassafras oil or camphor oil fractions, citronellal from citronella oil. These isolates may be converted chemically into other compounds, real synthetics, viz., citral into ionones, eugenol into vanillin, safrol into heliotropin, citronellal into citronellol, citronellyl acetate, hydroxycitronellal or synthetic menthol. Terpeneless and sesquiterpeneless oils have nothing to do with these isolates as the latter consist usually of only one well defined chemical substance, while the former are composed of several, often many, oxygenated compounds as present in the normal essential oil.

Because of the different composition, the deterpenation of each essential oil requires a special process. The general method is based upon two principles: (a) removal of the terpenes, sesquiterpenes and paraffins by fractional distillation *in vacuo* or (b) by extraction of the more soluble oxygenated compounds with dilute alcohol or other solvents. In many cases, especially with citrus oils, a combination of the two methods may be employed.

The commercial term "sesquiterpeneless" oils conventionally includes also the terpeneless oils. In some cases, especially when the content of sesquiterpenes in the natural oil is small, the two terms are employed synonymously. The trade designations and the names of the many brands on the market, however, are not always correct from the scientific point of view.

It would be more appropriate to name these products "Concentrated Oils," "Terpeneless Oils," and "Terpeneless *and* Sesquiterpeneless Oils."

"Concentrated oils" are those from which only a part of the hydrocarbons have been removed. This can be done by simple fractional distillation *in vacuo*. According to the process applied and the intended concentration, a wide range of concentrated oils, with different properties, may be obtained. Thus, we speak of a twofold lemon or orange oil, a fivefold oil, etc. "Terpeneless oils" are those from which all or most of the terpenes and waxes have been removed, usually by fractional distillation. "Terpeneless *and* sesquiterpeneless oils" are those from which the terpenes, the sesquiterpenes and the waxes have been eliminated. The common manufacturing practice is to distill off *in vacuo* first the terpenes, and then to extract the terpeneless oil with dilute alcohol, or other solvents, whereby the sesquiterpenes and waxes are eliminated; or, the sesquiterpenes and waxes may be removed by further fractionation of the terpeneless oil *in vacuo*. The resulting terpeneless *and* sesquiterpeneless oil represents the highest possible concentration of a natural essential oil.

The manufacture of these products requires that the operator be well acquainted with the chemical composition, especially with the boiling ranges of the various terpenes, sesquiterpenes and oxygenated compounds occurring in the natural oil which he expects to concentrate. The boiling range of terpenes varies in most cases from 150° to 180° at atmospheric pressure; that of sesquiterpenes from 240° to 280°. The boiling points of most oxygenated compounds (terpene alcohols, aldehydes, esters, etc.) lie between those of the terpenes and sesquiterpenes. Phenols, phenol ethers, and a few aromatic aldehydes form an exception, also the sesquiterpene alcohols, esters, etc., their boiling range falling into that of the sesquiterpenes or above.

As far as solubility in dilute alcohol is concerned, the terpenes are, in general, only sparingly soluble, the paraffins and sesquiterpenes practically insoluble. The oxygenated compounds, on the other hand, possess in general much better solubility: the alcohols, aldehydes, ketones, and phenols are most soluble, the esters and phenol ethers somewhat less soluble.

As pointed out, the terpenes may be removed by fractional distillation of the natural oil under reduced pressure. Most constituents of essential oils being deleteriously affected by heat, the distillation temperature must be kept as low as possible, which can be achieved with the aid of a good vacuum. For best results a well-constructed fractionation still as described in the section on "Distillation of Essential Oils" should be employed. It must be equipped with an efficient fractionation column.

It should be borne in mind that the terpenes cannot be removed quantitatively from a natural oil by mere fractional distillation; indeed, one of the greatest disadvantages of fractional distillation lies in the incomplete separation of the constituents, especially if their boiling points lie close together. A typical example is lemon oil which, aside from citral, contains also lower boiling aldehydes, such as octyl, nonyl and decyl aldehyde. If natural lemon oil is fractionated at 2 mm. pressure, the lower boiling terpenes should come over first, and theoretically the terpene fraction should contain no citral. However, even with a very efficient fractionation column, the aldehyde content of the terpene fraction will amount to about 1.0 per cent. The terpene fraction may be refractionated, but it will still retain small quantities of aldehydes; furthermore, repeated heating affects the flavor. Separation of the oxygenated compounds by chemical means is limited to certain cases only.

Repeated fractionation results in several intermediary fractions which consist of terpenes and a slight amount of oxygenated compounds, the latter increasing in proportion as the distillation temperature rises. Fractionation may be conducted in such a way that the residual oil is free from terpenes, but in this case the residual oil will be deprived also of those portions of the oxygenated constituents which have been carried over into the intermediary fractions. In order to recover these compounds, it will be necessary to refractionate the intermediary fractions, but, as said, prolonged heating is likely to have a deleterious effect upon the odor and especially the flavor of the fractions. Fractionation may be controlled by testing each fraction for solubility and for its rotatory power.

The elimination of the sesquiterpenes presents even more difficulties than that of the terpenes. In some cases the sesquiterpenes may be separated from the terpeneless oils by mere fractionation *in vacuo*, provided that the oils are not affected by the relatively high boiling temperature required for the distillation of sesquiterpenes (about 120°–140° at 10 mm. pressure) and by the partial overheating in the still which easily takes place. A vacuum of 3 to 5 mm. is desirable. In this case, too, the manufacturer must be familiar with the boiling points, at reduced pressure, of the various oil constituents. In some cases the differential in the boiling points of two compounds, as prevailing at atmospheric pressure, does not remain constant at reduced pressures; it may even be reversed. Too, every fraction should be tested for its rotatory power and for solubility in dilute alcohol, the insoluble ones to be rejected as containing mainly sesquiterpenes. Refractionation of the rejected fractions may be necessary. Even at a pressure of only 1 mm., a relatively high temperature is required to distill over the oxygenated compounds, most of them boiling between 90°–110°. Moreover, the temperature in the still itself will usually be about 10° and even 20° higher than the boiling point of the liquid, and intense local heating occurs especially along the walls of the still. All these factors tend to impart to the oil a note which the expert easily recognizes as "distilled" or slightly "burnt," as it does not occur in natural cold-pressed citrus oils, for example. Furthermore, the influence of heat seems to decompose the so-called "molecular compounds" which some authorities assume to occur in natural oils. It is a well-known fact that, upon aging, the odor of a perfume or flavor mixture changes and improves considerably. This may be caused by chemical reactions of functional groups—for example, by the interaction of alcohols and aldehydes which form acetals. Such compounds may exist in the natural oil and be decomposed upon heating and distilling.

Another method of removing the high boiling sesquiterpenes and waxes from the terpeneless oil consists in *steam* distilling the terpeneless oil at reduced pressure. This process is more gentle than dry distillation *in vacuo* and leaves the high boiling sesquiterpenes and waxes as residues in the still. In this case the distillate should be tested for solubility; any sesquiterpenes distilled over may be removed by treating the fractions with dilute alcohol. This method, however, has the inherent disadvantage that compared with dry vacuum distillation it takes much longer, especially in the case of oils containing a large percentage of high boiling compounds. Also, certain constituents of an oil are liable to dissolve in the distillation water—e.g., phenylethyl alcohol or eugenol. In this case the distillation water has to be returned into the still for cohobation.

In view of these inadequacies, some manufacturers remove all remaining terpenes and sesquiterpenes from concentrated oil by extracting the latter with dilute alcohol. The strength of the alcohol to be employed for this purpose depends primarily upon the solubility of the oxygenated compounds. Thus, the concentrated oil from which most of the terpenes and sesquiterpenes have been eliminated by fractionation *in vacuo* or by steam distillation under reduced pressure is shaken for some time with fifteen to twenty times its volume of dilute alcohol, for instance, with 60 per cent alcohol by volume; or the concentrated oil is first dissolved in the corresponding volume of strong alcohol and then the required amount of distilled water is gradually added with continuous stirring until the desired degree of alcohol dilution is reached. In both cases the turbid mixture should be cooled for a prolonged period and set aside until clarified. Thus, the oxygenated constituents dissolve in the dilute alcohol, while the terpenes and sesquiterpenes remain undissolved and (together with traces of oxygenated compounds) may be separated.

Because of the small differential in the specific gravity of the undissolved parts of the oil and that of the solution, emulsions may form and the separation of the two layers may require some time. In order to break the emulsion, small quantities of low boiling petrol ether are added, or the emulsion may be separated by centrifuging. The undissolved oil is repeatedly treated with dilute alcohol in order to extract any quantities of oxygenated compounds which it might still retain.

The clear solution of oxygenated compounds in dilute alcohol is then transferred into a still and the alcohol fractionated off at reduced pressure, until only oil and water remain in the still. The two layers of oil and water can easily be separated. The employment of an efficient condenser will prevent losses of alcohol. The recovered alcohol and the residual water may be used again for treating the next batch of oil.

The literature on essential oils contains many references to the preparing of terpeneless and sesquiterpeneless oils, one of the most comprehensive ones being the paper by Littlejohn.[46] As stated, no standard method has been adopted yet and every manufacturer uses his own process. Romeo[47] reported that terpeneless and sesquiterpeneless citrus oils are manufactured in Sicily by first removing the terpenes by fractional distillation. The sesquiterpenes are then eliminated from the terpeneless oil by extracting the oil with dilute alcohol, the strength of which should be somewhat lower than that in which the sesquiterpeneless oils must finally be soluble. The sesquiterpeneless oil is separated from the alcoholic solution by the addition of water or by distilling off the alcohol under reduced pressure. This constitutes the general method described previously and with some modifications it forms, today, the basis of most commercial processes.

A more novel method has been described and patented by van Dijck and Ruys.[48] In this process the natural oil is extracted by two solvents which are only partially soluble in one another—for instance, pentane and dilute methyl alcohol. The two solvents are made to flow, according to the countercurrent principle, through a horizontal glass cylinder and the oil is entered in the middle. The terpenes dissolve in the pentane phase, the oxygenated compounds in the methylalcohol phase. After separation of the two phases, the solvents are removed by distillation, only low temperatures being necessary. This, according to the inventors, is the principal advantage of their method, aside from the fact that high-grade terpeneless oils are obtained in almost quantitative yield. The principal difficulties of this process lie in the necessity of working with large volumes of solvents, furthermore, in the tendency toward formation of emulsions which, however, might be broken in some cases by the addition of 0.1 per cent of citric or tartaric acid.

After having discussed the various methods of manufacturing terpeneless and sesquiterpeneless oils, it might be advisable to add a few words about

[46] *Flavours* **3**, No. 4, August (1940), 7.

[47] *La deterpenazione delle essenze di agrumi. Estr. dagli Atti del II. Congresso Nazionale di Chimica pura ed applicata.* Palermo, May (1926). *Ber. Schimmel & Co.* (1928), 38.

[48] *Perfumery Essential Oil Record* **28** (1937), 91.

their concentration, as there exists a great deal of confusion regarding this point. The price lists and tables on the concentration of terpeneless and sesquiterpeneless oils issued by the various essential oil houses differ widely in regard to their concentration value. Yet, the theoretical concentration could be calculated only from the actual yield of terpeneless or sesquiterpeneless oil as obtained from a given weight of natural oil. However, the actual odor and flavor strength of two oils, although of the same theoretical concentration, may differ, concentration not being necessarily proportionate to odor and flavor strength. Let us assume, for instance, that 100 kg. of natural lemon oil are converted into terpeneless oil and that the yield is 8 kg. of terpeneless oil containing about 40 to 45 per cent of citral. (Some citral has been destroyed by distillation and, besides, the oxygenated compounds cannot be completely freed of terpenes.) In this case the actual concentration of the oil, but not necessarily of the flavor, is obviously twelve and one-half times.

It is difficult, if not impossible, to indicate general and definite limits for the physicochemical properties of concentrated, terpeneless and sesquiterpeneless oils because of the fact that these properties depend upon the degree of concentration and upon the relative proportions of oxygenated constituents originally present. Furthermore, every manufacturer has his own standards which are based upon his particular manufacturing process.

Littlejohn[49] listed the physicochemical properties of more than fifty terpeneless and sesquiterpeneless oils. Accordinging to this author, the specific gravity affords a valuable clue to the presence of any remaining terpenes. These hydrocarbons possess a low specific gravity and refractive index and their complete removal should raise the specific gravity and refractive index of the terpeneless oil relative to that of the original oil. The determination of the optical rotation, too, provides a good indication regarding the extent to which the terpenes have been eliminated.

By far the most important criterion for a terpeneless and sesquiterpeneless oil is its solubility in dilute alcohol, 70 per cent ethyl alcohol usually being employed for this purpose. A terpeneless oil should usually be soluble in 3 to 10 volumes of 70 per cent alcohol, while a sesquiterpeneless oil should be more soluble.

Aside from the determination of the physicochemical properties, it is advisable to test a terpeneless or sesquiterpeneless oil also for its content of oxygenated compounds, especially for alcohols, esters and aldehydes, which can be done by the usual analytical methods. Böcker[50] suggested a method of evaluating and examining terpeneless lemon oils which is based on treating the aldehyde-free oil with 51 per cent alcohol to remove all oxygenated

[49] *Flavours* 3, No. 4, August (1940), 7.
[50] *J. prakt. Chem.* [2], 89 (1914), 199; [2], 90 (1914), 393.

compounds, and on measuring the quantity of terpenes and sesquiterpenes left. For this purpose, 10 cc. of the oil is first treated with a solution of neutral sodium sulfite which removes the aldehydes. The remaining, unabsorbed oil is shaken with 100 times its volume of 51 per cent alcohol in a separatory funnel, cooled to about $-2°$ and left for a period of 6 hr. or more, until the liquids have completely separated when the lower layer can be removed. After washing the oil layer with a further quantity of 51 per cent alcohol, all undissolved oil is transferred to a burette tube and its volume carefully measured. From this amount the percentage of nonoxygenated constituents of the original oil can be determined. In order to obtain more exact results the terpeneless oil is first fractionated and the process applied to both the first and the last fraction. Böcker's method is not absolutely quantitative, as some terpenes will dissolve in the weak spirit, also because the transfer of the oils from the separatory funnel to the measuring burette always causes some loss.

The main advantage of the terpeneless and, especially of the sesquiterpeneless oils consists in their better solubility in dilute alcohol. The employment of these oils, therefore, effects a considerable saving of alcohol in the finished goods; odor and flavor of the oil are better utilized. A further advantage consists in the fact that, by the process of concentration, the oils are also freed of any products of decomposition or resinification which might result from improper handling or aging of the natural oils. Another merit of the terpeneless *and* sesquiterpeneless oils lies in their better stability. While natural citrus oils are apt to resinify, primarily due to polymerization of certain hydrocarbons, the concentrated oils are much more stable. Thus, they may be employed in powders, for the flavoring of gelatin desserts, for example, or for the scenting of bath salts.

The introduction of terpeneless and sesquiterpeneless oils on the market has met with some resistance. Several authorities contend that the elimination of terpenes and sesquiterpenes removes also a part of the characteristic odor and flavor of the natural oil. The application of heat undoubtedly has some effect on the delicate flavoring constituents of the oil and, if improperly prepared, concentrated oils may not display the freshness and bouquet of the original oil. Furthermore, the terpeneless and sesquiterpeneless oils contain a lower proportion of natural fixatives such as waxes and stearoptenes which contribute to the retaining of the flavor on the palate. Weighing the advantages against the disadvantages, the conclusion may be drawn that concentrated, terpeneless and sesquiterpeneless oils have their definite place in many formulas where highest possible concentration, solubility, and stability are required, but that they cannot replace the natural oils for all purposes.

SUGGESTED ADDITIONAL LITERATURE

"Concentrated Citrus Oils," by A. H. Bennett, *Perfumery Essential Oil Record* **25** (1934), 111.

"Preparation of Terpeneless Oils," by A. M. Burger, *Riechstoff Ind.* **13** (1938), 217. *Chem. Abstracts* **33** (1939), 1089.

"Removal of Terpenes from Essential Oils," by Pietro Leone, *Riv. ital. essenze profumi* **28** (1946), 5, 39, 82. *Chem. Abstracts* **40** (1946), 5201.

"Terpeneless and Sesquiterpeneless Oils," by Y. R. Naves, *Mfg. Chemist* **18**, No. 4 (1947), 173.

Chapter 4

THE EXAMINATION AND ANALYSIS OF ESSENTIAL OILS, SYNTHETICS, AND ISOLATES

BY

Edward E. Langenau

Note. All temperatures in this book are given in degrees centigrade unless otherwise noted.

CHAPTER 4

THE EXAMINATION AND ANALYSIS OF ESSENTIAL OILS, SYNTHETICS, AND ISOLATES

"The Examination and Analysis of Essential Oils, Synthetics, and Isolates" describes the commercial methods of testing and evaluating the raw materials of the essential oil industry. Most of these methods have been used in the New York and Clifton laboratories of Fritzsche Brothers, Inc., during the course of many years. They frequently represent standard official procedures or modifications of such procedures. Many highly specialized techniques which are of value in the scientific examination of essential oils have not been included because they are seldom used in a commercial laboratory.

Acknowledgment is due to the following standard reference works for much basic material used in the descriptions and discussions of these analytical methods:

"The United States Pharmacopoeias," United States Pharmacopoeial Conventions, Washington, D. C.
"The National Formularies," American Pharmaceutical Association, Washington, D. C.
"The British Pharmacopoeias," General Council of Medical Education and Registration of the United Kingdom, Constable & Co., London.
"Official and Tentative Methods of Analyses of the Association of Agricultural Chemists," Association of Agricultural Chemists, Washington, D. C.
E. Gildemeister and F. Hoffmann, "Die ätherischen Ole," 3d Ed., Schimmel & Co., Miltitz, 1929–1931.
S. Mulliken, "A Method for the Identification of Pure Organic Compounds," 1st Ed., John Wiley & Sons, Inc., New York, N. Y., 1904.
Berichte, Schimmel & Co., Miltitz.

I. INTRODUCTION

The essential oil chemist works in a highly specialized field requiring careful analytical ability, ingenuity, and a highly developed sense of smell and taste. He must always be on the alert for known adulterants and impurities and for new and hitherto untried adulterants. Above all, he must have sufficient chemical background and experience to be able to interpret the results of his analyses.

Crude adulteration of oils has lessened considerably, because of careful analytical control. Seldom does one encounter today the adulteration of lemon oil with turpentine, or the addition of acetanalide to vanillin. However, some adulterations are still in evidence, especially where strict analytical control is not maintained by government agencies and buyers. For example, at a comparatively recent date, the orange oils of French Guinea were so badly adulterated with kerosene and mineral oil fractions that the reputation of this oil suffered; government control entirely checked this gross adulteration. Ceylon citronella oils have been adulterated with mineral oil fractions for so long that the trade has almost accepted this as a necessary evil, trying to limit the amount of adulteration rather than to stop it altogether. Such crude adulterations usually offer no problem to the essential oil chemist. A routine analysis easily discloses such falsifying.

A much more dangerous and common type of adulteration is the addition of materials that do not materially affect the physicochemical properties of an oil. Often materials are added that are normal constituents of the oil: materials that are obtained as by-products, isolates from other oils, or synthetics. Such "sophistication" is much more difficult to detect and often may be suspected but proved only with great difficulty at best.

It is in cases such as these that a well-developed sense of smell and taste proves of immense value. Here it is important for the chemist to know what adulterants to expect. Organoleptic tests, in conjunction with physicochemical analyses, also are of great importance in evaluating the quality of unadulterated oils.

A discussion of the general procedure to be followed in examining an essential oil, isolate, or synthetic may prove of value.

A study of the odor, and in some cases the flavor, helps materially in detecting adulteration or "sophistication," and in judging quality. Comparison should be made with an oil of high quality, a "type" oil of known purity. A drop or two of the oil in question is placed upon a strip of blotting paper; the same amount of the pure "type" oil is placed upon a second strip; and the two held together at right angles by means of a clip.[1] The odor of the two oils should be studied carefully and compared at intervals. When first on blotters, addition of the more volatile adulterants is often discovered. Solvent "by-notes" may also be detected in the case of a product obtained by extraction. When the blotters have dried considerably, addition of the less volatile adulterants may often be detected—materials such as cedarwood and heavy camphor oil. The study of the odor and flavor often suggests the presence of adulterants which may be confirmed by special chemical or

[1] The wooden clips sold for drying photographic negatives prove very satisfactory.

physical tests. Or adulterants indicated by the analysis may be confirmed by a study of the odor and flavor. Moreover, organoleptic tests are probably the only satisfactory method, thus far developed, of detecting burnt, pyroligneous "by-notes" resulting from improper distillation, and of detecting slight spoilage in certain oils, as for example the citrus oils.

Also of great importance is the determination of the physical and chemical properties of a given oil. The specific gravity, the optical rotation, the solubility in dilute alcohol, and the refractive index should be determined for all oils and liquid isolates and synthetics as a matter of routine. Other special tests are also to be carried out, depending upon the material under consideration (e.g., ester content, total alcohol determination, congealing point, evaporation residue). For an optically inactive, crystalline solid, the best criterion of purity lies in the determination of the melting point.

Comparing these analytical figures with results of previous analyses and with data published in the literature, the chemist may obtain an indication of the purity and quality of the oil. Crude adulteration often is discovered at this point.

The relationship between the individual chemical and physical properties is often very revealing. Thus, the addition of orange terpenes to an orange oil will cause a lowering of the specific gravity, refractive index, and the evaporation residue, and a corresponding increase in the optical rotation; the addition of turpentine oil will cause a lowering of the optical rotation as well as of the other three properties.

Another factor to be considered is possible sources of adulteration or contamination. Benzaldehyde should be tested for chlorine, since a positive halogen test would indicate manufacture from benzyl chloride or insufficient purification. Low refractive index and specific gravity of a linaloe oil suggest adulteration with ethyl alcohol, a common form of adulteration for this oil.

The value of the relationship between each physical and chemical property, and between the analysis and the odor and flavor cannot be stressed too strongly.

The analytical figures obtained in as complex a material as an essential oil seldom represent actual percentages of single constituents. Thus, in the case of an ester determination, all saponifiable material is calculated as a certain ester, regardless of the fact that unquestionably other esters are present or that other constituents are capable of saponification. The figures obtained, however, are no less valuable for practical purposes. Nevertheless, it may be seen that in this field of chemistry it is of utmost importance that a procedure be rigidly followed in order to assure reproducible results that are of practical value.

The main purpose of the following discussion is to help standardize such analytical procedures so that chemists throughout the essential oil industry may obtain results that can be reproduced by other workers in this field as well as by chemists in related industries.

II. SAMPLING AND STORAGE

It is important that the sample used for analysis be representative of the entire contents of the container. Since most materials encountered by the analyst are homogeneous oils, sampling is not a difficult operation. However, a few cautions are noted to assure a representative sample.

Most essential oils are obtained by steam distillation with subsequent separation of the oil and water layers; therefore, shipments of oil frequently contain water. If the oil has a specific gravity of less than 1.0, the water will be found at the bottom of the drum. It is a wise precaution to test each drum for water by introducing a sampling thief made of a long glass tube with one end slightly constricted. With the other end of the glass tube securely closed (by pressing the thumb over the opening), the thief is introduced into the drum and lowered until the constricted end just touches the bottom of the drum; the thumb is removed to permit the oil and water (if present) to enter the tube; the thumb is replaced and the tube withdrawn (the thief should be held in a *vertical* position). The oil within the tube is permitted to drain into a flint glass bottle or graduate. Any water, sediment (e.g., dirt and rust) or precipitated waxes are readily discernible. For oils that have a specific gravity greater than 1.0, any water that is present will appear as a supernatant layer. Hence, in introducing the thief, the tube is not closed with the thumb and is lowered into the drum slowly.

The sample in the flint bottle or graduate should receive a cursory examination—color, clarity, viscosity, the presence or absence of sediments, separated waxes, and water, all should be observed and noted; finally the odor of the sample should be studied. Oils stored in drums for long periods will frequently show a slight musty "by-note" which rapidly disappears; this is a typical "drum-note" and does not reflect adversely on the quality of the oil. Freshly distilled oils frequently show a slight, sharp empyreumatic or burned note which disappears as the oil is aged. The "drum-note" and the "freshly distilled note" should disappear if the oil is permitted to stand in an open graduate overnight. If such notes do not disappear, the oils should be examined more carefully; it may be necessary to aerate the oil to remove persistent "by-notes." A small sample of the oil (about 50 cc.) should be treated by bubbling air through the oil for a period of several hours. It may be necessary to warm the air which is bubbled through the oil. Diagram 4.1 shows a convenient apparatus for carrying out such eration. If this treatment yields a satisfactory oil, the contents of the drum may be treated similarly.

Citrus oils frequently deposit large amounts of waxes. These are easily observed by the water testing technique described above. Another indication is the "feel" of the tube as it hits the bottom of the drum.

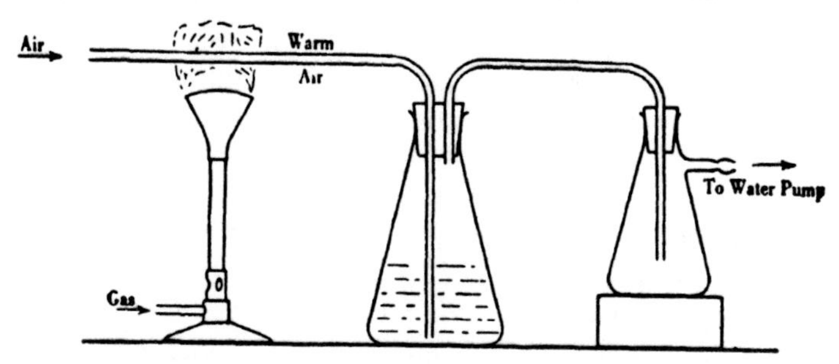

DIAGRAM 4.1. Apparatus for the aeration of essential oils.

If the drum shows no evidence of water or insignificant amounts of water and sediment, the sample drawn may be used for analysis. If water, wax, or sediment is present to a large extent, a fresh sample should be drawn from the supernatant oil.

When the sample is received by the chemist, it should be clarified and freed from sediment and separated waxes by decantation or by filtration if necessary. A small amount of dry sodium chloride placed in the folded filter will frequently aid in removing traces of water and will remove the haze from an oil. Treatment with clay or kieselguhr may be necessary to remove a haze caused by suspended materials. If the oil is very dark in color owing to the presence of heavy metals or other metallic impurities, the color may be lightened by shaking with tartaric acid and filtering (see p. 311). The color and appearance of the oil both before and after treatment should be described in the analytical report, as well as the treatment used.

Because of possible variation in the oils of a shipment of more than one drum, it is best to sample each drum of oil. However, if the oils are to be bulked (e.g., several drums are to be pumped into a tank to yield a uniform lot), an average or representative sample may be made based on the weights of the oil in the individual drums. The odor and general appearance of each drum should be examined before such a sample is made, to prevent the addition of a drum of poor quality to the tank. If this average sample is found to be of inferior quality or shows any abnormalities, then each drum must be resampled and examined individually.

Oils that congeal at temperatures normally encountered should be given special attention. These drums should be permitted to stand in a warm place and stirred occasionally (or heated in a steam room if the congealing point is high) until the last trace of solid material is dissolved. Anise oil may be taken as a typical example. During the cold weather a drum may be received in a frozen condition. Upon standing in a warm place, the anethole slowly melts until the drum is half solid, half liquid. The solid settles to the bottom. If a sample of the liquid portion is drawn, it will be deficient in anethole and may fail to meet the official requirements of "The United States Pharmacopoeia"; such a sample would not be representative of the drum. Synthetics and isolates such as anethole, benzyl benzoate and diphenyl oxide usually require a steam room to melt them completely to assure a representative sample. Upon chilling or long standing, some oils deposit small amounts of crystalline materials such as menthol, cedrol, or camphor; in these cases, the oil should be gently warmed and stirred to redissolve the crystalline material before a sample is drawn.

In sampling materials other than oils, certain precautions should be observed. For resinoids, oleoresins and balsams the drum or can should be stirred well with a flat stick to assure thorough mixing. A sample should not be drawn before the material is uniform. This is of importance for items such as styrax, which usually shows a separation of styrene, polystyrene and water.

The sampling of crude drugs offers considerable difficulties. "The United States Pharmacopoeia"[2] gives four methods for the sampling of vegetable drugs from original containers to obtain an "official," representative sample.

> *I.* "It is recommended that gross samples of vegetable or animal drugs in which the component parts are 1 cm. or less in any dimension, and all powdered or ground drugs, be taken by means of a sampler which removes a core from the top to the bottom of the container, not less than two cores being taken in opposite directions; that when the total weight of the drug to be sampled is less than 100 kilos (200 pounds) at least 250 g. shall constitute an *official sample.* When the total weight of the drug to be sampled is in excess of 100 kilos, repeated samples shall be taken by the above method, and according to the schedule given below, mixed and quartered, two of the diagonal quarters being rejected, the remaining two quarters being combined and carefully mixed, and again subjected to a quartering process in the same manner until two of the quarters weigh at least 250 g., which latter quarters shall constitute an *official sample.*
>
> *II.* "It is recommended that gross samples of vegetable drugs in which the component parts are over 1 cm. in any dimension be taken by hand. When the total weight of the drug to be sampled is less than 100 kilos, at least 500 g. shall constitute

[2] Thirteenth Revision, 710.

an *official sample*, and this shall be taken from different parts of the container or containers. When the total weight of the drug to be sampled is in excess of 100 kilos, repeated samples shall be taken by the above method and according to the schedule below, mixed and quartered, two of the diagonal quarters being rejected, and the remaining two quarters being combined and carefully mixed, and again subjected to a quartering process in the same manner until two of the quarters weigh not less than 500 g., which latter quarters shall constitute an *official sample*.

Schedule Recommended for Sampling

Number of Packages in Shipment	*Number of Packages To Be Sampled*
1 to 10	1 to 3
10 to 25	3 to 4
25 to 50	4 to 6
50 to 75	6 to 8
75 to 100	8 to 10

When over 100, the total number sampled should not be less than 10.

III. "When the total weight of a drug to be sampled is less than 10 kilos it is recommended that the above methods be followed but that somewhat smaller quantities be withdrawn, and in no case should the final *official sample* weigh less than 125 g.

IV. "In addition to the withdrawing of *official samples* according to methods *I*, *II*, and *III*, the *official sample* may consist of the total amount of a direct purchase made by Federal, State, or Municipal Food and Drug Act enforcement officials."

The sampling of a pure chemical which is a solid (e.g., vanillin) will now be considered. The well-known method of quartering[3] will assure a representative sample. Most manufacturers give an identifying number to each batch manufactured, and consequently it may be assumed that each batch is uniform so that a sample taken at random will be representative of the whole batch.

The final sample should be transferred to a bottle of light-resistant[4]

[3] Treadwell and Hall, "Analytical Chemistry," John Wiley & Sons, Inc., New York, Vol. II (1942), 45.

[4] "The United States Pharmacopoeia" (Twelfth Revision, 6) defines a light-resistant container as "a container which is opaque, or designed to prevent photochemical deterioration of the contents beyond the official limits of strength, quality, or purity, under customary conditions of handling, storage, shipment, or sale.

"Unless otherwise directed, a light-resistant container shall be composed of a substance which in a thickness of 2 mm. shall not transmit more than 10 per cent of the incident radiation of any wave length between 2900 and 4500 angstrom units.

"If the immediate container in its construction is less than 2 mm. in thickness, the same 10 per cent limit of light transmission shall apply.

"If the immediate container in its construction is not light-resistant, it must be provided with an opaque covering, be enclosed in an opaque covering or in an opaque container."

(The definition of the Thirteenth Revision, 5, is essentially the same.)

glass (amber, blue or green). The bottle should be well filled to prevent adverse action by the air and well stoppered with a sound *cork*. Screw caps should be used with caution, since the liners may contaminate the oil.[5] If a screw cap is to be employed it is well to stopper the bottle with a cork before using the screw cap.

If the shipment of oil is to be stored for any appreciable length of time, the precautions noted below should be observed:

1. The oils should be clarified and thoroughly dried.

2. The oils are best stored in glass containers in a cool[6] place protected from light and air. Half filled containers should be avoided. Storage in glass is frequently impractical; if drums or cans must be used, heavily galvanized or heavily tinned iron usually will prove satisfactory. Aluminum[7] and stainless steel can be used with some oils, but not universally.

3. Certain oils are much more susceptible to oxidation and polymerization than others; oils rich in terpenes (e.g., citrus oils) and oils containing large amounts of aldehydes (e.g., benzaldehyde) are readily affected. Some oils (e.g., vetiver, sandalwood and patchouly) show very good keeping qualities and may actually improve upon aging.

4. In general, the use of antioxidants for essential oils is not necessary *if the oils are properly and carefully stored.*[8]

III. DETERMINATION OF PHYSICAL PROPERTIES

1. SPECIFIC GRAVITY

Specific gravity is an important criterion of the quality and purity of an essential oil. Of all the physicochemical properties, the specific gravity has been reported most frequently in the literature. Values for essential oils vary between the limits of 0.696 and 1.188 at 15°;[9] in general, the gravity is less than 1.000. For each individual oil, however, the limits are much narrower and in most cases have been established during the course of years.

[5] The laboratories of Fritzsche Brothers, Inc., repeatedly have examined samples contaminated by such liners—the oils are frequently hazy, difficult to clarify, and show a strongly positive Halogen Test.

[6] Preferably at temperatures not exceeding 20°.

[7] Many aluminum containers are lacquer lined; often the oil will act as a partial solvent for the lacquer and introduce contaminating material.

[8] The importance of proper storage of oils is evidenced by the following observations: Italian lemon oils, stored under optimum conditions for four years, retained their fresh character; orange oils, stored experimentally under adverse conditions, spoiled within one week, developing the characteristic terbinthinate odor. The laboratories of Fritzsche Brothers, Inc., examined various oils that had been properly stored for more than fifty years: many of these oils showed no signs of spoilage whatsoever, in spite of the fact that no chemical antioxidants had been used.

[9] Gildemeister and Hoffman, "Die ätherischen Öle," 3d Ed., Vol. I, 699.

The specific gravity of an essential oil at 15°/15° may be defined as the ratio of the weight of a given volume of oil at 15° to the weight of an equal volume of water at 15°.[10]

For determination of this physical property, accuracy to at least the third decimal place is necessary. Therefore, hydrometers are practically worthless and should not be used. The Mohr-Westphal balance may be used but it has the disadvantage that relatively large amounts of oil are required for a determination. Other types of specific gravity balances have been developed which require less oil and which have proven satisfactory. Pycnometers offer the most convenient and rapid method for determining specific gravities. A conical shaped pycnometer having a volume of about 10 cc. with a ground-in thermometer and a capillary side tube with a ground glass cap proves very satisfactory (see Diagram 4.2).[11] Sprengel or Ostwald tubes give even more accurate results; if desired they may be used. However, a determination cannot be made as rapidly or as conveniently. Cleaning these tubes will prove considerably more difficult and time consuming. A small Sprengel tube or a Gay-Lussac specific gravity bottle having a capacity of about 2 cc. will often prove of value when only small amounts of oil are available. For routine analyses the conical pycnometer as described above is recommended.

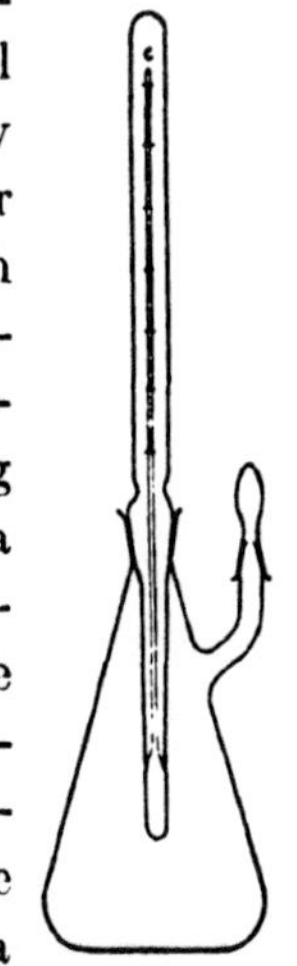

DIAGRAM 4.2. Pycnometer.

Procedure: Clean the pycnometer by filling it with a saturated solution of chromium trioxide in sulfuric acid and allow it to stand for at least 3 hr. Empty the pycnometer and rinse thoroughly with distilled water. Fill the pycnometer with recently boiled distilled water which has been cooled to a temperature of about 12° and place it in a water bath, previously cooled to 12°. Permit the temperature to rise slowly to 15°. Adjust the level of the water to the top of the capillary side arm, removing any excess with a blotter or cloth, and put the ground glass cap in place. Remove the pycnometer from the water bath, dry carefully with a clean cloth, permit it to stand for 30 min. and weigh accurately. Empty the pycnometer, rinse

[10] The density of a liquid is the weight of a unit volume. Thus, density may be expressed in pounds per cubic foot, or more frequently in grams per cubic centimeter. At 3.98° (the temperature of maximum density for pure water, free from air) 1 cc. of water weighs 0.999973 g.; furthermore, at this temperature 1 ml. of water weighs exactly 1 g. Since the coefficient of expansion of water is small, the density of a liquid expressed in grams per cubic centimeter corresponds closely to the specific gravity. However, the fundamental difference in the two concepts should be thoroughly understood.

[11] This pycnometer is similar to that described in A.S.T.M. Designation D 153 with the exception that the capacity is approximately 10 cc. instead of 50 cc.

several times with alcohol and finally with ether. Remove the ether fumes with the aid of an air blast and permit the pycnometer to dry thoroughly. Weigh accurately after standing 30 min. The "water equivalent" of the pycnometer may be found by subtracting the weight of the empty pycnometer from its weight when full.

Fill the clean, dried pycnometer with the oil previously cooled to a temperature of 12°. Following the same procedure as above, place the pycnometer in a water bath and permit it to warm slowly to 15°. Adjust the oil to the proper level, put the cap in place, and wipe the pycnometer dry. Accurately weigh after 30 min.

The weight of the oil contained in the pycnometer divided by the water equivalent gives the specific gravity of the oil at 15°/15° (in air).

For a given pycnometer the water equivalent need be determined only once; therefore, it is important that this determination be performed with great care and accuracy.

For scientific work or for cases where the gravity is in question, the determination should be carried out exactly as described above. However, for routine analyses it is permissible to determine the specific gravity of an oil at room temperature compared with water at 15° and then to reduce this value to a temperature of 15°/15° by use of a proper correction factor. Numerous workers have determined correction factors for various oils and have recommended a general value from 0.00042[12] to 0.00084[13] per degree centigrade. However, as Bosart[14] pointed out, it would be unsatisfactory to take the average figure obtained from a variety of oils and apply it to a particular oil, all the more so when there is a difference of opinion as to what that figure should be.

In the investigation carried out by Bosart, values were obtained which ranged from 0.00070 to 0.00099 per degree[15] for the forty-two essential oils he examined. For synthetics and isolates normally encountered by the essential oil chemist or perfumer, values ranged from 0.00067 to 0.00114 per degree.[16] Hence, it is unjustifiable to use an average correction factor if

[12] Harvey, *J. Soc. Chem. Ind.* **24** (1905), 717. *Ber. Schimmel & Co.*, October (1905), 87.

[13] Schreiner and Downer, *Pharm. Arch.* **4** (1901), 167. These authors determined the specific gravity of thirty-two essential oils at $\frac{15^\circ}{15^\circ}, \frac{20^\circ}{20^\circ}, \frac{25^\circ}{25^\circ}$; they recommended the use of the factor, 0.0064 per degree. Bosart has recalculated this figure in order to convert the specific gravity at $\frac{25^\circ}{15^\circ}$ to $\frac{15^\circ}{15^\circ}$, and has arrived at the factor 0.00084.

[14] *Ind. Eng. Chem., Ind. Ed.* **28** (1936), 867.

[15] *Ibid.*

[16] *Perfumery Essential Oil Record* **30** (1939), 145.

accurate data are to be obtained. A summary of Bosart's work is given in Table 4.1 and Table 4.2.

TABLE 4.1. VARIATIONS IN SPECIFIC GRAVITY PER DEGREE CENTIGRADE FOR ESSENTIAL OILS

Almond, Bitter	0.00089	Linaloe	0.00083
Anise	0.00082	Mace	0.00082
Bay	0.00085	Mirbane	0.00098
Bergamot	0 00081	Orange, Sweet	0.00078
Bois de Rose, Brazilian	0.00081	Origanum	0.00076
Cade	0.00074	Palmarosa	0.00073
Camphor	0.00081	Patchouly	0.00073
Cananga	0.00074	Pennyroyal	0.00078
Caraway	0.00078	Peppermint	0.00076
Cassia	0.00081	Petitgrain	0.00081
Cedarwood	0.00071	Pine	0.00079
Citronella, Ceylon	0.00081	Rosemary	0.00081
Citronella, Java	0.00093	Sandalwood, East Indian	0.00070
Clove	0.00085	Sassafras, Artificial	0.00087
Eucalyptus (*Eucalyptus globulus*)		Spearmint	0.00079
70 to 80%	0.00084	Spike	0.00082
Geranium, African	0.00076	Tansy	0.00080
Geranium, Bourbon	0.00076	Thyme	0.00079
Ho	0.00083	Vetiver	0.00071
Lavender	0.00082	Wintergreen (*Gaultheria*	
Lemon	0.00077	*procumbens*)	0.00099
Lemongrass	0.00079	Ylang Ylang	0.00073

The proper correction is to be added if the temperature, at which the determination was made, is above 15°; conversely to be subtracted if the temperature is below 15°. These correction factors may also prove of use for converting specific gravities given in the literature at temperatures other than 15° when compared with water at 15°.

It is customary to report specific gravities for essential oils at 15°/15°. For oils that are not liquid at this temperature the specific gravity is conveniently reported at some higher temperature, compared with water at 15°. Thus, the gravity of rose oils is often reported[17] at 30°/15°. "The United States Pharmacopoeia" and "The National Formulary" specify a temperature of 25°/25° for most essential oils. "The British Pharmacopoeia" specifies 15.5°/15.5°. In order to convert the specific gravities from 15°/15° to 25°/25°, the conversion factors given in Table 4.3 may be used.[18] These corrections are to be subtracted from the values determined at 15°/15°.[19]

[17] "The United States Pharmacopoeia," Thirteenth Revision, 456.

[18] *Ber. Schimmel & Co.* April (1906), 73.

[19] A more exact determination will result if the water equivalent of the pycnometer at 25° and the weight of the oil contained in the pycnometer at 25° are determined by the method described under "Procedure."

TABLE 4.2. VARIATIONS IN SPECIFIC GRAVITY PER DEGREE CENTIGRADE FOR ISOLATES AND SYNTHETICS

Acetal	0.00103	Ethyl Propionate	0.00111
Acetaldehyde	0.00129	Eugenol	0.00087
Acetophenone	0.00086	Geraniol	0.00071
Allyl Alcohol	0.00088	Geranyl Acetate	0.00085
Allyl Formate	0.00121	Glycerol	0.00062
n-Amyl Acetate	0.00094	Heliotropin	0.00093
α-Amyl Cinnamic Aldehyde	0.00076	Heptaldehyde	0.00086
n-Amyl Ether	0.00079	Heptyl Alcohol	0.00073
Amyl Salicylate	0.00085	Hydroxycitronellal	0.00077
Anisic Aldehyde	0.00085	Ionone	0.00076
Benzaldehyde	0.00089	Isoamyl Acetate	0.00097
Benzyl Acetate	0.00092	Isoamyl Formate	0.00097
Benzyl Alcohol	0.00076	Isoeugenol	0.00087
Benzyl Benzoate	0.00081	Isopulegol	0.00083
Benzyl Ether	0.00079	Isosafrole	0.00088
Bornyl Acetate	0.00086	Lauryl Alcohol	0.00067
Brombenzene	0.00134	*d*-Limonene	0.00077
Bromstyrene	0.00114	Linaloöl	0.00082
n-Butyl Acetate	0.00102	Linalyl Acetate	0.00084
n-Butyl Benzoate	0.00086	Methyl Acetate	0.00127
n-Butyl *n*-Butyrate	0.00093	Methyl Acetophenone	0.00081
n-Butyl Formate	0.00100	Methyl Anthranilate	0.00088
n-Butyl Lactate	0.00097	Methyl Benzoate	0.00095
n-Butyl Propionate	0.00099	Methyl *n*-Butyrate	0.00107
n-Butyl *d*-Tartrate	0.00091	Methyl *n*-Caproate	0.00094
Butyraldehyde	0.00105	Methyl Formate	0.00143
n-Caproic Acid	0.00087	Methyl Heptenone	0.00084
n-Caprylic Acid	0.00082	Methyl Nonyl Ketone	0.00076
Carvacrol	0.00076	Methyl Phenylacetate	0.00093
Carvone	0.00080	Methyl Phthalate	0.00093
Cinnamic Aldehyde	0.00080	Methyl Propionate	0.00118
Cinnamyl Alcohol	0.00074	Methyl Salicylate	0.00098
Citral	0.00080	Nitrobenzene	0.00098
Citronellal	0.00082	Octyl Alcohol	0.00068
Citronellol	0.00070	Phellandrene	0.00078
p-Cresyl Acetate	0.00093	Phenyl Acetate	0.00098
p-Cymene	0.00079	Phenylethyl Acetate	0.00090
Decyl Alcohol	0.00068	Phenylethyl Alcohol	0.00074
Diethyl Phthalate	0.00084	Phenylpropyl Alcohol	0.00073
Dipentene	0.00080	Pinene	0.00082
Diphenyl Methane	0.00078	*n*-Propyl Acetate	0.00110
Diphenyl Oxide	0.00085	*n*-Propyl Formate	0.00114
Ethyl Acetate	0.00120	Rhodinol	0.00071
Ethyl Benzoate	0.00092	Safrole	0.00089
Ethyl *n*-Butyrate	0.00103	Salicyl Aldehyde	0.00097
Ethyl Caproate (Tech.)	0.00092	Terpineol	0.00078
Ethyl Formate	0.00126	Terpinyl Acetate	0.00082
Ethyl *n*-Heptoate	0.00089	Valeric Acid	0.00091

TABLE 4.3. FACTORS FOR CONVERSION OF SPECIFIC GRAVITY FROM $\frac{15°}{15°}$ TO $\frac{25°}{25°}$

Oil	Factor	Oil	Factor
Almond, Bitter	0.0068	Pimenta	0.0068
Anise	0.0060	Rosemary	0.0066
Cajuput	0.0064	Sandalwood	0.0047
Caraway	0.0057	Sassafras	0.0067
Cassia, Rectified	0.0062	Savin	0.0058
Chenopodium	0.0063	Spearmint	0.0062
Clove	0.0065	Sweet Birch (*Betula lenta*)	0.0076
Copaiba	0.0054	Thyme	0.0061
Coriander	0.0067	Turpentine	0.0066
Cubeb	0.0055	Turpentine, Rectified	0.0065
Erigeron	0.0062	Wintergreen (*Gaultheria procumbens*)	0.0076
Eucalyptus (*Eucalyptus globulus*)	0.0063		
Fennel	0.0062		
Hedeoma	0.0060	*Synthetic*	
Juniper Berries	0.0062	Benzaldehyde	0.0069
Lavender	0.0067	Cinnamic Aldehyde	0.0056
Lemon	0.0058	Eucalyptol	0.0067
Mustard	0.0080	Eugenol	0.0066
Nutmeg	0.0065	Methyl Salicylate	0.0079
Orange, Sweet	0.0057	Safrole	0.0069
Peppermint	0.0054		

2. OPTICAL ROTATION

Most essential oils when placed in a beam of polarized light possess the property of rotating the plane of polarization to the right (dextrorotatory), or to the left (laevorotatory). The extent of the optical activity of an oil is determined by a polarimeter and is measured in degrees of rotation. Of the numerous types of polarimeters that are available, the most convenient for use with essential oils is probably the half-shadow instrument of the Lippich type.[20]

The angle of rotation is dependent upon the nature of the liquid, the length of the column through which the light passes, the wave length of the light used, and the temperature.

Both the degree of rotation and its direction are important as criteria of purity. In recording rotations it is customary to indicate the direction by the use of a plus sign (+) to indicate dextrorotation (rotation to the right,

[20] For a discussion of the theory involved, the reader is referred to H. Landolt: "The Optical Rotating Power of Organic Substances and Its Practical Application" (translated by J. H. Long), The Chemical Publishing Co., Easton, Pa. (1902). Landolt thoroughly covers the field of optical activity, including, *inter alia*, the causes of optical activity and inactivity, the theory and construction of the polarimeter, and the various types of instruments available.

i.e., clockwise) or a minus sign (−) to indicate laevorotation (rotation to the left, i.e., counterclockwise).

Since the scale reading for an optically active liquid is directly proportional to the length of the transmitting column of liquid, it is necessary to use a standard tube, 100 mm. long. If for any reason a longer or shorter tube is used, the rotation should be calculated for a tube of 100 mm. and reported as such. Rotations for essential oils given in the literature may be assumed to be for this standard tube unless a different length is specified.

It has become customary in polarimetric work to use sodium light. A suitable source may be obtained by placing large crystals of sodium chloride upon the grid of a Meeker burner or by wrapping a piece of asbestos, previously saturated in a strong salt solution, around the conventional Bunsen burner. By far the most convenient and satisfactory method of maintaining a constant light source is the use of a sodium vapor lamp. Such lamps, designed especially for use with polarimeters, are available.

Although "The United States Pharmacopoeia" and "The National Formulary" specify 25° as the official temperature for all optical rotations, nevertheless, a standard temperature of 20° is usually adopted for essential oils reported in the literature. For most essential oils the change in optical rotation with temperature variations normally encountered in the laboratory is very small; hence, in routine analyses the readings are usually taken at room temperature. No corrections for temperature variations are made except in the case of citrus oils which contain large amounts of highly active terpenes. The corrections to be used, per degree centigrade, are:

Orange Oil	13.2′
Lemon Oil	8.2′
Grapefruit Oil	13.2′

The proper correction is to be added if the reading is taken at a temperature higher than the desired temperature and, conversely, to be subtracted if the temperature of the reading is lower than the desired temperature.

In scientific work the temperature at which the rotation was determined should be specified. To adjust the temperature to standard, the polarimeter tubes may be immersed in a constant temperature bath. Use may also be made of special water jacketed tubes.

All determinations should be carried out in a dark room. Monochromatic sodium light should be employed.

a. **Liquids.**—The oil or liquid should be free from suspended material. Often oils are hazy owing to the presence of small amounts of water; such an oil should be dried with anhydrous sodium sulfate and filtered before a determination is attempted.

Procedure: Place the 100 mm. polarimeter tube containing the oil or liquid under examination in the trough of the instrument between the polarizer and analyzer. Slowly turn the analyzer until both halves of the field, viewed through the telescope, show equal intensities of illumination. At the proper setting, a small rotation to the right or to the left will immediately cause a pronounced inequality in the intensities of illumination of the two halves of the field.

Determine the direction of rotation. If the analyzer was turned counterclockwise from the zero position to obtain the final reading, the rotation is laevo (−); if clockwise, dextro (+).[21]

After the direction of rotation has been established, carefully readjust the analyzer until equal illumination of the two halves of the field is obtained. Adjust the eyepiece of the telescope to give a clear, sharp line between the two halves of the field. Determine the rotation by means of the protractor; read the degrees directly, and the minutes with the aid of either of the two fixed verniers; the movable magnifying glasses will aid in obtaining greater accuracy. A second reading should be taken; it should not differ by more than ±5′ from the previous reading.

Some oils are too dark in color for an accurate determination of the optical rotation when a 100 mm. tube is used. In such cases, a 50 mm. tube may be employed, or even a 25 mm. tube, if necessary. Since the rotation is reported for a 100 mm. tube, any experimental error will be multiplied by 2 for a 50 mm. tube, and by 4 for a 25 mm. tube. Conversely, if a clear, light colored oil is examined which is only slightly optically active, the use of a longer tube (200 mm.) may often prove of advantage; the value to be reported will be found by dividing the observed rotation by 2; any experimental error will also be halved.

b. **Solids.**—The optical activity of a solid is best determined in solution and expressed as specific rotation. The following formulas may be used:

$$[\alpha]_D^{t°} = \frac{100\alpha}{lpd} \tag{1}$$

$$[\alpha]_D^{t°} = \frac{100\alpha}{lc} \tag{2}$$

[21] Since most instruments are calibrated only to 180°, some confusion may exist as to the direction of rotation; this is especially true if the liquid is highly optically active. Thus, a reading of +100° may be reported mistakenly as a reading of −80°. If any doubt exists in the mind of the chemist, the determination should be repeated using a 50 mm. tube. In the example given above, a reading of +50° would be obtained, indicating that the correct value for a 100 mm. tube is +100°; the other possible reading with the smaller tube (that is, −130°) corresponds to a value of −260° for a 100 mm. tube: so high a value would be most unusual for an essential oil. The optical rotation of an essential oil seldom is greater than ±100°.

where: $[\alpha]_D^{t°}$ = specific rotation at temperature $t°$, using sodium light;
α = observed rotation in degrees of the solution at temperature $t°$, using sodium light;
l = length of polarimeter tube in decimeters;
d = specific gravity of the solution at the temperature $t°$;
p = concentration of the solution expressed as the number of grams of active substance in 100 g. of solution;
c = concentration of solution expressed as the number of grams of active substance in 100 cc. of solution.

Formula (2) is more convenient, since it does not require the determination of the specific gravity of the solution.

The experimental value for the specific rotation of a solid is dependent upon the concentration of the solution and upon the particular solvent employed; therefore, the concentration and solvent used should be given when the specific rotation of a solid is reported. The rotation should be determined as soon as possible after the solution has been prepared, so that any change that might result from mutarotation will be minimized.

The use of specific rotation for a complex mixture such as an essential oil is not recommended. For the sake of completeness, the following formula is given:

$$[\alpha]_D^{t°} = \frac{\alpha}{ld} \qquad (3)$$

Formula (3) applies to optically active liquids.

The symbol $[\alpha]_D^{t°}$ is reserved exclusively for specific rotation; optical rotation determined in a 100 mm. tube is indicated by $\alpha_D^{t°}$, the brackets being omitted. If no temperature is given, it may be assumed that the optical rotation was determined at room temperature.

3. REFRACTIVE INDEX

When a ray of light passes from a less dense to a more dense medium, it is bent or "refracted" toward the normal. If e represents the angle of refraction, and i the angle of incidence, according to the law of refraction,

$$\frac{\text{Sin } i}{\text{Sin } e} = \frac{N}{n}$$

where n is the index of refraction of the less dense, and N, the index of refraction of the more dense medium.

Refractometers offer a rapid and convenient method for the determination of this physical constant. Of the various types, the Pulfrich or the

Abbé refractometer proves very satisfactory.[22] The Abbé type, with a range of 1.3 to 1.7, is recommended for the routine analyses of essential oils, the accuracy of this instrument being sufficient for all practical work. The readings may be made directly from the scale without consulting conversion tables; only one or two drops of the oil are required for a determination; the temperature at which the reading is taken may be adjusted conveniently.

Procedure: Place the instrument in such a position that diffused daylight or some form of artificial light can readily be obtained for illumination. Circulate through the prisms a stream of water at 20°. Carefully clean the prisms of the instrument with alcohol and then with ether. To charge the instrument, open the double prism by means of the screw head and place a few drops of the sample on the prism, or, if preferred, open the prisms slightly by turning the screw head and pour a few drops of sample into the funnel-shaped aperture between the prisms. Close the prisms firmly by tightening the screw head. Allow the instrument to stand for a few minutes before the reading is made so that the sample and instrument will be at the same temperature. Move the alidade backward or forward until the field of vision is divided into a light and dark portion. The line dividing these portions is the "border line," and, as a rule, will not be a sharp line but a band of color. The colors are eliminated by rotating the screw head of the compensator until a sharp, colorless line is obtained. Adjust the border line so that it falls on the point of intersection of the cross hairs. Read the refractive index of the substance directly on the scale of the sector. A second reading should be taken a few minutes later to assure that temperature equilibrium has been attained.

Occasionally, the instrument should be checked by means of the quartz plate that accompanies it, using monobromnaphthalene, or if such a plate is not available, by means of distilled water at 20°; the refractive index of pure water at this temperature is 1.3330.

Great care should be exercised when determining refractive indexes during hot, humid weather, since moisture in the air may condense on the cooled prisms. This will result in a blurred and indistinct line of separation between the light and dark fields if the oil between the prisms does not dissolve the condensed moisture; if the oil dissolves the moisture, the dividing line will be sharp, but the observed index will be low.

[22] For a discussion of the theory involved and for a description of the instruments, the reader is referred to a standard text on physical chemistry, e.g., Findlay, "Introduction to Physical Chemistry," Longmans, Green & Co. (1933), 103; Daniels, Mathews, and Williams, "Experimental Physical Chemistry," McGraw-Hill Book Co., New York (1941), 44.

It has become the accepted procedure to report refractive indexes for essential oils at 20°, using a monochromatic sodium light source,[23] unless the material is a solid at that temperature. Thus, in the case of rose oil the refractive index is often given at 30°;[24] in the case of anethole, at 25°.[25]

Whenever possible, however, all observations should be made at 20°. The use of factors to reduce readings to 20° is not recommended. Various investigators, notably Bosart, have reported the change of refractive index with temperature for numerous oils. According to the findings of Bosart,[26] the values for the fifty-four oils examined lie between the limits of 0.00039 and 0.00049 per degree centigrade, and for the forty-seven synthetics and isolates between the limits of 0.00038 and 0.00054. A summary of Bosart's work is given in Table 4.4 and Table 4.5. These tables may be used con-

TABLE 4.4. CHANGE IN REFRACTIVE INDEX OF ESSENTIAL OILS

	Correction per Degree		Correction per Degree
Almond, Bitter	0.00049	Linaloe	0.00044
Anise	0.00049	Mace	0.00046
Bay Leaves	0.00047	Mawah	0.00041
Bergamot	0.00044	Mustard	0.00054
Bois de Rose	0.00044	Orange, Sweet	0.00045
Cajuput	0.00045	Origanum	0.00042
Camphor, Brown, s.g. 0.95–0.97	0.00043	Palmarosa	0.00040
Camphor, s.g. 1.020	0.00044	Patchouly	0.00042
Camphor, white	0.00045	Pennyroyal	0.00042
Cananga	0.00041	Peppermint	0.00040
Caraway	0.00044	Petitgrain	0.00044
Cassia	0.00048	Pimenta	0.00047
Cedarwood	0.00040	Pine	0.00042
Cinnamon, Ceylon	0.00048	Rosemary	0.00044
Citronella, Ceylon	0.00046	Sandalwood, E. I.	0.00039
Citronella, Java	0.00047	Sassafras, Art.	0.00045
Clove	0.00045	Savin	0.00044
Copaiba	0.00040	Spearmint	0.00043
Coriander	0.00047	Spike	0.00045
Erigeron	0.00046	Sweet Birch (*Betula lenta*)	0.00045
Eucalyptus (*Eucalyptus globulus*)	0.00044	Tansy	0.00042
Fennel	0.00047	Thyme, Red, 40–45%	0.00044
Geranium, African	0.00040	Turpentine	0.00046
Geranium, Bourbon	0.00040	Vetiver	0.00039
Ho	0.00043	Wintergreen (*Gaultheria procumbens*)	0.00045
Lavender	0.00043	Ylang Ylang	0.00042
Lemon	0.00046		
Lemongrass	0.00044		

[23] The Abbé refractometer is calibrated for the D-line of sodium vapor light.
[24] "The United States Pharmacopoeia," Thirteenth Revision, 456.
[25] "The National Formulary," Eighth Edition, 51.
[26] *Perfumery Essential Oil Record* 28 (1937), 95.

TABLE 4.5. CHANGE IN REFRACTIVE INDEX OF SYNTHETICS AND ISOLATES

	Correction per Degree		Correction per Degree
Acetophenone	0.00047	Hydroxycitronellal	0.00040
α-Amyl Cinnamic Aldehyde	0.00050	Ionone	0.00044
Amyl Salicylate	0.00012	Isoeugenol	0.00050
Anisic Aldehyde (Aubepine)	0.00046	Isopulegol	0.00045
Benzaldehyde	0.00047	Limonene	0.00045
Benzyl Acetate	0.00045	Linalool	0.00046
Bornyl Acetate	0.00043	Linalyl Acetate	0.00043
Bromstyrene	0.00054	Methyl Anthranilate	0.00048
Carvacrol, Tech.	0.00043	Methyl Benzoate	0.00048
Cinnamic Alcohol	0.00044	Methyl Heptenone	0.00046
Cinnamic Aldehyde	0.00052	Methyl Phenylacetate	0.00046
Citral	0.00045	Methyl Salicylate	0.00047
Citronellal	0.00044	Nitrobenzene (Mirbane Oil)	0.00049
Citronellol	0.00040	o-Nitrotoluene	0.00048
p-Cresyl Acetate	0.00046	Orange Terpenes	0.00046
p-Cymene	0.00049	Phenylethyl Acetate	0.00046
Diethyl Phthalate	0.00041	Phenylethyl Alcohol	0.00041
Diphenyl Oxide	0.00049	Phenyl Methyl Carbinyl Acetate	0.00046
Eucalyptol	0.00046	Phenylpropyl Alcohol	0.00038
Eugenol	0.00046	Rhodinol	0.00040
Geraniol	0.00041	Safrole	0.00045
Geranyl Acetate	0.00045	Terpineol	0.00044
Geranyl Butyrate	0.00043	Terpinyl Acetate	0.00041
Geranyl Formate	0.00043		

veniently to convert values reported in the literature at other than 20°. If an oil is encountered which is not listed in the table, the use of a correction factor of 0.00045 per degree will give approximately correct results. If the refractive index is reported at a temperature above 20°, the proper correction must be added; conversely, if reported at below 20°, the correction must be subtracted.

4. MOLECULAR REFRACTION

It is beyond the scope of this work to treat thoroughly of molecular refraction.[27] However, a brief discussion of the fundamental concepts involved may prove useful.

The index of refraction of a liquid varies with the temperature and with the wave length of the light. In order to obtain a constant which is independent of the temperature, Gladstone and Dale[28] introduced the use of "specific refractivity." Subsequently, Lorentz,[29] and Lorenz[30] independ-

[27] Reference may be made to the original papers of Eisenlohr, of Swientoslawski, and of Brühl and to any standard text on physical chemistry for further discussion of this interesting physical property.

[28] *Roy. Soc. London, Phil. Trans.* **148**, Part I (1858), 887.

[29] *Ann. Physik Chem.* N.S. 9 (1880), 641.

[30] *Ibid.* 11 (1880), 70.

ently deduced an expression for specific refractivity, based upon the electromagnetic theory of light, which shows considerably less variation than the empirical expression of Gladstone and Dale. In order to compare the refractivities of different liquids, the use of molecular refractivity (molecular refraction) is necessary. This constant is equal to the product of the molecular weight of a substance and its specific refractivity.

Using the Lorentz and Lorenz expression:

$$R = Mr = \left(\frac{n^2 - 1}{n^2 + 2}\right)\left(\frac{M}{d}\right)$$

where: R = the molecular refractivity;
M = the molecular weight;
r = the specific refractivity;
n = observed refractive index at temperature $t°$;
d = density at temperature $t°$.

The molecular refractivity has been found to be essentially additive. Hence, it is possible to calculate atomic refractivities for the different elements from a series of molecular refractivities of different compounds. By means of these atomic constants, the molecular refractivity of a pure chemical compound can be calculated as the sum of the atomic refractivities.

TABLE 4.6. ATOMIC REFRACTIONS FOR THE D-LINE*

	Eisenlohr (1910)	Conrady	Brühl-Conrady (1891)
CH_2	4.618		
C	2.418	2.495	2.501
H	1.100	1.051	1.051
O″	2.211	2.281	2.287
O<	1.643	1.679	1.683
O′	1.525	1.517	1.521
Cl	5.967	5.976	5.998
Br	8.865	8.900	8.927
I	13.900	14.120	14.120
Double bond between C-atoms	1.733	1.707	1.707
Triple bond between C-atoms	2.398		

* From Eisenlohr, "Spektrochemie organischer Verbindungen," Stuttgart, (1912), 44, 46, 48.

Investigation has shown, however, that the molecular refractivity is influenced by the presence of double and triple bonds, and also by the constitution of the molecule. Table 4.6 gives values for atomic refractivities for the D line of the solar spectrum (sodium light), 5893 angstrom units, calculated by different investigators. By use of these constants it is often possible to establish or confirm the chemical constitution of a pure chemical compound.

Certain anomalies have been observed. When double bonds are present in a conjugated position, the molecular refractivity will show in general a higher value than one would expect; this is known as optical exaltation. In some cases optical depression is also encountered. It is interesting to note that conjugated double bonds in a ring compound cause no exaltation or depression.

The application of molecular refraction is limited to pure individual chemical compounds; it becomes meaningless when applied to mixtures as complex as essential oils. Nevertheless, this constant has played a very important role in the elucidation of structure in the case of many individual constituents of essential oils after separation and purification.

5. SOLUBILITY

***a.* Solubility in Alcohol.**—Since most essential oils are only slightly soluble in water and are miscible with absolute alcohol, it is possible to determine the number of volumes of dilute alcohol required for the complete solubility of one volume of oil. The determination of such a solubility is a convenient and rapid aid in the evaluation of quality of an oil. In general, oils rich in oxygenated constituents[31] are more readily soluble in dilute alcohol than oils rich in terpenes.

Adulteration with relatively insoluble material will often greatly affect the solubility. Sometimes an actual separation of the adulterant may be observed. For example, adulteration of citronella oils (which are normally soluble in 80 per cent alcohol) with relatively large amounts of petroleum fractions will result in a poor solubility for the oil in 80 per cent alcohol and an actual separation of oily droplets of the adulterant. However, certain oils will show a normal separation in dilute alcohol. Expressed orange oil, for example, will separate natural waxes in 90 per cent alcohol. In alcohol of lower strength such an oil will separate a terpene fraction in addition to the waxes. Use of this fact sometimes is made in the preparation of terpeneless and sesquiterpeneless oils, concentrates and extracts.

The solubility of an oil may change with age. Polymerization is usually accompanied with a decrease in solubility; i.e., a stronger alcohol may be required to yield a clear solution. Such polymerization may be very rapid if the oil contains large amounts of easily resinified terpenes—e.g., juniper berry oil, bay oil. Improper storage may hasten polymerization; factors such as light, air, heat, and the presence of water, usually exert an

[31] However, the oxygenated constituents belonging to the sesquiterpene series are relatively insoluble; e.g., cedrol, santalol. Several other exceptions are also encountered; e.g., safrole, anethole.

unfavorable influence. Occasionally the solubility of an oil improves upon aging—e.g., oil of anise.[32]

Alcohols of the following strengths are customarily used in determining solubilities of essential oils:

50%-60%-70%-80%-90%-95% and occasionally 65% and 75%.

These are volume percentages at 15.56°/15.56°. In preparing dilute alcohols it is convenient to weigh the alcohol (95 per cent by volume) and the distilled water to give the proper volume percentage. Preparation in this manner is independent of temperature. The strength of the alcohol should be checked by determining the specific gravity at 15.56°/15.56°. Final adjustments may be made if necessary.

TABLE 4.7. PREPARATION OF DILUTE ALCOHOLS

Alcohol (% by volume)	Specific Gravity 15.56°/15.56°	95% Alcohol by volume, (g.)	Distilled Water (g.)
50	0.9342	460	540
60	0.9133	564	436
65	0.9019	619	381
70	0.8899	676	324
75	0.8771	734	266
80	0.8636	796	204
90	0.8336	927	73
95	0.8158	1000	0

Procedure: Introduce exactly 1 cc. of the oil into a 10 cc. glass-stoppered cylinder (calibrated to 0.1 cc.), and add slowly, in small portions, alcohol of proper strength. Shake the cylinder thoroughly after each addition. When a clear solution is first obtained, record the strength and the number of volumes of alcohol required. Continue the additions of alcohol until 10 cc. has been added. If opalescence or cloudiness occurs during these subsequent additions of alcohol, record the point at which this phenomenon occurs. In the event that a clear solution is not obtained at any point during the addition of the alcohol, repeat the determination, using an alcohol of higher strength.

Since the solubility is dependent upon the temperature, all determinations should be made at 20°. It should be noted, however, that "The United States Pharmacopoeia"[33] and "The National Formulary"[34] specify an official temperature of 25° for solubilities; "The British Pharmacopoeia,"[35] a tem-

[32] This is due to the presence of the difficultly soluble anethole, which yields upon oxidation the readily soluble anisic aldehyde.

[33] Thirteenth Revision, 8.

[34] Eighth Edition, 10.

[35] (1932), 9.

perature of 15.5°. The proper temperature may be maintained by frequent immersion of the cylinder in a water bath previously adjusted to the desired temperature.

If an oil is not clearly soluble in the dilute alcohols, it is advisable to describe more fully the appearance of the solubility test.

The following terms, which are relative and entirely empirical, are used in the laboratories of Fritzsche Brothers, Inc., to describe the appearance of the solution:

Clearly soluble	Opalescent
Slightly hazy	Slightly turbid
Hazy	Turbid
Slightly opalescent	Cloudy

A further term occasionally used is "fluorescent."

In the case of turbidity or cloudiness, record any separation of wax or oil that occurs, as well as the period of time required for such separation.

If an oil is soluble in a number of volumes of alcohol which is not a multiple of ½, report the solubility as being between the closest such limits. For example, if 2.7 volumes of 70 per cent alcohol were required to obtain a clear solution, and the solution remained clear upon further additions of 70 per cent alcohol until a total of 10 volumes had been added, the solubility would be recorded as:

"Clearly soluble in 2.5 to 3 volumes of 70 per cent alcohol and more, up to 10 volumes."

The behavior of the oil is best described by the following typical notations:

1. Clearly soluble in ____ volumes of ____ per cent alcohol and more, up to 10 volumes.
2. Clearly soluble in ____ volumes of ____ per cent alcohol; opalescent with more, up to 10 volumes.
3. Clearly soluble in ____ volumes of ____ per cent alcohol; opalescent to turbid with more, up to 10 volumes. No separation observed after 24 hr.
4. Clearly soluble in ____ volumes of ____ per cent alcohol and more, up to ____ volumes; opalescent in ____ volumes and more, up to 10 volumes.
5. Hazy in ____ volumes of ____ per cent alcohol; cloudy with more, up to 10 volumes. Oily separation observed after ____ hr.
6. Clearly soluble up to 10 volumes of ____ per cent alcohol.

***b.* Solubility in Nonalcoholic Media.**—Several solubility tests have been introduced for the rapid evaluation of oils. The following have proven valuable.

I. Carbon Disulfide Solubility for the Presence of Water.[36]—Oils rich in oxygenated constituents frequently contain dissolved water. This is particularly true in the case of oils containing large amounts of phenolic bodies—e.g., oil of bay. Such oils fail to give a clear solution when diluted with an equal volume of carbon disulfide or chloroform. This is the basis of a rapid test to ascertain whether or not an oil has been sufficiently dried.

II. Potassium Hydroxide Solubility for Phenol-Containing Oils.—Phenolic isolates and synthetics as well as oils consisting almost exclusively of phenolic bodies may be evaluated rapidly by dissolving 2 cc. of the oil in 20 or 25 cc. of a 1 N aqueous solution of potassium hydroxide[37] in a 25 cc. glass-stoppered, graduated cylinder. This test is particularly of value in the case of sweet birch and wintergreen oils. (See "Detection of Adullerants," p. 331.) It is well to examine critically the odor of the solution or any insoluble portion, whereby additions of foreign, odor-bearing substances may be detected.

Upon prolonged standing, the alkaline solution may saponify an ester group, if present. If the products of such a saponification are soluble in the alkaline solution, no separation will be observed—e.g., methyl salicylate. If the products are not completely soluble, a separation may occur—e.g., amyl salicylate.

Solutions of the alkali phenolates are frequently good solvents for other compounds; thus terpeneless bay oils containing about 90 per cent eugenol often form clear solutions with a 1 N potassium hydroxide solution. In this connection see the discussion under "Phenol Determination," p. 293.

III. Sodium Bisulfite Solubility for Aldehyde-Containing Oils.—Oils (such as oil of bitter almond, free from prussic acid), and synthetics (such as benzaldehyde, tolyl aldehyde, cinnamic aldehyde, and anisic aldehyde) and isolates (such as citral) may reveal impurities by their incomplete solution in dilute bisulfite solution. This test is usually carried out in a 25 cc. glass-stoppered, graduated cylinder: shake 1 cc. of the oil with 9 cc. of a freshly prepared saturated solution of sodium bisulfite and then add 10 cc. of water with further shaking. The odor of the resulting solution should be carefully examined. Because of the relative insolubility of certain bisulfite addition compounds, no general procedure is satisfactory for all aldehydes. Thus, some must be heated in a beaker of boiling water; and some require a larger amount of water to yield a clear solution. Each chemist soon

[36] "The National Formulary," Sixth Edition, 272.

[37] The potassium Hydroxide Test Solution of "The United States Pharmacopoeia" (13th Rev., 842) may be used; this is prepared by dissolving 6.5 g. of potassium hydroxide, A.R., in sufficient water to yield 100 cc. of solution. Since the potassium phenolates are more soluble than the corresponding sodium compounds, the use of potassium hydroxide is to be recommended.

develops his own techniques in testing these aldehydes; hence, specialized procedures have been omitted here.

6. CONGEALING POINT

The congealing point[38] offers a distinct advantage over the melting point and the titer, in the case of mixtures, such as essential oils.[39] In determining the congealing point, the oil is supercooled so that, upon congelation, immediate crystallization with liberation of heat occurs. This results in a rapid rise of temperature, which soon approaches a constant value and remains at this temperature for a period of time. This point is known as the "congealing point." With increasing percentage of crystalline material in an oil, the congealing point will approach a maximum.[40] Hence, this physical property is a good criterion of the percentage of such material. The congealing point is important in the evaluation of anise, sassafras and fennel oils.

> *Procedure:* Place about 10 cc. of the oil in a dry test tube of 18 to 20 mm. diameter. Cool in water or in a suitable freezing mixture, the temperature of which should be about 5° lower than the supposed congealing point of the liquid. To initiate congelation, rub the inner walls of the tube with a thermometer, or add a small amount of the substance previously solidified by excessive freezing. The thermometer should be rubbed quickly up and down in the mixture in order to cause a rapid congelation throughout, with its subsequent liberation of heat. The temperature should be read frequently; at first the rise of temperature is rapid, but soon approaches a constant value for a brief interval of time. This value is taken as the congealing point of the oil.
>
> The process described above should be repeated several times to assure obtaining the true congealing point.

The thermometer used should be calibrated in 0.1° units and should be accurately standardized. A thermometer covering the range of −5° to +50° is satisfactory for most determinations.

Before the oil is tested, it should be thoroughly dried with sodium sulfate, since the presence of small amounts of water will often materially lower the congealing point.

In the case of sassafras oils, it is well to initiate the congelation by the addition of a small piece of solid safrole since sassafras oil can be congealed only with great difficulty if no "seed" is used.

[38] The so-called "congealing point" of rose oil is not a true congealing point, but is determined by the same method as that used for titer determinations in fixed oils. (See "Special Tests and Procedures," p. 329.)

[39] The melting point is usually used for crystalline solids.

[40] This maximum will be the "congealing point" of the pure crystalline compound.

For a more exact determination of the congealing point, the test tube containing the supercooled oil may be insulated by means of an air jacket. This is frequently of particular importance when determining congealing points which are much below room temperature, as, for example, the congealing point of euclayptus oils. Gildemeister and Hoffmann recommend the use of the Beckman apparatus,[41] frequently used for the determination of molecular weights by the lowering of the freezing point. The use of a larger sample (up to 100 cc.) may make the congealing point sharper.

7. MELTING POINT

The importance of the determination of the melting point of a solid, crystalline material is obvious.

A brief but comprehensive discussion of the determination of melting points has been given by Shriner and Fuson,[42] from which much of the following is taken.

> *Procedure:* Heat a piece of 15 mm. glass tubing[43] in a flame until the glass is soft; then draw out into a thin walled capillary tube about 1 mm. in diameter. Cut into lengths of about 6 cm., and seal one end in a flame. Powder a small amount of the compound in a polished agate mortar and introduce some of the powder into the capillary tube. Hold the capillary tube vertically and gently rub with a file, which causes the powder to settle to the bottom; pack the material by tapping the tube on the desk. Fasten the tube to the thermometer by means of a rubber band (cut from a piece of rubber tubing) so that the sample is close to the mercury bulb (see Diagram 4.3). Place a heavy white mineral oil in the beaker and heat with a low flame. Clamp a cylindrical metal shield, open at the top and bottom, in the position as shown in Diagram 4.3 in order to protect the flame from drafts. Heat at a rate to cause a rise in temperature of about 1° or 2° per min. Stir the oil bath continuously. Note the temperature at which the compound starts to melt[44] and that at which it is entirely liquid; record these values as the melting point range. Note also the temperature recorded by the auxiliary thermometer (t_2); the bulb of this thermometer should be placed midway between the surface of the oil and the top of the mercury thread in t_1. Calculate the stem correction by means of the following formula:
>
> $$\text{Correction} = 0.000154N(t_1 - t_2)$$

[41] "Die ätherischen Öle," 3d Ed. Vol. I, 707.

[42] "The Systematic Identification of Organic Compounds," John Wiley & Sons, Inc., New York (1940), 85.

[43] Soft glass test tubes are very satisfactory since the walls are relatively thin.

[44] According to "The United States Pharmacopoeia," Thirteenth Rev., 668, the temperature at which the column of the sample is observed to collapse definitely against the side of the tube at any point is defined as the beginning of the melting. The temperature at which the material becomes liquid throughout is defined as the end of the melting.

where: N = number of degrees of mercury thread above the level of the oil bath;
t_1 = observed melting point;
t_2 = average temperature of the mercury thread.

This correction is to be added to the observed melting point.

It is often time saving to run a preliminary melting point, raising the temperature of the bath very rapidly. After the approximate melting point is known, a second determination is carried out raising the temperature rapidly until within 10° of the approximate value and then proceeding slowly as described above. A fresh sample of the compound should be used for each determination.

The thermometer should be calibrated by observing the melting points of several pure compounds such as the following:

Melting Point (corr.)	
0°	Ice
53	*p*-Dichlorobenzene
90	*m*-Dinitrobenzene
114	Acetanilide
121	Benzoic Acid
132	Urea
157	Salicylic Acid
187°	Hippuric Acid
200	Isatin
216	Anthracene
238	Carbanilide
257	Oxanilide
273	Anthraquinone
317	*N,N*-Diacetylbenzidine

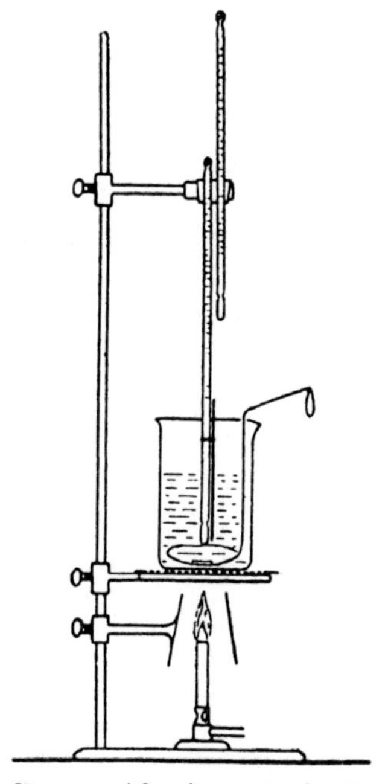

DIAGRAM 4.3. Apparatus for the determination of melting point.

If the same apparatus and thermometer are used in all melting point determinations, it is convenient to prepare a calibration curve. The observed melting point of the standard compound is plotted against the corrected value, and a curve is drawn through these points. In subsequent determinations the observed value is projected horizontally to the curve and then vertically down to give the corrected value. Such a calibration curve includes corrections for inaccuracies in the thermometer and stem correction.

The use of short stemmed, standardized Anschütz thermometers eliminates the need for an auxiliary thermometer and subsequent correction for emergent stem.

It is important to record the melting point range of a compound since this is a valuable index of purity. A large majority of pure organic compounds melt within a range of 0.5° or melt with decomposition over a narrow range of temperature (about 1°).

When determining the melting point of a solid that readily sublimes—e.g., borneol—certain precautions become necessary. The rate of heating of the oil bath should be increased considerably. The capillary should not be introduced into the hot oil until the temperature is within 10° to 20° of the expected melting point. The use of a sealed capillary may be necessary, i.e., a capillary that has both ends fused. The use of a Fisher-Johns or similar type apparatus is not recommended for materials that sublime readily.

Other types of melting point apparatus have proven satisfactory—e.g., the Fisher-Johns, Thiele, and Thiele-Dennis.

If a compound has a high melting point a Maquenne block may conveniently be used. It is claimed that the Dennis melting point apparatus is very satisfactory for compounds melting up to 300°.

Special types have been developed for determination of the melting point of waxes,[45] and the softening point of amorphous material.[46]

8. BOILING RANGE

In the case of isolates and synthetics, the determination of the boiling range is an important criterion of purity.

Procedure: Use the apparatus shown in Diagram 4.4. The bulb of the distilling flask should have a capacity of 50 cc. The neck of the flask above the side arm should be as short as possible. The bottom of the flask rests in a circular opening, 2.5 cm. in diameter, cut in a square piece of asbestos board having a thickness of about 3 mm.; this perforation should be slightly beveled on the upper edge to make it fit closely to the surface of the flask.[47] A wrapping of asbestos paper reaching to a point about 1 cm. above the side arm should be used to prevent condensation due to drafts.

Introduce 25 cc. of the sample into the flask by means of a pipette. Add a small clay chip. Insert the thermometer along the central axis of the flask with its bulb slightly below the side tube; attach a light auxiliary thermometer to the main thermometer to correct for stem exposure, the bulb of this second thermometer being placed half way from the cork to the top of the mercury column at the expected reading. (A short-stemmed thermometer of the Anschütz type having the proper

[45] A.S.T.M. Melting Point Apparatus, Designation D87.

[46] A.S.T.M. Softening Point Apparatus (Ring and Ball Method), Designation D36.

[47] This is to prevent upward leakage of hot gases from the flame and subsequent superheating.

range may be used; this will require no correction for stem exposure.) Distill at a uniform rate of about 0.5 cc. per min. until the level of the liquid remaining in the flask falls to the level of the asbestos diaphragm.

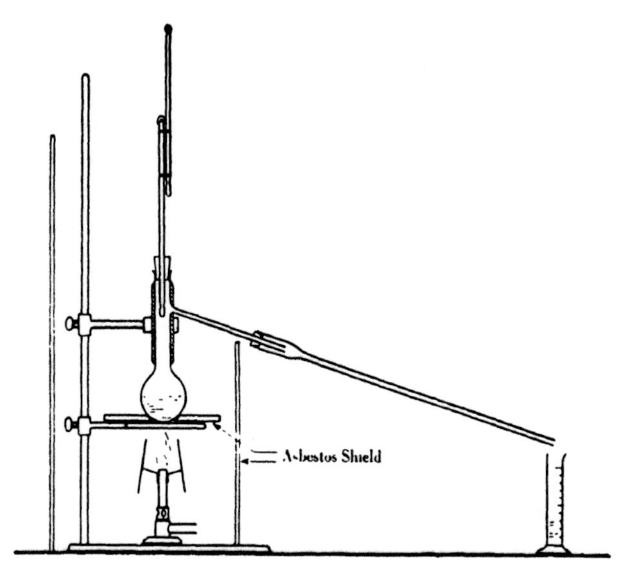

DIAGRAM 4.4. Apparatus for the determination of boiling range.

Since some time will elapse before the thermometer can acquire the temperature of the vapor, little significance can be attached to readings taken before the end of the first minute after the fall of the first drop of distillate from the side tube. Any readings taken after the liquid falls below the level of the asbestos board will be greatly influenced by superheating. In the case of pure compounds that boil without decomposition, the difference between the first and last significant readings should not amount to more than 1°.

The stem exposure correction may be found by the following formula:

$$\text{Correction} = 0.000154\ N(t_1 - t_2)$$

where: N = number of degrees of emergent stem;
t_1 = observed temperature of main thermometer;
t_2 = temperature of auxiliary thermometer.

This correction is to be added.

To reduce boiling points taken under pressures between 720 mm. and 780 mm. to their approximate values at 760 mm., apply a correction of 0.1° for every 2.7 mm. difference; the correction is to be added if the observed pressure is below 760 mm. and to be subtracted if above 760 mm.

The percentage of an essential oil which distills below a given temperature is frequently of importance in evaluating the oil; also, the percentage which distills between certain limits. However, it must be remembered that when fractionating an oil, the quantitative results of different observers will vary greatly; this is due to differences in the types of distilling flasks and condensers employed, to the distillation rates, and to the degree of superheating of the vapors.

Examination of the various fractions is of great importance; the determination of physical and chemical properties of these fractions and a study of the odor is frequently very revealing. Furthermore, suspected adulterants may be tested for chemically, and if present identified by derivatives.

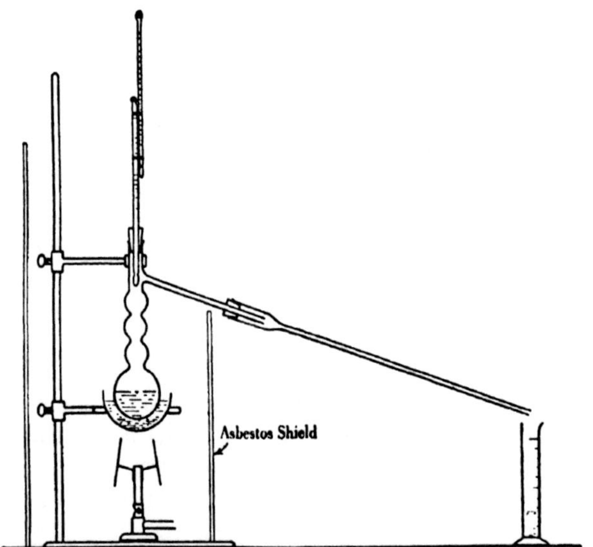

DIAGRAM 4.5. Apparatus for the determination of boiling range.

Only through experience will the chemist know whether or not it is better to distill at atmospheric pressure or under vacuum. In general, for the collection of first fractions it is better to distill at atmospheric pressure. Usually it is more advantageous to separate fractions according to the temperature, measuring the volumes collected; occasionally it is desirable to collect definite amounts, noting the temperatures at which these fractions are obtained.

For fractionations at normal pressure the following technique will generally give satisfactory results. The procedure as described is intended primarily for the distillation of turpentine oil and for the removal and collection of the first 10 per cent of citrus oils.

Procedure: See Diagram 4.5. Place 50 cc. of the oil in a 100 cc. three-bulb Ladenburg flask of approximately the following dimensions: the lower or main bulb 6 cm. in diameter, with the smaller condensing bulbs 3.5 cm., 3.0 cm., and 2.5 cm., respectively, in diameter; the distance from the bottom of the flask to the side arm, 20 cm. Support the flask in a hemispherical metal oil bath, 4 in. in diameter, containing a suitable heating medium such as glycerin, cottonseed oil, or high boiling mineral oil. Attach a Pyrex straight-tube condenser, 22 in. long, having a water cooled jacket,[48] and fitted with an adapter which is long enough to extend into the graduated cylinder used as a receiver. Use a short-stemmed thermometer of the Anschütz type or a long-stemmed thermometer with an auxiliary thermometer for stem correction. Add a few small clay chips. Heat the bath with a Bunsen burner protected from drafts by a chimney. Fasten a large sheet of asbestos board vertically to act as a shield for the flame, bath and flask. Distill the oil at a uniform rate of 1 drop per sec. until the required distillate is obtained.

9. EVAPORATION RESIDUE

An important criterion of purity is the evaporation residue; i.e., the percentage of the oil which is not volatile at 100°.

A determination of the evaporation residue is of special value in the case of the citrus oils; a low value for an expressed oil suggests the possibility of the addition of terpenes, or other volatile constituents; a high value may indicate the addition of foreign material, such as, rosin, fixed oils, or high boiling sesquiterpenes. In the case of rectified oils such as turpentine, a high value may indicate improper or lack of rectification, or polymerization due to age or improper storage. In the case of certain solids, such as camphor, thymol, or menthol, a high evaporation residue will indicate insufficient purification.

It is important to study the odor of an oil as it volatilizes during the heating. Often "by-notes" of foreign low boiling adulterants or contaminants may be discovered. The odor of the final residue while still hot should also be carefully studied for the addition of high boiling adulterants, such as cedarwood.

The consistency of the residue, both when hot and cool, and the color sometimes indicates the presence of particular adulterants. For example, an orange oil, which has a brittle residue instead of the usual soft waxy residue, should be carefully investigated for rosin.

Acid numbers and saponification numbers may be determined on suspicious residues: rosin usually raises the acid number considerably; fixed oils raise the saponification number.

[48] For oils containing mostly high boiling constituents (such as cassia and bay), use an air-cooled condenser.

The fact that essential oils are complex mixtures makes an exact determination of the nonvolatile residue very difficult. "Constant weight" cannot conveniently be attained because of the fact that waxes and other high boiling nonvolatile material tend to retain or "fix" some of the lower boiling constituents. "The United States Pharmacopoeia" defines constant weight as the value obtained when "two consecutive weighings do not differ by more than 0.1 per cent, the second weighing following an additional hour of drying."[49] Even after "constant weight," according to this definition, has been attained, further prolonged heating will give much lower results. Hence, a certain standardization of technique becomes necessary.

Procedure: Weigh accurately (to the closest milligram) a well cleaned Pyrex evaporating dish that has been permitted to stand in a desiccator for 30 min. To this tared dish add the requisite amount of oil or solid (weighed to the closest centigram) and heat on a steam bath for the prescribed length of time. Then permit the evaporating dish to cool to room temperature in the desiccator and weigh (to the closest milligram). Calculate the nonvolatile residue obtained, the so-called "evaporation residue," and express as a percentage of the original oil.

	Size of Sample (*g.*)	*Period of Heating* (*hr.*)
Oil Bergamot	5	5
Oil Grapefruit	5	6
Oil Lemon	5	4½
Oil Limes, Expressed	5	6
Oil Mace	3	8
Oil Mandarin	5	5
Oil Nutmeg	3	8
Oil Orange	5	4½
Oil Tangerine	5	5
Oil Turpentine	5	4½
Oleoresin Capsicum, Oleoresin Ginger	2	4
Camphor	2	4
Copaiba	0.5	6
Menthol	2	2
Styrax	2	2
Thymol	2	4
Floral Waters	100	After last of liquid has evaporated, heat for an additional hour.

It is well to bear in mind that the size, shape, and composition of the evaporating dish employed in such a determination, as well as the size of sample and time of heating, will influence the analytical result obtained.

[49] Twelfth Revision, 3. The Thirteenth Revision, 7, limits the difference to not more than 0.05%.

Flat bottom evaporating dishes of Pyrex glass are very satisfactory; they offer the further advantage of more easily permitting an observation of the color and opacity of the residue. Conventional Pyrex evaporation dishes, 80 mm. in diameter and 45 mm. deep, are to be recommended. The use of such dishes tends to minimize the formation of polymerization products in most cases.

Certain exceptional products will require special treatments, however; evaporation residues on such materials as diacetyl are meaningless because of the rapid formation of polymerization products unless the determinations are carried out in vacuum with the application of little or no heat.

In evaluating oleoresins, evaporation residues should also be determined. Here it is best to express the results as "loss of weight on heating." The analytical results obtained will include the loss of volatile solvent as well as the loss of part of the naturally occurring essential oil. An abnormally high value often indicates the incomplete removal of the volatile solvent used in the manufacture of the oleoresin.

10. FLASH POINT

The flash point may prove useful in the evaluation of an essential oil. Unfortunately insufficient data exist to use this property as a criterion of quality for normal, unadulterated oils. However, the flash point has value as an indication of adulteration: additions of adulterants such as alcohol and low boiling mineral spirits will greatly lower the flash point.

Occasionally it is necessary to determine the flash point of a synthetic, solvent, or a mixture because of shipping regulations.

Several types of instruments are available for the determination; e.g., the Pensky-Martin closed tester,[50] the Tag closed tester,[51] the Cleaveland[52] and the Tag open cup testers. The Tag open cup tester is simple, inexpensive, and entirely satisfactory for use in the essential oil industry. The procedure described below is intended primarily for this instrument (see Diagram 4.6).

> *Procedure:*[53] Fill the metal bath with water of about 60° F. (15.6° C.) temperature,[54] leaving room for displacement by the glass oil cup which is placed in the water bath. Suspend the thermometer in a vertical position so that the bottom of the bulb is about $\frac{1}{4}$ in. from the bottom of the glass cup and so that

[50] A description of the instrument and a detailed procedure for its use may be found in A.S.T.M. Designation D93–42.

[51] A.S.T.M. Designation D56–36.

[52] A.S.T.M. Designation D92–33.

[53] This procedure is based on the directions for using the Tag Open Fire Tester supplied by the Tagliabue Manufacturing Co., Brooklyn, N. Y.

[54] It is customary in the United States to report flash points in degrees Fahrenheit.

the thermometer is suspended half way between the center and the back of the glass cup. Fill the glass cup with the oil to be tested in such a manner that the top of the meniscus is exactly at the filling line at room temperature (i.e., $\frac{1}{8}$ in. from the upper edge of the cup). Be sure that there is no oil on the outside of the cup or on its upper level edge; use soft paper to clean the cup in preference to a cloth. Remove any air bubbles from the

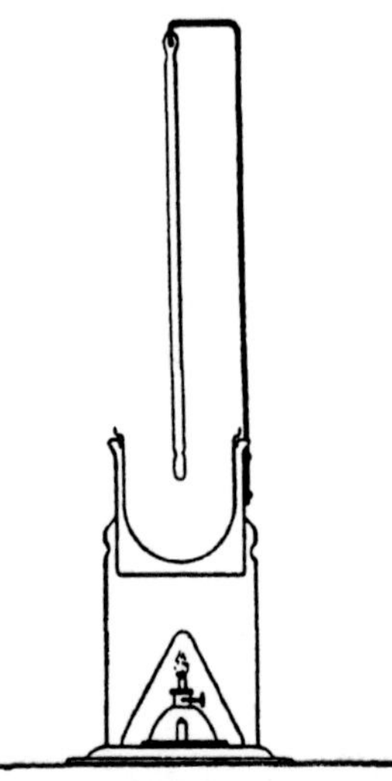

DIAGRAM 4.6. Tag open cup tester for the determination of flash point.

surface of the oil. Adjust the horizontal flashing taper guide wire in place. The instrument should stand level and should be protected from drafts. It is desirable that the room be darkened sufficiently so that the flash may be readily discernible. Avoid breathing over the surface of the oil. Heat the water bath with a small burner so that it will raise the temperature of the oil at a rate not faster than 2° F. (1.1° C.) per min. without removing the burner during the whole operation. Adjust the test flame on the flashing taper so that it is the same size as the metal bead mounted on the instrument. Apply this test flame to the oil at 5° F. (2.8° C.) intervals: hold the flashing taper in a horizontal position and draw it across the guide wire quickly and without pause from left to right. (The time of passage of the test flame across the cup should be approximately 1 sec.) The first or initial flash[55] is called the "flash point." Continue heating and testing the oil until the surface ignites and continues to burn until quickly blown out with a mouth-open breath. This burning point temperature is called the "fire test" or "fire point." Repeat the determination and try for a flash at the proper trial temperatures indicated in Table 4.8.

[55] The true initial flash should not be confused with a bluish halo that sometimes surrounds the test flame

TABLE 4.8. TRIAL TEMPERATURE TABLE FOR FLASH POINTS
(All Temperatures in ° F.)

For Oils Expected to Have a Fire Test of	Try for Flash						
	First at	Then at					
110	85	90	95	100	105	108	110
115	90	95	100	105	110	113	115
120	95	100	105	110	115	118	120
125	100	105	110	115	120	123	125
130	100	105	110	115	120	125	130
135	105	110	115	120	125	130	135
140	110	115	120	125	130	135	140
145	115	120	125	130	135	140	145
150	120	125	130	135	140	145	150

IV. DETERMINATION OF CHEMICAL PROPERTIES

1. DETERMINATION OF ACIDS

Most essential oils contain only small amounts of free acids. Consequently the acid content is usually reported as an acid number rather than as a percentage calculated as a specific acid.

The acid number of an oil is defined as the number of milligrams of potassium hydroxide required to neutralize the free acids in 1 g. of oil.

In determining the acid number, dilute alkali must be employed since many of the esters (e.g., the formates) normally present in essential oils are capable of saponification even in the cold in the presence of strong alkalies. Moreover, phenols will react with the alkali hydroxides, making it necessary to use special indicators (such as phenol red) for oils containing large amounts of phenolic bodies; this is particularly true in the case of the salicylates.

The acid number of an oil often increases as the oil ages, especially if the oil is improperly stored; processes such as oxidation of aldehydes and hydrolysis of esters increase the acid number. Oils which have been thoroughly dried and which are protected from air and light show little change in the amount of free acids.

Procedure: Weigh accurately about 2.5 g. of the oil into a 100 cc. saponification flask. Add 15 cc. of neutral 95 per cent alcohol and 3 drops of a 1% phenolphthalein solution. Titrate the free acids with a standardized 0.1 N aqueous sodium hydroxide solution, adding the alkali dropwise at a uniform rate of about 30 drops per min. The contents of the flask must be continually agitated. The first appearance of a red coloration that does not fade within 10 sec. is considered the end point.

If the determination requires more than 10 cc. of alkali, it should be repeated using a 1 g. sample of the oil; if more than 10 cc. of alkali is still required, then a 1 g. sample is titrated with 0.5 N aqueous sodium hydroxide solution.

The acid number is calculated by means of the following formulas:

$$\text{Acid number} = \frac{5.61\ (\text{no. of cc. of } 0.1\text{ N NaOH})}{\text{wt. of sample in g.}}$$

$$= \frac{28.05\ (\text{no. of cc. of } 0.5\text{ N NaOH})}{\text{wt. of sample in g.}}$$

TABLE 4.9. MOLECULAR WEIGHTS* OF ACIDS

Acids	*Molecular Wt.*
Monobasic Acids	
Acetic	60.05
Anisic	152.14
Anthranilic	137.13
Benzoic	122.12
Butyric	88.10
Capric	172.26
Caproic	116.16
Caprylic	144.21
Cinnamic	148.15
Formic	46.03
Furoic	112.08
Lactic	90.08
Lauric	200.31
Methyl Anthranilic	151.16
Myristic	228.37
Oenanthic	130.18
Oleic	282.46
Pelargonic	158.24
Phenylacetic	136.14
Phenylpropionic	150.17
Propionic	74.08
Pyruvic	88.06
Salicylic	138.12
Stearic	284.47
Tiglic	100.11
Undecylenic	184.27
Undecylic	186.29
Valeric	102.13
Dibasic Acids	
Malonic	104.06
Phthalic	166.13
Sebacic	202.25
Succinic	118.09
Tartaric	150.09
Tribasic Acids	
Citric	192.12

* All molecular weights have been calculated from the values of the International Atomic Weights adopted by the Committee on Atomic Weights in 1938.

For oils containing large amounts of free acid (e.g., orris oil), the free acid content may be expressed as a percentage, calculated as a specific acid. In such cases it is well to use a 0.5 N alcoholic sodium hydroxide solution.

$$\text{Free acid content (percentage)} = \frac{ma}{20w}$$

$$\text{Free acid content (percentage)} = \frac{mb}{100w}$$

where: m = molecular weight of the acid;
a = number of cc. of 0.5 N alkali used for neutralization;
b = number of cc. of 0.1 N alkali used for neutralization;
w = weight of sample in grams.

If the acid is dibasic, the result must be divided by 2; if tribasic, by 3.

In Table 4.9 are listed the molecular weights of those acids frequently encountered by the essential oil chemist.

2. DETERMINATION OF ESTERS

a. Determination by Saponification with Heat.—The determination of the ester content is of great importance in the evaluation of many essential oils. Since most esters which occur as normal constituents of essential oils are esters of monobasic acids, the process of saponification may be represented by the following reaction:

$$RCOOR' + NaOH \rightarrow RCOONa + R'OH$$

where R and R′ may be an aliphatic, aromatic, or alicyclic radical (R may also be a hydrogen atom).

Procedure: Into a 100 cc. alkali-resistant saponification flask weigh accurately about 1.5 g. of the oil. Add 5 cc. of neutral 95% alcohol and 3 drops of a 1% alcoholic solution of phenolphthalein, and neutralize the free acids with standardized 0.1 N aqueous sodium hydroxide solution.[56] Then add 10 cc. of 0.5 N alcoholic sodium hydroxide solution, measured accurately from a pipette or a burette. Attach a glass, air-cooled condenser to the flask, 1 m. in length and about 1 cm. in diameter, and reflux the contents of the flask for 1 hr. on a steam bath. Remove and permit to cool at room temperature for 15 min. Titrate the excess alkali with standardized 0.5 N aqueous hydrochloric acid. A further addition of a few drops of phenolphthalein solution may be necessary at this point.

[56] This usually requires not more than 5 drops of the 0.1 N alkali.

In order to determine the amount of alkali consumed, carry out a blank determination, observing the same conditions but omitting the oil. The difference in the amounts of acid used in titrating the actual determination and the blank gives the amount of alkali used for the saponification of the esters. The blank should require an excess of about 100 per cent over the amount used in the determination. If insufficient excess is used, results will be obtained which are too low.

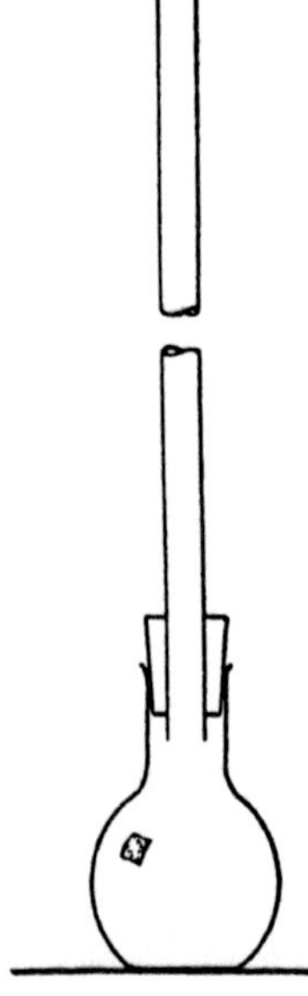

DIAGRAM 4.7. Saponification flask.

It is well to use saponification flasks (see Diagram 4.7) made of "Jena Glass" or of the special alkali-resistant glass recently made available by Corning Glass Company. These flasks minimize the amount of alkali consumed by the action of the sodium hydroxide on the glass itself. More accurate results are thus obtained. This is of importance when the ester determination requires more than 1 hr. of refluxing, as, for example, in the case of the isovalerates.

The alcoholic 0.5 N sodium hydroxide solution is best prepared by adding 11.5 g. of metallic sodium of analytical grade to 1 liter of 95 per cent ethyl alcohol. (If larger amounts of solution are to be prepared, use 43.5 g. of sodium for each gallon of alcohol.) The sodium should be added slowly, a few small pieces at a time. After weighing out the sodium and cutting it into small pieces, it should be protected from atmospheric moisture until it is used by immersion in low boiling petroleum ether. After the required amount of sodium has been added, the solution is set aside for several days to permit any carbonate to settle; it is filtered into the reagent reservoir and permitted to stand for a few days before it is used. A clear, water white solution is thus obtained. The 0.5 N hydrochloric acid may be prepared by diluting 85 cc. of concentrated acid to 2 liters; it should then be carefully standardized.

Calculation of Results.—The ester content may be calculated from the following formula:

$$\text{Percentage of ester} = \frac{am}{20s}$$

where: a = number of cc. of 0.5 N sodium hydroxide used in the saponification;

m = molecular weight of the ester;

s = weight f the sample in grams.

This formula assumes that the ester is monobasic; for esters of dibasic acids (e.g., dimethyl phthalate) and dihydroxy alcohols (e.g., glycol diacetate), the ester content is divided by 2; for tribasic acids (e.g., triethyl citrate) and trihydroxy alcohols (e.g., triacetin), by 3.[57]

The ester may also be expressed by the ester number, which is defined as the number of milligrams of potassium hydroxide required to saponify the esters present in 1 g. of oil. The use of the ester number is especially convenient when the ester present in the oil is unknown, since a knowledge of the molecular weight of the ester is not required.

$$\text{Ester number} = \frac{28.05a}{s}$$

Ester numbers are frequently used for oils which contain very small amounts of ester; e.g., oil of black pepper and oil of cubeb. A high ester number in such cases is usually indicative of adulteration.

The ester number may readily be converted to an ester content, expressed as a weight percentage, by the following formula if the acid radical of the ester is monobasic:

$$\text{Percentage of ester} = \frac{m\ (\text{ester no.})}{561.04}$$

If the acid is dibasic, the result must be divided by 2; if tribasic, by 3. Also, if the alcohol radical contains two hydroxy groups, the result[58] must be divided by 2; if three hydroxy groups, by 3.

In Table 4.10 are listed the molecular weights of those esters which are frequently encountered.

Modification of the General Procedure.—Certain esters are not completely saponified in a period of 1 hr. by the procedure described above. Notable exceptions are the salicylates which should be refluxed for 2 hr.; terpinyl acetate, 2 hr.; menthyl acetate, 2 hr.; isovalerates, 6 hr. Certain esters of sesquiterpene alcohols require 2 hr. or more—e.g., cedryl acetate, 4 hr. A solution of potassium hydroxide in a high boiling solvent (such as the monoethyl ether of ethylene glycol) has been recommended[59] for the determination of difficultly saponifiable esters. Such a solution also permits of rapid saponification (ca. 15 min.) of other esters. Since such high temperatures may have an adverse effect upon some of the constituents of an essential oil, this method should be applied with caution.

[57] This is based on the assumption that the ester is neutral in the case of di- and tribasic acids, and that all alcoholic groups have been esterified in the case of esters of di- and trihydroxy alcohols.

[58] This is based on the assumption that the ester is neutral in the case of di- and tribasic acids, and that all alcoholic groups have been esterified in the case of esters of di- and trihydroxy alcohols.

[59] Steet, *Analyst* 61 (1936), 687.

TABLE 4.10. MOLECULAR WEIGHTS* OF ESTERS

Esters	Molecular Wt.	Esters	Molecular Wt.
Esters of Monobasic Acids		Cinnamyl Acetate	176.21
Allyl Salicylate	178.18	Cinnamyl Benzoate	238.27
Amyl Acetate	130.18	Cinnamyl Butyrate	204.26
Amyl Anisate	222.28	Cinnamyl Cinnamate	264.31
Amyl Benzoate	192.25	Cinnamyl Formate	162.18
Amyl Butyrate	158.24	Cinnamyl Propionate	190.23
Amyl Caproate	186.29	Cinnamyl Valerate	218.29
Amyl Caprylate	214.34	Citronellyl Acetate	198.30
Amyl Cinnamate	218.29	Citronellyl Benzoate	260.36
Amyl Formate	116.16	Citronellyl Butyrate	226.35
Amyl Furoate	182.21	Citronellyl Caproate	254.40
Amyl Heptine Carbonate	210.31	Citronellyl Cinnamate	286.40
Amyl Laurate	270.45	Citronellyl Formate	184.27
Amyl Myristate	298.50	Citronellyl Propionate	212.32
Amyl Oenanthate	200.31	Citronellyl Valerate	240.38
Amyl Phenylacetate	206.28	Cresyl Acetate	150.17
Amyl Propionate	144.21	Cresyl Butyrate	178.22
Amyl Pyruvate	158.19	Cresyl Cinnamate	238.27
Amyl Salicylate	208.25	Cresyl Phenylacetate	226.26
Amyl Undecylate	256.42	Cyclohexanyl Acetate	142.19
Amyl Undecylenate	254.40	Cyclohexanyl Butyrate	170.25
Amyl Valerate	172.26	Decyl Acetate	200.31
Anisyl Acetate	180.20	Decyl Formate	186.29
Anisyl Formate	166.17	Dimethyl Benzyl Carbinyl Acetate	192.25
Benzyl Acetate	150.17	Ethyl Acetate	88.10
Benzyl Benzoate	212.24	Ethyl Amyl Carbinyl Acetate	172.26
Benzyl Butyrate	178.22	Ethyl Anisate	180.20
Benzyl Cinnamate	238.27	Ethyl Anthranilate	165.19
Benzyl Formate	136.14	Ethyl Benzoate	150.17
Benzyl Heptine Carbonate	216.27	Ethyl Butyrate	116.16
Benzyl Phenylacetate	226.26	Ethyl Caprate	200.31
Benzyl Propionate	164.20	Ethyl Caproate	144.21
Benzyl Salicylate	228.24	Ethyl Caprylate	172.26
Benzyl Valerate	192.25	Ethyl Cinnamate	176.21
Bornyl Acetate	196.28	Ethyl Decine Carbonate	210.31
Butyl Acetate	116.16	Ethyl Formate	74.08
Butyl Benzoate	178.22	Ethyl Furoate	140.13
Butyl Butyrate	144.21	Ethyl Heptine Carbonate	168.23
Butyl Formate	102.13	Ethyl Hexyl Carbinyl Acetate	186.29
Butyl Furoate	168.19	Ethyl Lactate	118.13
Butyl Lactate	146.18	Ethyl Methyl Phenyl Glycidate	206.23
Butyl Phenylacetate	192.25	Ethyl Myristate	256.42
Butyl Propionate	130.18	Ethyl Octine Carbonate	182.26
Butyl Salicylate	194.22	Ethyl Oenanthate	158.24
Butyl Stearate	340.58	Ethyl Oleate	310.51
Butyl Undecylenate	240.38	Ethyl Pelargonate	186.29
Butyl Valerate	158.24	Ethyl Phenylacetate	164.20
Cedryl Acetate	264.40		

* All molecular weights have been calculated from the values of the International Atomic Weights adopted by the Committee on Atomic Weights in 1938.

TABLE 4.10.—*Continued*

Esters	Molecular Wt.	Esters	Molecular Wt.
Ethyl Propionate	102.13	Methyl Formate	60.05
Ethyl Pyruvate	116.11	Methyl Furoate	126.11
Ethyl Salicylate	166.17	Methyl Heptine Carbonate	154.20
Ethyl Undecylate	214.34	Methyl Laurate	214.34
Ethyl Undecylenate	212.32	Methyl Methyl Anthranilate	165.19
Ethyl Valerate	130.18	Methyl Myristate	242.39
Geranyl Acetate	196.28	Methyl Octine Carbonate	168.23
Geranyl Benzoate	258.35	Methyl Oenanthate	144.21
Geranyl Butyrate	224.33	Methyl Pelargonate	172.26
Geranyl Formate	182.26	Methyl Phenylacetate	150.17
Geranyl Phenylacetate	272.37	Methyl Phenylpropionate	164.20
Geranyl Propionate	210.31	Methyl Propionate	88.10
Geranyl Tiglate	236.34	Methyl Salicylate	152.14
Geranyl Valerate	238.36	Methyl Valerate	116.16
Guaiyl Acetate	264.40	Neryl Acetate	196.28
Guaiyl Phenylacetate	340.49	Neryl Butyrate	224.33
Heptyl Acetate	158.24	Neryl Formate	182.26
Heptyl Caproate	214.34	Neryl Phenylacetate	272.37
Heptyl Formate	144.21	Neryl Propionate	210.31
Heptyl Oenanthate	228.37	Neryl Valerate	238.36
Heptyl Propionate	172.26	Nonyl Acetate	186.29
Heptyl Valerate	200.31	Nonyl Butyrate	214.34
Hexyl Acetate	144.21	Nonyl Lactone	156.22
Hexyl Butyrate	172.26	Octyl Acetate	172.26
Hexyl Formate	130.18	Octyl Benzoate	234.33
Hexyl Valerate	186.29	Octyl Butyrate	200.31
Isopulegyl Acetate	196.28	Octyl Formate	158.24
Isopulegyl Formate	182.26	Octyl Oenanthate	242.39
Linalyl Acetate	196.28	Octyl Propionate	186.29
Linalyl Anthranilate	273.36	Octyl Valerate	214.34
Linalyl Benzoate	258.35	Phenyl Benzoate	198.21
Linalyl Butyrate	224.33	Phenylethyl Acetate	164.20
Linalyl Cinnamate	284.38	Phenylethyl Anthranilate	241.28
Linalyl Formate	182.26	Phenylethyl Benzoate	226.26
Linalyl Phenylacetate	272.37	Phenylethyl Butyrate	192.25
Linalyl Propionate	210.31	Phenylethyl Cinnamate	252.30
Linalyl Valerate	238.36	Phenylethyl Dimethyl Carbinyl Acetate	206.28
Menthyl Acetate	198.30	Phenylethyl Formate	150.17
Menthyl Salicylate	276.36	Phenylethyl Phenylacetate	240.29
Menthyl Valerate	240.38	Phenylethyl Propionate	178.22
Methyl Acetate	74.08	Phenylethyl Salicylate	242.26
Methyl Anisate	166.17	Phenylethyl Valerate	206.28
Methyl Anthranilate	151.16	Phenyl Methyl Carbinyl Acetate	164.20
Methyl Benzoate	136.14	Phenylpropyl Acetate	178.22
Methyl Butyrate	102.13	Phenylpropyl Butyrate	206.28
Methyl Caprate	186.29	Phenylpropyl Cinnamate	266.33
Methyl Caproate	130.18	Phenylpropyl Formate	164.20
Methyl Caprylate	158.24	Phenylpropyl Propionate	192.25
Methyl Cinnamate	162.18	Phenylpropyl Valerate	220.30
Methyl Decine Carbonate	196.28		

TABLE 4.10.—*Continued*

Esters	Molecular Wt.	Esters	Molecular Wt.
Propyl Acetate	102.13	Vetivenyl Butyrate	290.43
Propyl Butyrate	130.18	Vetivenyl Formate	248.35
Propyl Formate	88.10	Vetivenyl Propionate	276.41
Propyl Propionate	116.16	Vetivenyl Valerate	304.46
Propyl Valerate	144.21		
Rhodinyl Acetate	198.30	*Esters of Dibasic Acids*	
Rhodinyl Benzoate	260.36	Diamyl Phthalate	306.39
Rhodinyl Butyrate	226.35	Dibenzyl Succinate	298.33
Rhodinyl Formate	184.27	Dibutyl Phthalate	278.34
Rhodinyl Phenylacetate	274.39	Dibutyl Tartrate	262.30
Rhodinyl Propionate	212.32	Diethyl Phthalate	222.23
Rhodinyl Valerate	240.38	Diethyl Malonate	160.17
Santalyl Acetate	262.38	Diethyl Sebacate	258.35
Terpinyl Acetate	196.28	Diethyl Succinate	174.19
Terpinyl Anthranilate	273.36	Dimethyl Malonate	132.11
Terpinyl Butyrate	224.33	Dimethyl Phthalate	194.18
Terpinyl Cinnamate	284.38		
Terpinyl Formate	182.26	*Esters of Tribasic Acids*	
Terpinyl Propionate	210.31	Triethyl Citrate	276.28
Terpinyl Valerate	238.36	Trimethyl Citrate	234.20
Thujyl Acetate	196.28		
Undecalactone	184.27	*Esters of Trihydroxy Alcohols*	
Vetivenyl Acetate	262.38	Triacetin	218.20

In the case of salicylates, benzoates, and phthalates, an addition of 5 cc of water should be made before the ester is heated on the steam bath to prevent the separation of the sodium salts of the acids during the saponification.

If the oil contains large amounts of free acids, these should be determined separately by the procedure described under "Determination of Acids." The saponification number, representing the sum of the acid number and the ester number, is then determined for the oil using the general procedure described above, except that the free acids are not neutralized before the addition of the 0.5 N alkali.

In the case of oils containing large amounts of esters (e.g., oil of wintergreen), or esters of low-molecular weight (e.g., methyl formate), or esters of dibasic or tribasic acids, it becomes necessary to vary the size of the sample and the amount of alkali employed. If 10 cc. of alkali is insufficient, 20 cc. may be used. For synthetic esters, it is often necessary to decrease the size of the sample; usually 1 g. (of the pure synthetic) is used and 20 cc. of alkali. In the case of esters of low molecular weight or esters of polybasic acids, a 0.5 g. sample and 20 cc. of alkali may be required.

Relatively small samples are also required in the case of certain darkly colored oils. It may also be necessary to dilute the saponified oil with alcohol in order to ascertain the end point of the titration, and to use a spotplate. The use of thymolphthalein (in place of phenolphthalein as an indicator) has been suggested[60] for determinations involving red or brown solutions, such as result during the saponification of oleoresins. Thymolphthalein changes from a deep blue to colorless in the range pH 9.3 to pH 10.5.

The determination of the ester content by saponification will not yield satisfactory results if the oil contains appreciable amounts of aldehydes, unless the aldehydes are removed and the residual oil saponified.

It has been reported that certain phenols also may interfere with the ester determination.[61] In addition to esters, lactones may be determined quantitatively by saponification.

b. **Determination by Saponification in the Cold.**—As an analytical procedure, saponification in the cold is not generally applicable. In most cases, long periods of time are necessary to complete the process; furthermore, side reactions frequently occur which give rise to inconsistent and deceptive results.

Saponification in the cold has a definite value for the determination of those esters which are very easily saponified; this is particularly true for certain formates. Thus, cold saponification is used in the analysis of geranium oils to determine the amount of "actual formate," since the standard procedure for the determination of esters with a reflux period of 1 hr. saponifies not only the geranyl formate but also other esters including geranyl tiglate.

For the determination of geranyl formate in geranium oils, the following procedure has given satisfactory results.

> *Procedure:* Into a 100 cc. saponification flask, weigh accurately about 1.5 g. of the oil. Add 5 cc. of neutral alcohol and 3 drops of a 1% alcoholic solution of phenolphthalein, and neutralize the free acids quickly with standardized 0.1 N aqueous sodium hydroxide solution. Add 10 cc. of 0.5 N alcoholic sodium hydroxide solution, measured accurately from a burette or pipette, and titrate the excess alkali *immediately* with standardized 0.5 N aqueous hydrochloric acid. Calculate the ester content as geranyl formate in the usual manner.

In the case of pure synthetic formates, it is advisable to add 5 cc. of water to the flask in order to dissolve the sodium formate which otherwise may precipitate out of solution.

3. DETERMINATION OF ALCOHOLS

a. **Determination by Acetylation.**—The alcoholic constituents of an essential oil are determined by acetylation; i.e., the oil is acetylized with

[60] Saxl, *Chemist-Analyst* **32** (1943), 11.
[61] Gildemeister and Hoffmann, "Die ätherischen Öle," 3d Ed., Vol. I, 797.

acetic anhydride and the ester content of the resulting oil is determined; from this value the percentage of alcohol in the original oil may be calculated.

The basic chemical processes involved in this determination may be summarized by the following equations:

$$\left.\begin{matrix} R1 \\ R2 \\ R3 \end{matrix}\right\} C{-}OH + O\begin{matrix} \diagup O{=}CCH_3 \\ \\ \diagdown O{=}CCH_3 \end{matrix} \longrightarrow \left.\begin{matrix} R1 \\ R2 \\ R3 \end{matrix}\right\} COOCCH_3 + CH_3COOH$$

$$\left.\begin{matrix} R1 \\ R2 \\ R3 \end{matrix}\right\} COOCCH_3 + NaOH \longrightarrow \left.\begin{matrix} R1 \\ R2 \\ R3 \end{matrix}\right\} COH + CH_3COONa$$

where R1, R2 and R3 may be a hydrogen atom, an aliphatic, aromatic or alicyclic radical.

For this determination, a special acetylation flask of approximately 100 cc. capacity is employed. This flask is equipped with an air-cooled condenser attached to the flask by means of a ground glass joint (see Diagram 4.8). A condenser 1 m. in length is to be preferred in order to prevent the loss of volatile constituents.

Procedure: Introduce into a 100 cc. acetylation flask 10 cc. of the oil (measured from a graduated cylinder), 10 cc. of acetic anhydride (similarly measured) and 2.0 g. of anhydrous sodium acetate. Attach the air condenser, and boil the contents of the flask *gently* for exactly 1 hr. on a sand bath suitably heated by an open Bunsen flame or an electric hot-plate. Permit the flask to cool for 15 min. and introduce 50 cc. of distilled water through the top of the condenser. Heat the flask on a steam bath for 15 min. with frequent shaking to destroy the excess of acetic anhydride. Transfer the contents of the flask to a separatory funnel and rinse the flask with two 10 cc. portions of distilled water; add these rinsings to the separatory funnel. Shake thoroughly to assure good contact of the aqueous layer with the oil. When the liquids have separated completely, reject the aqueous layer and wash the remaining oil repeatedly with 100 cc. portions of saturated salt solution, until the washings are neutral to litmus; this usually requires three washings. Dry the resulting oil with anhydrous sodium sulfate and filter. (If the oil has been washed properly, not more than 0.2 cc. of 0.1 N aqueous sodium hydroxide solution should be required per gram of acetylized oil in order to neutralize the remaining trace of acetic acid.)

Saponify the acetylized oil, using the procedure described under "The Determination of Esters," p. 265.

In order to secure accurate and reproducible results it is important to use exactly 2.0 g. of sodium acetate and to reflux the mixture for exactly 1 hr. A notable exception occurs in the case of citronella oils, which require a reflux period of 2 hr.

Calculation of Results.—If the original oil contains a negligible quantity of saponifiable constituents, the free alcohol may be calculated by the following formula:

$$\text{Percentage of alcohol in the original oil} = \frac{am}{20(s - 0.021a)}$$

where: a = number of cc. of 0.5 N sodium hydroxide solution required for the saponification of the acetylized oil;
s = weight of acetylized oil in grams used in the saponification;
m = molecular weight of the alcohol.

For oils which have not been thoroughly investigated and whose alcoholic constituents are not well known, it is frequently more convenient to report the result as an ester number after acetylation.

$$\text{Ester number after acetylation} = \frac{28.05a}{s}$$

The ester number after acetylation is numerically equal to the number of milligrams of potassium hydroxide required to saponify the esters present in 1 g. of the acetylized oil.

If the original oil contains an appreciable amount of esters (as indicated by the ester number), the percentage of free alcohol may be estimated by the following formula:

$$\text{Percentage of free alcohol in the original oil} = \frac{dm}{561.04 - 0.42d}$$

where d = (ester number after acetylation − ester number). Although this expression is not mathematically precise, nevertheless it is sufficiently accurate for all practical work and has been used traditionally by essential oil chemists.

DIAGRAM 4.8. Acetylation flask.

For the evaluation of essential oils, it is often desirable to know the percentage of total alcohol; i.e., the percentage of free alcohol plus the percentage of alcohol combined as ester present in the original unacetylized oil.

$$\text{Percentage of total alcohol in the original oil} = \left[\frac{am}{20(s - 0.021a)}\right]\left[1 - \frac{42.04e}{100(m + 42.04)}\right]$$

where e = ester content in per cent. This formula assumes that all of the esterified alcohol present in the original oil is combined as the acetate.

All formulas in this chapter that calculate the result of an acetylation as a *percentage* actually refer to all constituents which are capable of acetylation under the experimental conditions, *calculated as a specific alcohol.* Thus for example, the "total alcohol" in citronella oils includes not only the geraniol, free and as ester, but also all other acetylizable constituents and their esters, such as, borneol, citronellol, sesquiterpene alcohols, and the aldehyde citronellal, *all calculated as geraniol.* These formulas further assume that the alcohol is a monohydroxy compound. Table 4.11 gives the molecular weights of alcohols frequently encountered in the analysis of essential oils.

TABLE 4.11. MOLECULAR WEIGHTS* OF ALCOHOLS

Alcohols	*Molecular Wt.*
Amyrol	222.36
Anisyl Alcohol	138.16
Benzyl Alcohol	108.13
Borneol	154.25
Cedrenol	220.34
Cedrol	222.36
Cinnamyl Alcohol	134.17
Citronellol	156.26
Costol	220.34
Cyclohexanol	100.16
Decyl Alcohol	158.28
Duodecyl Alcohol	186.33
Elemol	222.36
Farnesol	222.36
Fenchyl Alcohol	154.25
Geraniol	154.25
Guaiol	222.36
Isoborneol	154.25
Isopulegol	154.25
Linaloöl	154.25
Menthol	156.26
Nerol	154.25
Nerolidol	222.36
Nonyl Alcohol	144.25
Octyl Alcohol	130.23
Phenylethyl Alcohol	122.16
Phenylpropyl Alcohol	136.19
Rhodinol	156.26
Santalol	220.34
Terpineol	154.25
Thujyl Alcohol	154.25
Undecyl Alcohol	172.30
Vetivenol	220.34

* All molecular weights have been calculated from the values of the International Atomic Weights adopted by the Committee on Atomic Weights in 1938.

Limitations and Modifications of the General Procedure.—As mentioned above, acetic anhydride employed under the experimental conditions described in the "Procedure" will react with certain compounds found in essential oils other than alcohols. Phenols will be quantitatively converted into the acetates. Certain aldehydes and ketones are partially acetylated and partially destroyed, or are converted to other compounds which are capable of acetylation.

Furthermore, some tertiary alcohols are not quantitatively converted to the acetate by this process of acetylation; the most important alcohols in this class are terpineol and linaloöl.

***b.* Determination of Primary Alcohols.**—Phthalic anhydride reacts with primary alcohols forming an acid phthalic ester.

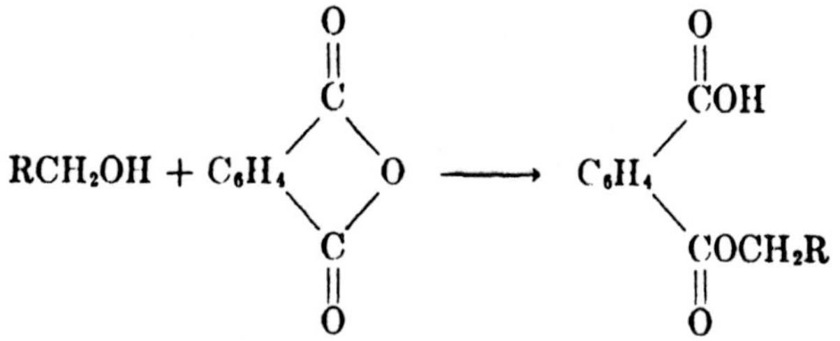

$$RCH_2OH + C_6H_4 \begin{matrix} \diagup C(=O) \diagdown \\ \\ \diagdown C(=O) \diagup \end{matrix} O \longrightarrow C_6H_4 \begin{matrix} \diagup \overset{O}{\overset{\|}{C}}OH \\ \\ \diagdown \underset{O}{\underset{\|}{C}}OCH_2R \end{matrix}$$

Under the experimental conditions described below, this reaction takes place readily at a temperature of about 100° in the case of primary alcohols; for secondary alcohols, the time required for reflux is greatly increased; for tertiary alcohols, no appreciable reaction occurs.

It is important that the phthalic anhydride does not contain free phthalic acid. This may be ascertained conveniently by shaking 1 g. of the anhydride with 10 cc. of benzene and warming to 40°; a clear solution indicates the absence of appreciable amounts of phthalic acid.

Procedure:[62] Into a 100 cc. acetylation flask introduce about 2 g. of powdered phthalic anhydride, accurately weighed, and about 2 g. of the oil, accurately weighed. Add 2 cc. of benzene, measured from a graduated cylinder. Heat the flask on a steam bath with frequent shaking for 2 hr. Then permit the flask to cool for 30 min. Add 60 cc. of 0.5 N aqueous potassium hydroxide solution, accurately measured from a pipette or burette. Stopper the acetylation flask with a ground glass stopper and shake thoroughly for 10 min. Titrate the excess of alkali with standardized 0.5 N hydrochloric acid, using 3 drops of a 1 per cent phenolphthalein solution as indicator.

Run a blank determination omitting the oil, and from this calculate the amount of alkali which would be required for the weight of phthalic anhydride used in the actual determination.

Calculate the percentage of primary alcohol by the following formula:

$$\text{Percentage of primary alcohol} = \frac{m(b - a)}{20w}$$

[62] *Ber. Schimmel & Co.*, October (1912), 39.

where: m = the molecular weight of the primary alcohol;
b = the calculated number of cc. of 0.5 N potassium hydroxide required for the amount of phthalic anhydride used in the determination;
a = the number of cc. of 0.5 N potassium hydroxide consumed in the determination;
w = weight of oil in grams.

c. **Determination of Tertiary Terpene Alcohols.**—Most tertiary alcohols suffer partial or complete breakdown and dehydration when treated with acetic anhydride. In the event that an oil contains a large percentage of such easily dehydrated alcohols, special techniques are required.

I. The Method of Glichitch.[63]—The Glichitch method of formylation for the estimation of easily dehydrated alcohols has been successfully employed for the determination of linaloöl and terpineol.

Procedure: Introduce 15 cc. of aceto-formic acid reagent in a 125 cc. glass-stoppered Erlenmeyer flask. Cool in an ice bath and add slowly 10 cc. of the oil to be tested. Allow the mixture to stand for not less than 72 hr. at room temperature. The ice in the bath should not be renewed. At the end of this interval pour the contents of the flask into a separatory funnel. Shake well with 50 cc. of ice cold water and allow to stand for 2 hr. Separate the oil and wash successively with 50 cc. of cold water, 50 cc. of a 5% sodium bicarbonate solution, and then with two 50 cc. portions of water. Separate the oil and dry with anhydrous sodium sulfate. Filter and saponify by refluxing with 0.5 N alcoholic sodium hydroxide. Calculate the alcohols in the usual way on the assumption that they are present as formates.

Preparation of the Aceto-Formic Reagent.—To 2 volumes of acetic anhydride, previously cooled to at least 0°, add slowly 1 volume of 100 per cent formic acid.[64] Mix thoroughly and then heat to 50° for 15 minutes and immediately cool in an ice bath.

II. The Method of Boulez.[65]—The Boulez method of acetylation makes use of a diluent in order to lessen the dehydrating effect of acetic anhydride. The period of acetylation, however, must be prolonged. This gives satisfactory results for linaloöl and terpineol if the prescribed conditions are rigidly followed.

The original method suggested oil of turpentine as a diluent in the ratio of 1 part of the oil under examination to 5 parts of oil of turpentine.[66] The

[63] *Bull. soc. chim.* [4] 33 (1923), 1284.

[64] It is very important to use a highly purified formic acid of substantially 100 per cent strength. The usual A.R. grade of formic acid (specific gravity = 1.20; HCOOH = approximately 87 per cent) is useless for the preparation of this reagent.

[65] *Bull. soc. chim.* [4] 1 (1907), 117.

[66] Boulez later suggested an even greater dilution—namely, 1 g. oil to 25 cc. of xylene. *Bull. soc. chim.* [4] 35 (1924), 419.

chemists of Schimmel & Company[67] modified the procedure by substituting xylene as a diluent in the ratio of 1:4. The period of acetylation is very important; for terpineol 5 hr. are required, longer or shorter periods give low values; for linaloöl, 7 hr.

This modified procedure gives reproducible data. Great care must be exercised during this determination since any error introduced will be multiplied by 5 in the final result.

III. Dehydration Methods.—Dehydration methods are based upon the catalytic decomposition of tertiary alcohols and the splitting off of water. The amount of water obtained is determined from which the percentage of tertiary alcohol may be calculated.

Such a method has been described by Ikeda and Takeda,[68] using zinc chloride. A very satisfactory dehydration catalyst is iodine. Additions of approximately 0.5 per cent of catalyst to the oil will prove sufficient. Such dehydration methods offer the advantage that only tertiary alcohols are determined, primary and secondary alcohol being unaffected. This is an advantage not found in the other methods described here. Hydroxy ketones and hydroxy aldehydes will interfere in this procedure, since both split off water under the experimental conditions.

A convenient method for the determination makes use of the distillation trap of Sterling and Bidwell.

> *Procedure:* Dry the oil thoroughly by permitting it to stand overnight in contact with anhydrous sodium sulfate. Into a 1 liter, round bottom flask, introduce a sufficiently large sample, accurately weighed, to yield about 5 cc. of water upon dehydration of the tertiary alcohol. Add 0.5% of solid iodine as catalyst and 500 cc. of xylene. Connect the flask to a standard Sterling and Bidwell water-trap; attach a water-cooled, straight tube condenser. Heat the flask by means of an oil bath. Proceed as in the "Determination of Water Content," p. 323. Measure the amount of collected water and calculate the percentage of tertiary alcohol.

This method does not yield highly accurate results, but is a convenient method for the determination of the tertiary alcohol content.

IV. Acetyl Chloride-Dimethyl Aniline Method.—This method, originally described by Fiore,[69] gives exceptionally concordant and satisfactory results in the case of linaloöl and linaloöl-containing oils. It has been carefully evaluated by the members of the Essential Oil Association of the U. S. A. and adopted by that body. Preliminary experiments with terpineol and

[67] *Ber. Schimmel & Co.*, April (1907), 128.
[68] *J. Chem. Soc. Japan* **57** (1936), 442. *Chem. Abstracts* **30** (1936), 5907.
[69] *News Capsule* (Essential Oil Association of U.S.A.), Vol. 1, No. 15 (1943).

other tertiary terpene alcohols indicate that this may prove to be a valuable method for many tertiary alcohols.

The method is described below in the final form in which it was accepted by the Essential Oil Association for the determination of linaloöl.

Procedure: Ten cc. of linaloöl or essential oil containing linaloöl, previously dried with sodium sulfate, is introduced into a 125 cc. glass-stoppered Erlenmeyer flask cooled with ice and water. To the cooled oil is added 20 cc. dimethyl aniline (monomethyl free) and the contents thoroughly mixed, then 8 cc. acetyl chloride (reagent grade) and 5 cc. of acetic anhydride are added, the anhydride serving as a solvent to prevent crystallization of the reaction mass. The mixture is cooled for a few minutes and permitted to stand at room temperature for one-half hour after which time the flask is immersed in a water bath maintained at 40° C. ± 1° for three hours. At the end of this time the acetylated oil is washed three times with 75 cc. of ice water, then with successive washes of 25 cc. of 5% sulfuric acid until the separated acid layer fails to liberate any dimethyl aniline with an excess of caustic. After removal of the dimethyl aniline, the acetylated oil is washed with 10 cc. of 10% sodium carbonate solution and then finally washed neutral with water.

The oil is separated, dried over anhydrous sodium sulfate and the ester number determined in the usual manner. The linaloöl content can thus be obtained directly from saponification tables or by substitution in the following formula:

$$\text{Percentage of linaloöl} = \frac{\text{cc. } N/2 \text{ KOH} \times 154.14}{20\,(\text{wt. sample} - \text{cc. } N/2 \text{ KOH} \times 0.021)}$$

As this test is further to be used for other oils containing linaloöl, besides linaloöl itself, a correction factor is necessary with oils containing significant amount of esters. For such oils, the following standard formula is recommended:

$$\text{Percentage of total linaloöl} = \frac{A \times 77.07}{B - (A \times 0.021)} \times (1 - (E \times 0.0021))$$

where: A = cc. half normal alkali required for saponification;
B = weight of sample;
E = per cent of esters calculated as linalyl acetate in the original oil.

***d.* Determination of Citronellol by Formylation.**—Most terpene alcohols are dehydrated by strong formic acid, giving rise to nonsaponifiable terpenes. A notable exception is citronellol which is converted almost quantitatively to the corresponding formate. This results in a convenient and satisfactory method for the determination of citronellol in the presence of geraniol and linaloöl.

Formylation has become a standard procedure for the determination of citronellol in rose oils. The procedure to be followed is identical to that described under "Determination by Acetylation," p. 272, with the exception that the 10 cc. of acetic anhydride is replaced with 20 cc. of 100 per cent formic acid, and the anhydrous sodium acetate is omitted.

Place in the flask short pieces of glass tubing to permit heat transfer throughout the mixture. This is particularly important if the oil contains a high percentage of geraniol, since the dehydration which results may dilute the formic acid sufficiently to cause the formation of two layers in the flask. Should this occur, there will be some danger of the lower layer becoming overheated and violently throwing out the contents of the flask through the air condenser. A small clay chip should also be placed in the flask to help prevent such overheating.

The percentage of alcohol (citronellol) in the original oil may be calculated from the amount of alkali consumed in the subsequent saponification.

$$\text{Percentage of alcohol in the original oil} = \frac{am}{20(s - 0.014a)}$$

where: a = number of cc. of 0.5 N sodium hydroxide solution required for the saponification of the formylated oil;
m = molecular weight of the alcohol;
s = weight of formylated oil in grams.

4. DETERMINATION OF ALDEHYDES AND KETONES

Of the many procedures which have been suggested for the determination of aldehydes and ketones, only four general methods have attained practical significance. These are the bisulfite method, the neutral sulfite method, the phenylhydrazine method, and the hydroxylamine methods.

***a.* Bisulfite Method.**—The bisulfite method is an absorption process based upon the general reaction:[70]

$$RCHO + NaHSO_3 \longrightarrow RCH\begin{matrix} OH \\ SO_3Na \end{matrix}$$

Upon shaking a measured quantity of oil with a hot aqueous solution of sodium bisulfite, an addition compound[71] forms which is generally water soluble and which dissolves in the hot bisulfite solution; the nonaldehyde

[70] There exists some question as to the linkage of the $-SO_3Na$ group to the C atom of the carbonyl group; this linkage may occur through the S atom or possibly through the O atom.

[71] In many cases, this addition compound is a water soluble sulfonate instead of (or in addition to) the normal bisulfite addition compound of the carbonyl group. See p. 282.

portion of the oil separates as an oily layer which can be measured conveniently in the graduated neck of a *cassia* flask.

These special flasks have been known traditionally as cassia flasks because they were first used for the determination of the cinnamic aldehyde content of cassia oil. They have a large bulbous body with a long thin neck graduated in divisions of 0.1 cc. The two types (having the dimensions shown in Diagram 4.9) have proved most useful in the laboratory. The larger flask with a capacity of 150 cc. and a thin neck graduated to contain 6 cc. is very satisfactory for the determination of aldehydes and ketones. The smaller flask with a capacity of 100 cc. and a neck graduated to contain 10 cc. may also be used for such determinations, although the accuracy will suffer somewhat. Furthermore, the capacity of these smaller flasks does not permit as thorough and as intimate contact of the oil and solution when the flask is shaken; if such a flask is used, the shaking should be thorough and prolonged. In general, the use of these smaller flasks is not recommended for the determination of aldehydes and ketones, unless the oil contains less than 40 per cent of reactive carbonyl compounds.

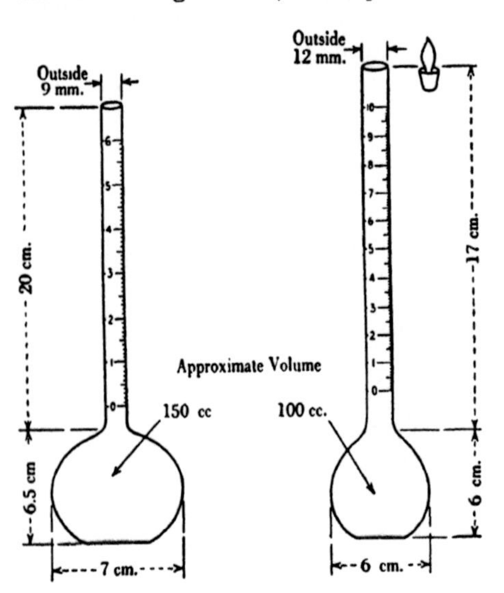

DIAGRAM 4.9. Cassia flasks.

The bisulfite method is perhaps the most convenient and simple of the four general methods. As such, it is frequently used in the trade because it requires no standardized solutions, analytical balance, or special skill.

This method has proved satisfactory for the estimation of cinnamic aldehyde in cassia oil, of benzaldehyde in bitter almond oil, of citronellal in *Eucalyptus citriodora* oil; it is the commercially accepted method for the determination of citral in lemongrass oil.[72]

The bisulfite method suffers from certain disadvantages inherent in the absorption process. Water-soluble adulterants analyze as apparent aldehyde. The time required for a determination is usually at least 1 hr. The

[72] The neutral sulfite method gives a more accurate value of the true citral content; in addition to citral, the bisulfite method determines other carbonyl constituents which occur as natural constituents of lemongrass oils (e.g., part of the methyl heptenone). The values obtained by the bisulfite method are generally about 4 per cent higher than those obtained by the neutral sulfite method for the normal lemongrass oils of commerce.

results obtained are volume percentages. These methods are applicable only to oils containing large amounts of aldehydes or ketones. Water-soluble sulfonates may be formed from noncarbonyl compounds having double bonds; these will interfere with the accuracy of the analytical results.[73]

The bisulfite method suffers from further disadvantages. There is no definite indication when all of the aldehyde has completely reacted. Although satisfactory for most aldehydes, the method is not suitable for the determination of such ketones as carvone, thujone, pulegone, menthone, fenchone, or camphor.

Procedure:[74] Into a 150 cc. cassia flask, having a thin neck graduated in 0.1 cc. divisions, introduce 75 cc. of a *freshly prepared*, saturated, aqueous solution of sodium bisulfite,[75] measured from a graduated cylinder. Pipette exactly 10 cc. of the oil into the flask. Upon thorough shaking, a semisolid mass frequently will result. Immerse the flask in a beaker of boiling water and occasionally shake until the solid addition compound has gone completely into solution. Shake the flask repeatedly to assure complete reaction of the aldehyde with the bisulfite solution. A further addition of 25 cc. of bisulfite solution is made, and the flask is again repeatedly shaken. After standing undisturbed in the beaker of boiling water for 10 min. to permit the unreacted oil to rise to the surface, add sufficient sodium bisulfite solution to force the unreacted oil into the neck of the flask. Any droplets of oil adhering to the sides are made to rise into the neck by gently tapping the flask, and by rotating it rapidly between the palms of the hands. After cooling the flask to room temperature, measure the amount of unreacted oil. The aldehyde content may then be calculated by means of the following formula:

Percentage of aldehyde = 10(10 − no. of cc. of unreacted oil).

As mentioned above, this result is a volume percentage. It may be converted into a weight percentage if the specific gravity of the original

[73] In this connection, see Dodge, *Am. Perfumer*, May (1940), 41. According to this authority only small amounts of unsaturated alcohols will dissolve if the solution of $NaHSO_3$ is stronger than molar (10.4 per cent).

[74] Variations of this procedure have been suggested by other authorities. Gildemeister and Hoffmann ("Die ätherischen Öle," 3d Ed., Vol. I, 739) suggest the use of a 30 per cent aqueous solution of sodium acid sulfite which does not contain too much free sulfurous acid; if necessary the solution should be neutralized with sodium carbonate. It has been the experience of the laboratories of Fritzsche Brothers, Inc., that a freshly prepared solution of $NaHSO_3$ made with Analytical Grade of reagent does not contain sufficient free H_2SO_3 to interfere with the reaction; the separation of the noncarbonyl portion of the oil is sharper and more complete if a saturated solution of $NaHSO_3$ is employed instead of a 30 per cent solution.

[75] At room temperature, this will be approximately a 40 per cent (wt./vol.) solution.

oil and of the aldehyde is known:

$$\text{Percentage by weight} = (\% \text{ by volume}) \left(\frac{d_{15°}^{15°} \text{ of aldehyde or ketone}}{d_{15°}^{15°} \text{ of oil}} \right)$$

After cooling to room temperature, a small amount of the bisulfite add tion compound will often precipitate out of solution, sometimes forming a the surface where the oil and aqueous layers meet; this renders an exac reading difficult. The addition of a few drops of water (added with medicine dropper in such a way that the water runs down along the insid of the neck of the flask), which will remain temporarily on top of the bisu fite solution, gives a sharp separation of the oil and aqueous layers. If th oil contains heavy metals, these should be removed before the determinatio by shaking the oil thoroughly with a small amount (about 1 per cent) c powdered tartaric acid and filtering; a sharper separation of the noncar bonyl layer will then result.

The procedure described above will prove satisfactory for those alde hydes which form water-soluble sulfonates in addition to the normal bisulfit addition compound—e.g., citral, citronellal,[76] cinnamic aldehyde.

For aldehydes which form only the normal addition compound (e.g compounds which have no double bonds other than those present in th carbonyl group or benzene ring) but which form water-soluble bisulfite add tion compounds, the procedure must be modified. For the determination c phenylpropyl aldehyde,[77] benzaldehyde,[78] and anisic aldehyde, use a 10 cc sample and only 50 cc. of the saturated bisulfite solution. The norma addition compound which forms usually will not dissolve in the saturate bisulfite solution even after heating; consequently the flask should be fille by the addition of 25 cc. portions of *water*[79] (instead of bisulfite solution After each addition, the flask should be thoroughly shaken and then im mersed in the boiling water for a period of about 5 min. The additio compound slowly dissolves and the nonreacting oily layer is driven into th neck of the flask and measured. Upon cooling and standing, some of th

[76] In the determination of citronellal, the addition compound will often separate upc cooling; hence the reading should be taken as soon as the *neck* of the flask has cooled t room temperature.

[77] In the determination of phenylpropyl aldehyde, considerable amounts of the additic compound separate upon cooling; however, a reading is possible.

[78] Use is made of the poor solubility of the benzaldehyde addition compound in *saturat* $NaHSO_3$ solution for the detection of benzaldehyde in cinnamic aldehyde; cinnamic ald hyde forms a sulfonate which dissolves completely in saturated $NaHSO_3$ solution. Henc the separation of a solid addition compound upon cooling the contents of the flask to roo temperature is indicative of the presence of benzaldehyde.

[79] Gildemeister and Hoffmann ("Die ätherischen Öle," 3d Ed., Vol. I, 740) also recon mend additions of water instead of $NaHSO_3$ solution for the determination of benzald hyde, anisic aldehyde, and phenylacetaldehyde.

addition compound may settle out of solution. However, a reading usually may be obtained.

In general, this modified procedure will not be satisfactory for the determination of decyl aldehyde,[80] cuminic aldehyde,[81] methyl heptenone,[82] or phenylacetaldehyde which has polymerized.[83]

b. **Neutral Sulfite Method.**—This is also an absorption method. Using a neutral sufite solution, sodium hydroxide is liberated as the reaction proceeds; this must be periodically neutralized with acid to permit the reaction[84] to go to completion.

$$RCHO + Na_2SO_3 + H_2O \longrightarrow RCH\begin{matrix} OH \\ SO_3Na \end{matrix} + NaOH$$

Although this method suffers from the disadvantages of an absorption process, nevertheless it offers certain advantages over the use of the bisulfite technique. Through the use of phenolphthalein, the exact end point of the reaction may be determined. Furthermore, some ketones react with neutral sulfite completely, so that this method may be used for their determination; this is specifically of importance for the determination of carvone in spearmint, dill, and caraway oils, of pulegone in pennyroyal oil, and of piperitone in eucalyptus oils. Carvone reacts smoothly requiring about 1 hr. for the determination The reaction with piperitone and with pulegone is very slow: only a 5 cc. sample should be used and the flask should be heated in a bath of *vigorously boiling* water.

> *Procedure:* Into a 150 cc. cassia flask, having a thin neck graduated in 0.1 cc. divisions, introduce 75 cc. of a freshly prepared, saturated, aqueous solution[85] of sodium sulfite, measured from a graduated cylinder. Add a few drops of a 1 per cent alcoholic phenolphthalein solution and neutralize the free alkali with a 50 per cent (by volume) aqueous acetic acid solution. Then pipette exactly 10 cc. of the oil into the flask and shake thoroughly. Immerse the flask in a beaker of boiling water and shake repeatedly. Neutralize the mixture from time to time

[80] Upon cooling, the entire contents of the flask will solidify making a reading difficult.

[81] The addition compound formed is not sufficiently soluble even when the flask is heated.

[82] The reaction with methyl heptenone is incomplete under the condition of the determination.

[83] The nonaldehyde portion settles to the *bottom* of the flask. Reclaire (*Perfumery Essential Oil Record* 12 (1921), 341) recommends the use of a special flask for this determination. The hydroxylamine method (see p. 285) will prove entirely satisfactory.

[84] See footnotes 71 and 72, p. 279.

[85] At room temperature this will be approximately a 30 per cent (wt./vol.) solution.

with the 50 per cent acetic acid.[86] Continue this procedure until no further pink color appears upon the addition of a few more drops of phenolphthalein solution. Permit the flask to remain in the boiling water for an additional 15 min. to assure complete reaction. Then add sufficient neutralized sodium sulfite solution to raise the lower limit of the oily layer within the graduated portion of the neck. Any droplets of oil adhering to the sides are made to rise into the neck by gently tapping the flask, and by rotating it rapidly between the palms of the hands. After cooling the flask to room temperature, measure the amount of unreacted oil. The aldehyde content may then be calculated by means of the formula given under bisulfite method.

The neutral sulfite method is the official method of "The United States Pharmacopoeia" for the determination of cinnamic aldehyde in cassia oil,[87] for carvone in spearmint oil,[88] and of "The National Formulary" for carvone in caraway oil.[89] It proves satisfactory for the determination of citral in lemongrass oils,[90] the reaction being very rapid.

As in the case of the bisulfite method, oils containing heavy metals should be treated with tartaric acid before a determination is attempted (see p. 311).

c. **Phenylhydrazine Method.**—The phenylhydrazine method is seldom used today. It attained importance as the first practical method for the assay of citral in lemon oil.[91] The official method of "The United States Pharmacopoeia," Tenth Revision, is included here, since commercial contracts occasionally specify that aldehydes be determined by the phenylhydrazine method.

An accurately measured amount of an alcoholic solution of freshly distilled phenylhydrazine is added to a weighed amount of the oil. The excess of phenylhydrazine is titrated with hydrochloric acid. A blank is run simultaneously, and from the difference in the amounts of standardized hydrochloric acid required for the blank and the determination, the percentage of aldehyde is calculated.

$$RCHO + C_6H_5NNH_2 \rightarrow C_6H_5NN{=}CHR + H_2O$$

[86] "The United States Pharmacopoeia," Thirteenth Rev., 132, suggests neutralization with a 30 per cent $NaHSO_3$ solution. However, the volume of solution frequently becomes too great to permit thorough shaking.

[87] Thirteenth Revision, 132.

[88] Thirteenth Revision, 510.

[89] Eighth Edition, 121.

[90] See footnote 72, p. 280.

[91] This method was first proposed by Kleber, *Am. Perfumer* 6 (1912), 284.

The method, as described, is suitable for the determination of aldehydes in the citrus oils.

> *Procedure:*[92] Place about 15 cc. of oil of lemon in a tared, 250 cc. Erlenmeyer flask, and weigh accurately. Add 10 cc. of an alcoholic solution of phenylhydrazine[93] (1 in 10) (not darker in color than pale yellow), and allow it to stand for 30 min. at room temperature. Then add 3 drops of a 0.1% aqueous solution of methyl orange, and neutralize the liquid by the addition of half-normal hydrochloric acid. If difficulty is experienced in determining the end point of the reaction, continue the titration until the liquid is distinctly acid, transfer it to a separatory funnel, and after the layers have separated draw off the alcoholic portion. Wash the oil remaining in the funnel with distilled water, adding the washings to the alcoholic solution, and titrate the latter with half-normal sodium hydroxide. Carry out a blank test identical with the foregoing, omitting the oil of lemon, and note the amount of half-normal hydrochloric acid consumed. Subtract the number of cc. of half-normal sodium hydroxide from the number of cc. of half-normal hydrochloric acid consumed in the test containing the oil of lemon, and this result from the number of cc. of half-normal hydrochloric acid consumed in the test without the oil of lemon. Each cc. of this difference corresponds to 0.07609 g. of aldehydes calculated as citral.

In the case of orange oils, the aldehyde is usually calculated as decyl; the factor then used is 0.07813. In the case of grapefruit oils, the aldehydes are frequently calculated as an equal mixture of octyl and decyl; the factor then used is 0.07112.

The results obtained in the above method represent percentages by weight.

***d.* Hydroxylamine Methods.**—Two important techniques have been developed, both based upon the use of hydroxylamine for the determination of aldehydes and ketones. The first makes use of a solution of hydroxylamine hydrochloride and the subsequent neutralization with standardized alkali of the hydrochloric acid liberated by the reaction. The second technique makes use of a solution of hydroxylamine (i.e., a solution of the hydrochloride with substantially all of the combined hydrochloric acid previously neutralized with alkali); after the reaction with the aldehyde or ketone, the mixture is titrated with standardized acid. The latter procedure is known as the Stillman-Reed method. Both modifications are based

[92] "The United States Pharmacopoeia," Tenth Revision, 260.

[93] The phenylhydrazine solution should be measured accurately from a pipette or burette.

upon the fundamental reaction:

$$RCHO + NH_2OH \cdot HCl \longrightarrow RCH{=}NOH + H_2O + HCl$$

$$\begin{matrix} R \\ & \diagdown \\ & & C{=}O + NH_2OH \cdot HCl \longrightarrow \\ & \diagup \\ R' \end{matrix} \quad \begin{matrix} R \\ & \diagdown \\ & & C{=}NOH + H_2O + HCl \\ & \diagup \\ R' \end{matrix}$$

The hydroxylamine methods offer many advantages over the absorption processes. Relatively small amounts of the oil are required for a determination. The reaction of hydroxylamine with aldehydes is rapid, shortening the time required for a determination. Water-soluble adulterants which do not contain a carbonyl group do not analyze as apparent aldehyde or ketone. The methods have proved satisfactory for the determination of certain ketones (such as menthone and thujone) which cannot be determined conveniently by the absorption procedures. In fact, hydroxylamine will react with practically all aldehydes and most ketones encountered by the essential oil chemist. Furthermore, these hydroxylamine methods prove exceptionally applicable to oils which contain only small amounts of aldehydes or ketones (e.g., lemon oils), and to oils containing large amounts of free acids (e.g., orris oils). The solutions used for the standard procedure are stable and can be kept for many months; however, the Stillman-Reed solution deteriorates rapidly and is best prepared when needed.

The hydroxylamine methods have certain disadvantages not inherent in absorption techniques. It must be remembered that the calculation of results involves the molecular weight of the aldehyde or ketone, giving percentages by weight; hence adulterations with carbonyl compounds of lower molecular weight give apparent percentages which are too high. If more than one aldehyde or ketone is present in an oil, all are calculated as a specific carbonyl compound. Since the reaction of hydroxylamine is quite universal, it is difficult to determine an individual component. Nor can the carbonyl and noncarbonyl portions be separated conveniently and studied individually.

> *Standard Procedure:*[94] Into a 100 cc. saponification flask weigh accurately the requisite amount of oil or synthetic and add 35 cc. of 0.5 N hydroxylamine hydrochloride solution, measured from a graduated cylinder. Permit the flask to stand at room temperature for the proper length of time and titrate the liberated hydrochloric acid with standardized 0.5 N alcoholic sodium hydroxide. The titration is continued until the original greenish shade of the hydroxylamine solution is obtained. A

[94] This is essentially the procedure described in "The United States Pharmacopoeia," Twelfth Revision, 314, for the determination of benzaldehyde in bitter almond oil.

second flask containing 35 cc. of hydroxylamine hydrochloride solution may be used as a blank to assure a more accurate color match.[95]

$$\text{Percentage of aldehyde or ketone} = \frac{am}{20s}$$

where: a = number of cc. of 0.5 N sodium hydroxide used for neutralization;
m = molecular weight of the aldehyde or ketone;
s = weight of sample in grams.

Preparation of 0.5 N Hydroxylamine Hydrochloride Solution: Dissolve 275 g. of recrystallized hydroxylamine hydrochloride[96] in 300 cc. of distilled water; warm to a temperature of 65° on a steam bath to yield a clear solution. Add this solution slowly to 2 gal. of 95% alcohol, and mix thoroughly. Then add 125 cc. of a 0.1% solution of bromphenol blue indicator in 50% alcohol, and sufficient 0.5 N alcoholic sodium hydroxide solution to change the yellow color of the solution to a greenish shade; this usually requires about 20 to 25 cc. of the alkali. The proper degree of neutralization is attained when 35 cc. of the solution shows a distinct greenish shade which changes to a distinct yellow upon the addition of 1 drop of 0.5 N hydrochloric acid. A stable solution of hydroxylamine hydrochloride is thus obtained which is approximately 0.5 N; an exact adjustment is unnecessary.

For lesser quantities of solution, dissolve 34.75 g. of recrystallized hydroxylamine hydrochloride in 40 cc. of distilled water and make up to 1 liter with 95% alcohol; add 15 cc. of the bromphenol blue solution and neutralize.

The proper size of sample and the proper length of time to give complete reaction and the molecular weights of the most frequently encountered aldehydes and ketones are given in Table 4.12.

Stillman-Reed Procedure:[97] Proceed as directed under the standard procedure but add 75 cc. of hydroxylamine solution, measured accurately by means of a burette or pipette. At the same time run a blank determination. After standing the required length of time, titrate with standardized 0.5 N hydrochloric acid to a green-yellow end point. Care should be taken to titrate both the blank and the sample to the same end point. Calculate the percentage of aldehyde or ketone as described above.

[95] In the case of very darkly colored oils, the size of sample should be greatly reduced and the end point determined with the aid of a spotplate. This is particularly important in the case of oils which have a greenish color—e.g., wormwood oils.

[96] The hydroxylamine hydrochloride offered by Commercial Solvents Corp. under the name of "hydroxylammonium chloride" is sufficiently pure, after recrystallization from water, for the preparation of this solution.

[97] *Perfumery Essential Oil Record* 23 (1932), 278.

TABLE 4.12. MOLECULAR WEIGHTS* AND REACTION TIME OF ALDEHYDES AND KETONES.

PART I

Carbonyl Compound	Molecular Wt.	Size of Sample (g)	Reaction Time
Acetaldehyde	44.05	0.5	immediate
Acetophenone	120.14	1.0	15 min.
α-Amyl Cinnamic Aldehyde	202.29	1.0	24 hr.
Anisic Aldehyde	136.14	1.0	15 min.
Benzaldehyde	106.12	1.0	immediate
Benzophenone†	182.21		
Benzylidene Acetone	146.18	1.0	15 min.
Butyraldehyde	72.10	1.0	15 min.
Camphor†	152.00		
Carvone	150.21	0.5	24 hr.
Cinnamic Aldehyde	132.15	1.0	15 min.
Citral	152.23	1.0	15 min.
Citronellal‡	154.25	1.0	15 min.
Cuminic Aldehyde	148.20	1.0	15 min.
Decyl Aldehyde	156.26	1.0	30 min.
Dodecyl Aldehyde	184.31	1.0	15 min.
Ethyl Amyl Ketone	128.21	1.0	15 min.
Fenchone†	152.23		
Furfural	96.08	1.0	15 min.
Heliotropin	150.13	1.0	15 min.
Heptyl Aldehyde	114.18	1.0	15 min.
Hexyl Aldehyde	100.16	1.0	15 min.
Hydrotropic Aldehyde	134.17	1.0	15 min.
Ionone	192.29	0.5	24 hr.
Irone	192.29	0.5	1 hr.
Isovaleric Aldehyde	86.13	1.0	15 min.
Menthone	154.25	0.5	24 hr.
Methyl Acetophenone	134.17	1.0	15 min.
Methyl Amyl Ketone	114.18	1.0	15 min.
Methyl Heptenone	126.19	1.0	24 hr.
Methyl Heptyl Ketone	142.24	1.0	15 min.
Methyl Hexyl Ketone	128.21	1.0	15 min.

* All molecular weights have been calculated from the values of the International Atomic Weights adopted by the Committee on Atomic Weights in 1938.

† Because of the slow reaction rate with hydroxylamine this method is not satisfactory for these ketones.

‡ Low values are obtained for this isolate if the usual hydroxylamine hydrochloride technique is used. According to Dodge, fairly satisfactory results may be obtained if the solution is well cooled and the titration carried out at low temperatures (−10°). *Am. Perfumer*, May (1940), 43.

TABLE 4.12 (*Cont.*).

PART I (*Cont.*)

Carbonyl Compound	Molecular Wt.	Size of Sample (g)	Reaction Time
Methyl Nonyl Ketone	170.29	1.0	15 min.
o-Methoxyacetophenone	150.17	1.0	24 hr.
Nonyl Aldehyde	142.24	1.0	15 min.
Octyl Aldehyde	128.21	1.0	15 min.
Perillic Aldehyde	150.21	1.0	15 min.
Phenylacetaldehyde	120.14	1.0	30 min.
Phenylpropyl Aldehyde	134.17	1.0	15 min.
Piperitone†	152.23		
Pulegone†	152.23		
Salicyl Aldehyde	122.12	1.0	15 min.
Thujone	152.23	0.5	24 hr.
Tolyl Aldehyde	120.14	1.0	15 min.
Umbellulone†	150.21		
Undecyl Aldehyde	152.14	1.0	15 min.
Vanillin	152.14	1.0	15 min.
Valeric Aldehyde	86.13	1.0	15 min.

PART II

Oil	Main Carbonyl Compound Present		Size of Sample (g)	Reaction Time
	Name	Molecular Wt.		
Almond, Bitter	Benzaldehyde	106.12	1.0	immediate
Caraway	Carvone	150.21	1.0	24 hr.
Cassia	Cinnamic Aldehyde	132.15	1.0	15 min.
Cedar Leaf	Thujone	152.23	1.0	24 hr.
Cherry Laurel	Benzaldehyde	106.12	1.0	immediate
Cinnamon	Cinnamic Aldehyde	132.15	1.0	15 min.
Citronella, Ceylon	Citronellal	154.25	2.5	15 min.
Citronella, Java	Citronellal	154.25	1.0	15 min.
Cumin	Cuminic Aldehyde	148.20	1.0	15 min.
Dill	Carvone	150.21	1.0	24 hr.
Geranium	Menthone	154.25	0.5	24 hr.
Grapefruit	Decyl Aldehyde*	156.26	5.0	30 min.

* Occasionally the carbonyl component of grapefruit oil is reported as a mixture of equal parts of octyl and decyl aldehydes; if this is desired, use 142.24 as an average of the molecular weights.

TABLE 4.12 (*Cont.*).

PART II (*Cont.*)

Oil	Main Carbonyl Compound Present		Size of Sample (g.)	Reaction Time
	Name	Molecular Wt.		
Lemon	Citral	152.23	5.0	15 min.
Lemon Concentrates, Terpeneless and Sesquiterpeneless	Citral	152.23	1.0	15 min.
Lemongrass	Citral	152.23	1.0	15 min.
Limes, Distilled	Citral†	152.23	5.0	15 min.
Limes, Expressed	Citral	152.23	5.0	15 min.
Mandarin	Decyl Aldehyde	156.26	5.0	30 min.
Orange	Decyl Aldehyde	156.26	5.0	30 min.
Orange Concentrates, Terpeneless and Sesquiterpeneless	Decyl Aldehyde	156.26	1.0	30 min.
Orris‡	Irone	192.29	1.0	1 hr.
Pennyroyal§	Pulegone	152.23	0.5	about 72 hr.
Peppermint	Menthone	154.25	0.5	24 hr.
Rue	Methyl Nonyl Ketone	170.29	1.0	15 min.
Sage, Dalmatian	Thujone	152.23	1.0	24 hr.
Spearmint	Carvone	150.21	1.0	24 hr.
Tansy	Thujone	152.23	0.5	24 hr.
Wormwood‖	Thujone	152.23	0.25	24 hr.

† Very little citral is actually present in distilled lime oil. The carbonyl components consist mainly of octyl aldehyde, decyl aldehyde, dodecyl aldehyde, and an unidentified aldehyde. However, it is customary to report the aldehyde content as citral.

‡ It is well to titrate the reaction mixture at the end of 1 hr. and then at the end of 24 hr: any appreciable difference in the two values indicates the presence of other carbonyl compounds, most likely one of the ionones.

§ Because of the slow reaction rate of hydroxylamine with pulegone, this method is not satisfactory for oil pennyroyal; use the neutral sulfite method.

‖ The dark color of wormwood oils with their natural greenish tint makes the determination of thujone quite difficult. The use of a very small sample and the use of a spot-plate to judge the end point is recommended. However, the accuracy of the determination suffers thereby; nevertheless an accuracy of ±5 per cent can be obtained.

Preparation of 0.5 N Hydroxylamine Solution: Dissolve 20 g. of recrystallized hydroxylamine hydrochloride in 40 cc. of water and dilute to 400 cc. with 95% alcohol. To this solution, in a 1 liter beaker, add, with stirring, 300 cc. of 0.5 N alcoholic potassium hydroxide and 2.5 cc. of a 0.4% bromphenol blue solution in 50% alcohol. Permit the solution to stand for 30 min. and filter. This solution cannot be stored for any appreciable period. A blank must always be run since the solution tends to deteriorate slowly.

In conclusion it might be well to point out that each of the general methods for the determination of aldehydes and ketones has its place in the analysis of essential oils. Thus, absorption methods permit of the easy separation of the noncarbonyl portion of the oil and of the separation of some aldehydes and ketones by regeneration from the bisulfite addition compound with strong alkali. It then becomes possible to study the odor and other properties of these individual portions and to detect more readily adulteration of the original oil. A comparison of the results from the absorption methods and from the hydroxylamine method is frequently very revealing; large differences may be indicative of adulteration with water-soluble constituents or additions of carbonyl compounds of low molecular weight.

From a consideration of the limitations of each method, it should be obvious that it is of utmost importance *always to record the method used when reporting an analytical result.*

5. DETERMINATION OF PHENOLS

Phenols react with the alkali hydroxides, giving rise to water-soluble phenolates. This is the basis of the classical method[98] for the estimation of phenols in essential oils. Since the potassium salts of many phenols are more soluble than the corresponding sodium salts, the use of potassium hydroxide is preferred.

It must be remembered that, in addition to the phenols, any alkali-soluble material (e.g., acids) will also go into solution as well as any water-soluble constituents or water-soluble adulterants (e.g., alcohol). This will give rise to erroneous results: the apparent phenol content will be too high. Further, an aqueous solution of alkali phenolates is a much better solvent for the nonphenolic portion of an oil than is the alkali solution itself; specifically this is important in the case of terpeneless bay oils.

When the determination has been completed, it often proves of value to separate the nonphenolic portion and to study its odor. The alkaline solution of the phenolates may be freed of traces of oil by washing with ether;

[98] The procedure was first applied by Gildemeister for the determination of phenols in thyme oil.

the phenols may then be regenerated by the addition of dilute sulfuric acid (1:3), extracted with ether, and obtained in a pure state by evaporating off the ether. (The separated ether layer should be dried with anhydrous sodium sulfate before this final evaporation.) The presence of foreign phenolic bodies frequently may be detected by this technique.

Modifications of the general procedure become necessary in the case of certain specific oils. Such modifications are noted below.

General Procedure: Into a well cleaned 150 cc. cassia flask, having a long, thin neck graduated in 0.1 cc. divisions, introduce 10 cc. of the oil, measured from a pipette. Add 75 cc. of an aqueous 1 N potassium hydroxide solution,[99] measured from a graduated cylinder. Stopper and shake thoroughly for exactly 5 min. Permit to stand undisturbed for 1 hr., after which the undissolved oil is forced into the neck by the addition of more potassium hydroxide solution. The alkaline solution must be added carefully to avoid disturbing the layer of separated oil. (This addition may conveniently be made by clamping the flask at a slight angle on a ring stand; above the flask is placed a ring to hold a separatory funnel containing the solution of alkali which is permitted to flow down along the inside of the neck of the cassia flask very slowly. If the flow of the alkali is adjusted to about 1 drop per sec., a clean separation of the oil is usually obtained. (See Diagram 4.10.) In order to make any droplets of oil adhering to the sides of the flask rise into the neck, gently tap or revolve the flask rapidly between the palms of the hands. Measure the quantity of oil that does not dissolve in the alkali. The phenol content, expressed as a volume/volume percentage, is calculated from the following formula:

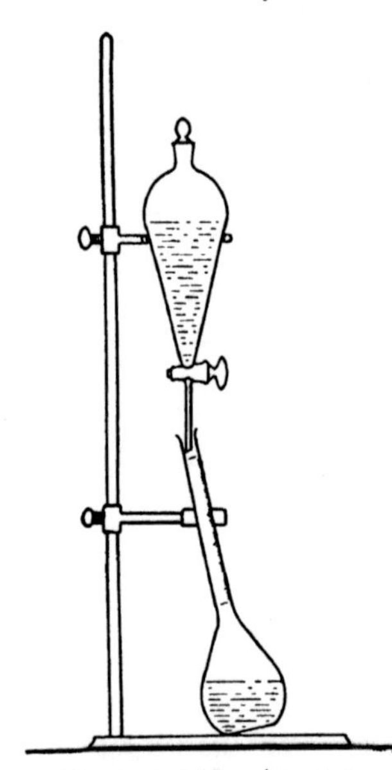

DIAGRAM 4.10. Apparatus for Phenol determination.

$$\text{Percentage of phenol} = 10(10 - \text{no. of cc. of undissolved oil})$$

[99] Gildemeister and Hoffmann, "Die ätherischen Öle," 3d Ed., Vol. I, 753, recommended the use of a 5 per cent solution of either NaOH or KOH be employed for thymol and carvacrol-containing oils; a 3 per cent solution for eugenol-containing oils. The official methods of the "United States Pharmacopoeia" and the "National Formulary" require the use of KOH T.S. solution (1 N).

Oils containing large amounts of heavy metals may not give a sharp separation of the nonphenolic oily layer and the alkaline solution in the neck of the flask. Such oils should be thoroughly shaken with a small amount (about 1 per cent) of powdered tartaric acid and filtered to remove the interfering metals before the determination of phenols is attempted.

Modification of the General Procedure:

I. Clove Oils.—Since clove oils contain aceteugenol in addition to free eugenol and since both constituents contribute to the value of the oil it is customary to saponify the former and report the total phenol content as eugenol. The general procedure is modified as follows:

After thoroughly shaking the oil and alkali for 5 min. in the cold, heat the flask on a steam bath for 10 min. Occasionally shake the flask during this heating to insure complete saponification. Immediately after removal of the flask from the steam bath add a further quantity of alkali in order to drive the unreacted oil into the neck of the flask. It is necessary to make this addition while the content of the flask is still hot since the nonphenolic portion may partially solidify.

II. Pimenta Oils.—The procedure described above for clove oils is also used for the determination of the phenol content of pimenta oils.

III. Terpeneless Bay Oils.—Because of the solvent effect of the potassium eugenolate upon the nonphenolic constituents of a terpeneless bay oil, the whole oil will go completely into solution if a 1 N solution of potassium hydroxide is used in this determination. Therefore, it becomes necessary to reduce the strength of the alkali to 3% and to use 125 cc. of this dilute alkaline solution for shaking out the phenols.

IV. Cinnamon Oils.—These oils offer some difficulty to the analyst. The formation of a troublesome emulsion and a very poor separation of the oil and the aqueous layers results because of the similarity of the gravity of the oil and the gravity of the solution. It is for this reason that a 3% solution cannot satisfactorily be used. Shaking for too long a period gives rise to results that are much too high. The following procedure, if followed exactly, will give results that can easily be duplicated, and which represent approximately the true eugenol content:

To 50 cc. of a 1 N potassium hydroxide solution in a cassia flask add 5 cc. of the cinnamon oil. Shake well for exactly 3 min. and let stand for 10 min. Fill the flask with potassium hydroxide solution, using the ring stand technique described under the general procedure. If the determination has been carried out carefully, the residual oil will rise into the neck in an unbroken column.

V. Thyme and Origanum Oils.—The phenolic constituents of thyme and origanum oils consists mainly of thymol and carvacrol. The separation of the phenolic constituents is an aid in the evaluation of these oils, since oils containing predominently thymol are generally considered of superior quality. Thymol is easily crystallized; carvacrol is a liquid at temperatures above 2°.

The separation and examination of the phenolic portion may conveniently be carried out after the determination of the phenol content.

Pour the contents of the cassia flask (used in the assay) into a separatory funnel and permit the nonphenolic portion to separate. Filter the aqueous layer through filter paper previously wetted with water. Transfer this filtered solution to a separatory funnel and acidify with dilute hydrochloric acid (1:3) until the mixture is strongly acid to litmus. Add 50 cc. of ether and shake thoroughly. Separate the ether layer, dry with anhydrous sodium sulfate and filter. Evaporate the ether *cautiously* on a steam bath and pour the liberated phenols into a test tube, and permit it to stand at room temperature for 30 min. If the phenols consist primarily of thymol, a crystalline mass results. If no crystals form after 30 min., cool to 5° by means of an ice bath. Rub the side of the test tube with a thermometer or glass rod and add a small crystal of thymol to initiate crystallization: if no crystals form after 30 min. the absence of an appreciable amount of thymol may be assumed.

In determining phenol contents it is well to remember that water-soluble constituents may be added to increase the apparent phenol contents. Alcohol and certain glycols are such adulterants which may be occasionally encountered. If the relationship between the specific gravity and the phenol content appears abnormal, the oil should be investigated further for the presence of possible adulterants.

6. DETERMINATION OF CINEOLE

Of the numerous methods that have been proposed for the determination of the cineole content of essential oils, the method of Kleber and von Rechenberg,[100] the method of Cocking,[101] and that of Scammell[102] (modified by Baker and Smith[103]) have proved the most valuable.

According to the Kleber and von Rechenberg method, the congealing point of the oil itself is determined, from which the cineole content may be determined by reference to a table or graph. The presence of oxygenated constituents other than cineole has little effect upon the values obtained; an accuracy of about ±1 per cent may be obtained. "The United States Pharmacopoeia"[104] makes use of this method to establish a minimum of 70

[100] *J. prakt. Chem.* [2] 101 (1920), 171.
[101] *Analyst* 52 (1927), 276.
[102] British Patent 14138 (1894).
[103] "Eucalypts and Their Essential Oils," 2nd Ed., Sydney, 1921, 364.
[104] Thirteenth Revision, 217.

per cent cineole in official eucalyptus oils. The main criticism of the Kleber and von Rechenberg method is the inconvenience of working at greatly reduced temperatures. The exact determination of a congealing point at a temperature much below 0° often presents difficulty.

Procedure: Place about 10 cc. of the oil in a heavy walled tube which is preferably equipped with an air or vacuum jacket. Immerse the tube in a mixture of ice and salt, or in a cooling bath of solid carbon dioxide in acetone. The true solidification point is determined; i.e., the temperature at which the crystals of cineole first appear as the oil is cooled, and at which the crystals disappear as the temperature is permitted to rise. Several determinations of the solidifying point should be made in order to obtain an exact reading. The percentage of cineole, corresponding to this temperature can be determined directly by reference to Table 4.13.

TABLE 4.13. DETERMINATION OF EUCALYPTOL CONTENT BY CONGEALING POINT

Temperature (° C)	Eucalyptol Content (%)	Temperature (° C.)	Eucalyptol Content (%)
1.2	100.0	− 9.0	80.3
1.0	99.4	−10.0	78.5
0.0	97.3	−11.0	76.5
−1.0	95.3	−12.0	75.3
−2.0	93.4	−13.0	73.7
−3.0	91.5	−14.0	72.2
−4.0	89.6	−15.0	70.6
−5.0	87.5	−16.0	69.2
−6.0	85.7	−17.0	67.5
−7.0	83.7	−18.0	66.2
−8.0	82.0	−19.0	64.8

The *o*-cresol method of Cocking offers certain advantages. Since the congealing point is well above room temperature, the determination is greatly simplified. Results are easily reproducible. According to Cocking, the accuracy is approximately ±3 per cent in the case of eucalyptus and cajuput oils. The *o*-cresol method has been accepted as the official method of "The British Pharmacopoeia."[105] The procedure as given below is essentially the official method of "The British Pharmacopoeia."

Procedure: Into a stout walled test tube, about 15 mm. in diameter and 80 mm. in length, place 3 g. (accurately weighed) of the oil, previously dried with anhydrous sodium sulfate,[106]

[105] (1932), 584.

[106] The use of calcium chloride for drying the oil as suggested by "The British Pharmacopoeia" is not to be recommended; anhydrous sodium sulfate is much to be preferred, eliminating the possibility of formation of addition products with primary alcohols that may be present in the oil.

together with 2.1 g. of melted *o*-cresol. The *o*-cresol used must be pure and dry, with a freezing point not below 30°. It is hygroscopic and should be stored in a small well-stoppered bottle, because the presence of moisture may lower the results.

Insert a thermometer, graduated in fifths of a degree, and stir the mixture well in order to induce crystallization; note the highest reading of the thermometer. Warm the tube gently until the contents are thoroughly melted, and insert the tube through a bored cork into a widemouthed bottle which is to act as an air jacket. The thermometer should be suspended from a ring stand in such a way that it does not touch the walls of the inner tube. Allow the mixture to cool slowly until crystallization commences, or until the temperature has fallen to the point previously noted. Stir the contents of the tube vigorously with the thermometer, rubbing the latter on the sides of the tube with an up and down motion in order to induce rapid crystallization. Continue the stirring and rubbing as long as the temperature rises. Take the highest point as the freezing point. Repeat this procedure until two readings agreeing within 0.1° are obtained. The percentage of cineole in the oil can be computed from Table 4.14.

It should be noted that in both methods relatively high cineole contents give more accurate analytical results. Therefore, if the cineole content of the oil is low, the determination is best carried out on a mixture of equal parts (by weight) of oil and pure cineole, (melting point, 1.2° or higher). In the Kleber and von Rechenberg method, if the cineole content is less than 65 per cent, this modified procedure should be followed; in the *o*-cresol method, of Cocking, if less than 50 per cent. The cineole content of the original oil may then be calculated by means of the following formula:

$$\text{Percentage of cineole in original oil} = 2 \times (\% \text{ of cineole in mixture} - 50)$$

The phosphoric acid method of Scammell as modified by Baker and Smith is based on the formation of a solid, loose molecular compound of cineole and phosphoric acid from which the cineole may be regenerated by the addition of water. The procedure recommended by these authors is given below:

Procedure: Place 10 cc. of the oil in a 50 cc. beaker or other suitable vessel and cool thoroughly in a bath of ice and salt. Slowly add 4 cc. of phosphoric acid,[107] a few drops at a time, mixing the acid and oil thoroughly between each addition by careful stirring. After all of the acid has been added, permit the mixture to remain in the bath for 5 min. to insure complete formation of the cineole-phosphoric acid addition compound. Then add 10 cc. of petroleum ether (boiling below 50°) which

[107] If the cineole content is below 30 per cent, add only 3 cc. of phosphoric acid.

TABLE 4.14. PERCENTAGE OF CINEOLE

Temp	0.0	0.1	0.2	0.3	0.4	0.5	0.6	0.7	0.8	0.9
24	45.6	45.7	45.9	46.0	46.1	46.3	46.4	46.5	46.6	46.8
25	46.9	47.0	47.2	47.3	47.4	47.6	47.7	47.8	47.9	48.1
26	48.2	48.3	48.5	48.6	48.7	48.9	49.0	49.1	49.2	49.4
27	49.5	49.6	49.8	49.9	50.0	50.2	50.3	50.4	50.5	50.7
28	50.8	50.9	51.1	51.2	51.3	51.5	51.6	51.7	51.8	52.0
29	52.1	52.2	52.4	52.5	52.6	52.8	52.9	53.0	53.1	53.3
30	53.4	53.5	53.7	53.8	53.9	54.1	54.2	54.3	54.4	54.6
31	54.7	54.8	55.0	55.1	55.2	55.4	55.5	55.6	55.7	55.9
32	56.0	56.1	56.3	56.4	56.5	56.7	56.8	56.9	57.0	57.2
33	57.3	57.4	57.6	57.7	57.8	58.0	58.1	58.2	58.3	58.5
34	58.6	58.7	58.9	59.0	59.1	59.3	59.4	59.5	59.6	59.8
35	59.9	60.0	60.2	60.3	60.4	60.6	60.7	60.8	60.9	61.1
36	61.2	61.3	61.5	61.6	61.7	61.9	62.0	62.1	62.2	62.4
37	62.5	62.6	62.8	62.9	63.0	63.2	63.3	63.4	63.5	63.7
38	63.8	63.9	64.1	64.2	64.4	64.5	64.6	64.8	64.9	65.1
39	65.2	65.4	65.5	65.7	65.8	66.0	66.2	66.3	66.5	66.6
40	66.8	67.0	67.2	67.3	67.5	67.7	67.9	68.1	68.2	68.4
41	68.6	68.8	69.0	69.2	69.4	69.6	69.7	69.9	70.1	70.3
42	70.5	70.7	70.9	71.0	71.2	71.4	71.6	71.8	71.9	72.1
43	72.3	72.5	72.7	72.9	73.1	73.3	73.4	73.6	73.8	74.0
44	74.2	74.4	74.6	74.8	75.0	75.2	75.3	75.5	75.7	75.9
45	76.1	76.3	76.5	76.7	76.9	77.1	77.2	77.4	77.6	77.8
46	78.0	78.2	78.4	78.6	78.8	79.0	79.2	79.4	79.6	79.8
47	80.0	80.2	80.4	80.6	80.8	81.1	81.3	81.5	81.7	81.9
48	82.1	82.3	82.5	82.7	82.9	83.2	83.4	83.6	83.8	84.0
49	84.2	84.4	84.6	84.8	85.0	85.3	85.5	85.7	85.9	86.1
50	86.3	86.6	86.8	87.1	87.3	87.6	87.8	88.1	88.3	88.6
51	88.8	89.1	89.3	89.6	89.8	90.1	90.3	90.6	90.8	91.1
52	91.3	91.6	91.8	92.1	92.3	92.6	92.8	93.1	93.3	93.6
53	93.8	94.1	94.3	94.6	94.8	95.1	95.3	95.6	95.8	96.1
54	96.3	96.6	96.9	97.2	97.5	97.8	98.1	98.4	98.7	99.0
55	99.3	99.7	100.0							

has previously been well cooled in the ice bath and incorporate well with the aid of a flat ended rod. Immediately transfer the mixture to a cooled Buchner funnel, 5 cm. in diameter. Filter off the noncombined portion rapidly with the aid of a water pump. Transfer the cake from the Buchner funnel to a piece of fine calico and spread the cake with a spatula so that it covers an area of about 6 × 8 cm. Fold over the calico into a pad and place between several layers of absorbent paper. Press well for 3 min. Break up the cake on a glazed tile with a spatula and transfer to a cassia flask. Decompose the cineole-phosphoric acid addition compound with warm water and force the liberated cineole into the neck of the flask by the further addition of water. After the separation is complete and the contents of the flask have cooled to room temperature measure the amount of cineole.

If the cineole content is found to be above 60 per cent, repeat the determination using a sample of the oil diluted with freshly distilled pinene or turpentine oil: three volumes of oil plus one volume of pinene. Make the necessary correction in calculating the percentage of cineole.

This test gives satisfactory results with oils containing as little as 20 per cent cineole and as high as 100 per cent; in the latter case, the oil must be previously diluted as described. It is important to add the acid slowly, to have the mixture very cold and to cool the petroleum ether thoroughly before adding.

Other methods for the quantitative determination of cineole have been suggested. In their monumental work on the eucalypts Baker and Smith[108] give a brief criticism of many of these methods.

7. DETERMINATION OF ASCARIDOLE

The analysis of wormseed oils offers some difficulty because of the lack of a satisfactory method for the determination of the active principle, ascaridole.

"The United States Pharmacopoeia"[109] based the official determination on the fact that ascaridole is soluble in dilute acetic acid. The technique consists of shaking a measured volume of the oil in a cassia flask with 60 per cent acetic acid, determining the volume of undissolved oil, and calculating the ascaridole content by difference. This is a method developed by Nelson.[110] Although this procedure represents one of the simplest determinations for ascaridole, it suffers from the fact that the analytical values are far from accurate. It has been found that normal oils containing as much as 70 to 80 per cent ascaridole (as indicated by solubility, gravity, distillation, and other methods for the determination of ascaridole) often analyze as low as 55 per cent by the Nelson method. Furthermore, Reindollar[111] has shown that "hi-test" oils containing large amounts of ascaridole give results by this method that are too high. A further disadvantage of the Nelson method lies in the fact that the determination is by no means specific

[108] "The Eucalypts," 2nd Ed. (1920), Sydney, Australia, p. 357.
[109] Eleventh Revision, 251.
[110] *J. Am. Pharm. Assocn.* 10 (1921), 836.
[111] *Ibid.* 28 (1939), 591.

for ascaridole; additions of cineole or cineole-containing oils analyze as ascaridole. This is also true of many other oxygenated compounds, such as terpineol.

> *Procedure:* Place 10 cc. of oil wormseed, measured from a pipette, in a 100 cc. cassia flask. Add 50 cc. of a solution of acetic acid made by diluting 60 cc. of glacial acetic acid with distilled water to measure 100 cc. Shake the mixture well for 5 min. Add sufficient of the acetic acid solution to raise the lower limit of the oily layer within the graduated portion of the neck and allow the liquids to separate, rotating the flask from time to time. Note the volume of the oily layer.
>
> Percentage of ascaridole = 10(10 − cc. of unreacted oil)

In order to overcome these difficulties, "The United States Pharmacopoeia"[112] later abandoned the Nelson method and then substituted the Cocking and Hymas[113] procedure, previously official in "The British Pharmacopoeia."[114] The procedure given below is based upon this official method:

> *Assay:* Place about 2.5 g. of oil chenopodium, accurately weighed, in a 50 cc. volumetric flask, fill to the mark with 90% acetic acid, mix well, and transfer a portion of this freshly prepared solution to a burette, graduated in 20ths of a cc. Into a glass-stoppered Erlenmeyer flask measure, from graduated cylinders, 3 cc. of a solution of potassium iodide (prepared by dissolving 8.3 g. of potassium iodide in sufficient distilled water to make 10 cc. of solution), 5 cc. of concentrated hydrochloric acid, and 10 cc. of glacial acetic acid. Immerse the flask in a freezing mixture until the temperature is reduced to −3°, add quickly about 5 cc. of the acetic acid solution of the oil, mix it with the cooled reagent as rapidly as possible, and observe the volume drawn from the burette after 2 min. (to allow for draining). Set the stoppered flask aside at a temperature between 5° and 10° for exactly 5 min.; then, without diluting, titrate the liberated iodine with tenth-normal sodium thiosulfate. At the same time, conduct a blank test, but dilute the reagent with 20 cc. of distilled water before titrating the liberated iodine. The difference between the two titrations represents the iodine liberated by ascaridole. Each cc. of tenth-normal sodium thiosulfate is equivalent to 0.00665 g. of $C_{10}H_{16}O_2$.[115]

If all conditions as outlined are rigidly followed, this method gives analytical results that are reproducible and relatively accurate for normal

[112] Twelfth Revision, 319. (Now official in "The National Formulary," Eighth Ed., 136.)

[113] *Analyst* 55 (1930), 180.

[114] (1932), 303.

[115] Pauly (see *Pharm. Arch.* 7 (1936), 9) contends that the empirical factor, 0.00665, is actually too high, since he obtained a value of 110 per cent for a redistilled ascaridole fraction using this factor. Experiments carried out in the laboratories of Fritzsche Brothers, Inc., appear to confirm this fact.

American oils. If, however, the ascaridole content is abnormally low, the results will not be sufficiently accurate.

Since this method is based on the oxidation of potassium iodide by the peroxide, ascaridole, and the subsequent determination of the amount of free iodine, additions of oxygenated constituents should not increase the apparent ascaridole content unless the added compound is capable of oxidizing the potassium iodide under the experimental conditions. Furthermore, since the liberated iodine is capable of being absorbed by unsaturates in the oil, it is very important to maintain the low temperatures as indicated in the procedure, to keep this secondary reaction at a minimum.

The tentative method for the determination of ascaridole as outlined in the Methods of Analysis of the Association of Official Agricultural Chemists[116] proves too cumbersome for rapid commercial analyses and control. This method, developed by Paget,[117] involves the reduction of ascaridole by titanium trichloride. It requires that the solution be protected from atmospheric oxygen; this entails storage of the solution under an atmosphere of hydrogen, and titrations under an atmosphere of carbon dioxide. According to the experiments of Reindollar,[118] the results obtained by this method are fairly concordant with those obtained by the method of Cocking and Hymas. The Association of Official Agricultural Chemists' method is given below, exactly as it appears in this official work:

> *Procedure:* "Weigh 1 ml of the oil in 100 ml volumetric flask and dilute to volume with alcohol. Place 50 ml of the $TiCl_3$ soln in Erlenmeyer flask through which current of CO_2 is passing. Fit flask with Bunsen valve, add 10 ml of diluted soln of the oil, close flask (with the Bunsen valve), and heat contents almost to boiling for 2 min. (Prolonged heating has no effect if contents are not boiled vigorously.) If pale violet color of the $TiCl_3$ disappears, add more reagent to insure excess. (Formation of a white precipitate does not interfere with determination.) Add 1 ml of 5% NH_4CNS soln and titrate back excess of $TiCl_3$ with the $FeNH_4(SO_4)_2$ soln in CO_2 atmosphere until faint, permanent, brownish red color is obtained.
>
> "Subtract quantity of $FeNH_4(SO_4)_2$ soln used, expressed in equivalent mg of $TiCl_3$, from number of mg of $TiCl_3$ taken. Difference is number of mg of $TiCl_3$ oxidized by oil taken. Convert mg of $TiCl_3$ oxidized into ascaridole by dividing by factor 1.284 (1 g of ascaridole is reduced by 1.284 g of $TiCl_3$).
>
> "Example: 0.9600 g of oil was made up to 100 ml and 10 ml aliquot was heated with 50 ml of the $TiCl_3$ soln (1 ml containing 0.0034 g of $TiCl_3$). It then required 5.9 ml of the reagent, each ml equivalent to 0.01545 g to $TiCl_3$, to back titrate. Grams of

[116] 6th Ed., 735.
[117] *Analyst* 51 (1926), 170.
[118] *J. Am. Pharm. Assocn.* 28 (1939), 589.

$TiCl_3$ oxidized is numerically equal to $(50 \times 0.0034) - (5.9 \times 0.01545)$, or 0.07885. Weight of oil in the aliquot was 0.0960 g. Hence percentage of ascaridole $= \frac{0.07885 \times 100}{0.096 \times 1.284} = 72.1\%$.

"a) *Standard ferric ammonium sulfate soln.*—Dissolve 39.214 g of pure, crystallized $Fe(NH_4)_2(SO_4)_2 \cdot 6H_2O$ in 200 ml of H_2O in liter flask, add 30 ml of H_2SO_4, and mix well. Weigh exactly 3.16 g of $KMnO_4$, dissolve in 200 ml of warm H_2O, and slowly add to soln in the flask, with stirring. ($KMnO_4$ soln should be just sufficient to oxidize the ferrous salt, but it is well to add the last few ml in small portions.) Cool soln and dilute to 1 liter with H_2O.

"b) *Standard titanium trichloride soln.*—Add 100 ml of commercial 15–20% $TiCl_3$ soln to 200 ml of HCl, boil 1 min., cool, and dilute to 4500 ml with H_2O. Place soln in container with H atmosphere provision and allow to stand 2 days for absorption of residual O. Preserve the $TiCl_3$ soln in an atmosphere of H (Chap. 21, Fig. 27), taking care to have all joints air-tight, and covering stoppers (preferably countersunk) with suitable wax. Standardize by titrating 20 ml of the $FeNH_4(SO_4)_2$ soln against the $TiCl_3$ soln in a protective stream of CO_2, using 1 ml of 5% NH_4CNS soln as indicator. 1 ml of 0.1 N $FeNH_4(SO_4)_2$ = 0.01545 g of $TiCl_3$."

The determination of the ascaridole content by a distillation technique is not to be recommended for routine analyses, since the ascaridole is so unstable that the oil is apt to decompose with explosive violence should the temperature not be carefully controlled.

Dodge[119] has suggested the use of a solution of sodium bisulfite to determine the ascaridole content of wormseed oils. The difficulty of determining the exact end point and the length of time required for a determination militates against the use of this technique.

In spite of its recognized deficiencies, the method of Cocking and Hymas, at the present time official in "The National Formulary," Eighth Ed., and "The British Pharmacopoeia," is probably the most useful for commercial analytical control laboratories.

8. DETERMINATION OF CAMPHOR

In order to determine the camphor content of essential oils (which contain no other carbonyl compounds) the gravimetric determination proposed by Aschan[120] may be employed.

Procedure: Introduce about 1 g. of the oil, accurately weighed, into a test tube and dissolve the oil in 2 g. of glacial

[119] *Drug & Cosmetic Ind.* **46** (1940), 414.
[120] *Finska Apoth. Tidskrift* (1925), 49. *Chem. Abstracts* **20** (1926), 1775.

acetic acid. Add 1 g. of semicarbazide hydrochloride and 1.5 g. of freshly fused anhydrous potassium acetate. Triturate thoroughly with a glass rod, stopper the tube with a plug of absorbent cotton and heat three hours in a water bath at 70°. Cool the mixture, add 10 to 15 cc. of water, stir thoroughly and transfer the precipitate quantitatively to a tared 4 to 5 cm. filter. Wash with water until all water soluble matter is removed, air dry, wash with petroleum ether and dry in air to constant weight. Determine the weight of semicarbazone from the increase in the weight of the filter. Calculate the content of camphor in the original oil by means of the following formula:

$$\text{Percentage of camphor} = \frac{72.7p}{s}$$

where: p = weight of semicarbazone in grams;
s = weight of oil in grams.

9. DETERMINATION OF METHYL ANTHRANILATE

The procedure described below is based on the classical method of Hesse and Zeitschel.[121] It depends on the actual separation of the ester from the volatile oil by the formation of the ether insoluble sulfate.

Procedure: Dissolve about 25 to 100 g. of oil in twice the volume of anhydrous ether. Cool the solution well in a freezing mixture, the temperature being reduced to at least 0°. Add, with constant stirring, a solution of 1 volume of concentrated sulfuric acid in 5 volumes of anhydrous ether until no further precipitate forms. Collect the precipitate in a small, well-cooled Büchner funnel and wash with dry, cold ether until odorless. Dissolve this precipitate in water with the aid of alcohol if necessary, and titrate with 0.5 N sodium hydroxide. Calculate the ester content by means of the following formula:

$$\text{Percentage of methyl anthranilate} = \frac{3.775a}{s}$$

where: a = cc. of alkali required;
s = original weight of oil taken in grams.

To this solution add an excess of 0.5 N sodium hydroxide and heat the mixture on a steam bath for 30 min. Titrate the free alkali which is unconsumed with 0.5 N hydrochloric acid. Calculate again the ester content by means of the following formula:

$$\text{Percentage of methyl anthranilate} = \frac{7.55b}{s}$$

where: b = cc. of alkali consumed in the saponification;
s = original weight of oil taken in grams.

[121] *Ber.* **34** (1901), 296.

If the ester is exclusively methyl anthranilate, *a* should be twice as large as *b*.

The procedure as described will determine all basic constituents which form ether insoluble sulfates (e.g., the methyl ester of methyl anthranilic acid) in addition to methyl anthranilate.

For the determination of methyl anthranilate in the presence of methyl N-methyl anthranilate, Erdmann[122] has suggested a procedure based on the diazotization of methyl anthranilate, a primary aromatic amine. The ester is washed out of the oil with dilute sulfuric or hydrochloric acid, and the acid solution treated with a 5 per cent sodium nitrite solution to diazotize the amine. The solution is then titrated with an alkaline solution of β-naphthol. (This solution is prepared by dissolving 0.5 g. β-naphthol in 0.5 cc. of sodium hydroxide, at least 30 per cent, and adding a solution of 15 g. of sodium carbonate in 150 cc. of water.) The azo dye thereby formed is insoluble and precipitates out. The titration is continued until no further precipitation occurs.

A combination of the method of Hesse and Zeitschel and that of Erdmann can be used to determine the percentage of methyl anthranilate and of methyl N-methyl anthranilate.

10. DETERMINATION OF ALLYL ISOTHIOCYANATE

Several methods, both volumetric and gravimetric, for the determination of allyl isothiocyanate in mustard oils have been suggested in the literature.[123] The most satisfactory one is a modification of the official procedure of "The United States Pharmacopoeia."[124] This method is based upon the reaction of the isothiocyanate radical with silver nitrate. The excess silver nitrate solution is determined by titration with standardized ammonium thiocyanate solution in the presence of ferric ion; the titration must be continued until the red color of ferric thiocyanate is first observed.

$$C_3H_5NCS + NH_4OH + AgNO_3 \rightarrow AgNCS + C_3H_5OH + NH_4NO_3$$

Procedure: Dilute about 4 cc. of the oil, accurately weighed, with sufficient alcohol to make exactly 100 cc. of solution. Pipette 5 cc. of this solution into a 100 cc. mustard oil flask (see Diagram 4.11) and add 50 cc. of 0.1 N silver nitrate solution and 5 cc. of 10% ammonia solution. Connect the flask to an air-cooled reflux condenser, 1 m. long, and heat on a steam bath for 1 hr. Allow the liquid to cool to room temperature and then add sufficient distilled water to fill the flask to the 100 cc. mark. Mix well and filter through a dry filter. Reject

[122] *Ber.* 35 (1902), 24.
[123] See "Suggested Additional Literature," p. 359.
[124] Twelfth Revision, 46. (Now official in "The National Formulary," Eighth Ed., 29.)

the first 10 cc. of filtrate. Transfer 50 cc. of the subsequent filtrate by means of a pipette into a 100 cc. saponification flask; add about 5 cc. of concentrated nitric acid and 2 cc. of an 8% solution of ferric ammonium sulfate. Titrate the excess of silver nitrate solution with standardized 0.1 N ammonium thiocyanate. Carry out a blank determination simultaneously, using 5 cc. of alcohol and the same quantities of reagents but omitting the oil. Calculate the percentage of allyl isothiocyanate; each cc. of 0.1 N silver nitrate is equivalent to 0.004958 g. of allyl isothiocyanate.

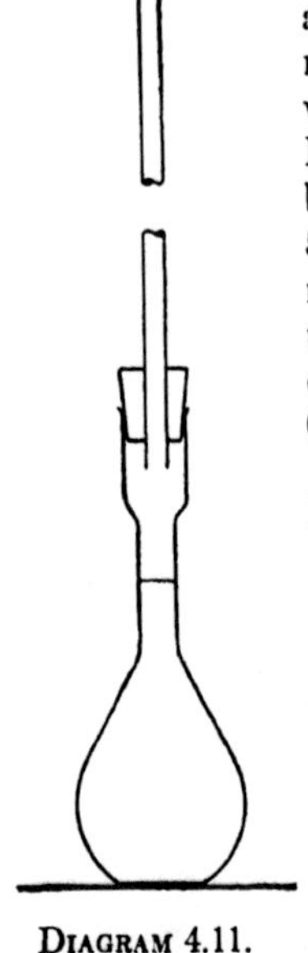

DIAGRAM 4.11. Mustard oil flask.

Percentage of allyl isothiocyanate

$$= \frac{19.832\,(b - a)}{w}$$

where: b = cc. of ammonium thiocyanate solution required for the blank;
a = cc. of ammonium thiocyanate solution required for the determination;
w = weight of oil used for the original dilution.

The blank should require about 24 to 25 cc. of ammonium thiocyanate; the actual determination, about 4 to 5 cc. if a 4 g. sample of oil is employed.

11. DETERMINATION OF HYDROGEN CYANIDE

Hydrogen cyanide occurs in the distillates of a number of plants. It plays an important part in the medicinal value of oil of bitter almond and oil of cherry laurel. The presence of hydrogen cyanide can be ascertained qualitatively by means of the Prussian blue test.

Procedure: To 1 cc. of the oil in a test tube add 1 cc. of distilled water, a few drops of a 10% aqueous sodium hydroxide solution and a few drops of a 10% ferrous sulfate solution.[125] Shake thoroughly and acidify with dilute hydrochloric acid. The precipitate of ferrous and ferric hydroxides dissolves, and in the presence of hydrogen cyanide the characteristic precipitate of Prussian blue appears.

In order to determine quantitatively the amount of hydrogen cyanide in an oil the titrimetric method of "The United States Pharmacopoeia"[126]

[125] Such a solution of ferrous sulfate always contains a small amount of ferric salt which is necessary for this reaction.

[126] Twelfth Revision, 314. (Now official in "The National Formulary," Eighth Ed., 31.)

has proved satisfactory. This method is based upon precipitation of the cyanide by silver nitrate solution.. The end point of the reaction can be determined by the red color of silver chromate. Part of the hydrogen cyanide found in oil of bitter almond and oil of cherry laurel is bound with benzaldehyde in the form of cyanhydrin; in order to liberate this hydrogen cyanide, a small amount of freshly precipitated magnesium hydroxide is added.

Procedure: Dissolve 0.75 g. of magnesium sulfate in 45 cc. of distilled water. Add 5 cc. of 0.5 N sodium hydroxide solution and 2 drops of a 10% solution of potassium chromate, and titrate the solution with 0.1 N silver nitrate solution to the production of a permanent reddish tint; this requires but a few drops of the silver nitrate solution. Pour this mixture into a 100 cc. Erlenmeyer flask containing 0.5 g. of oil of bitter almond accurately weighed. Mix well and titrate again with 0.1 N silver nitrate solution until a red tint, which does not disappear upon shaking, is produced. Conduct this titration as rapidly as possible. Calculate the hydrogen cyanide content by means of the following formula:

$$\text{Percentage of hydrogen cyanide} = \frac{0.2702a}{s}$$

where: a = number of cc. of 0.1 N silver nitrate required;
s = weight of sample in grams.

12. DETERMINATION OF IODINE NUMBER

The iodine number of a fat or oil represents the number of grams of iodine capable of being absorbed under certain fixed conditions by 100 g. of the substance. It is an indication of the degree of unsaturation in the fatty acid radical of the glycerides.

The use of iodine numbers for the evaluation of essential oils has never attained practical significance. This is due primarily to the unpredictable behavior of such oils in the presence of iodine solutions. It has been shown frequently that the iodine numbers of many essential oils vary with the size of the sample as well as with the period of contact with the reagent. Furthermore, the results do not correspond with the theoretical values expected.

In the case of fixed oils the iodine number is an important criterion of purity. Since the essential oil chemist occasionally is faced with the evaluation of such fixed oils as persic oil, sweet almond oil, olive oil, castor oil, and sesame oil, a procedure for the determination of iodine numbers is included here.

Procedure:[127] Introduce about 0.25 g. of the oil, accurately weighed,[128] into a glass stoppered Erlenmeyer flask of 250 cc. capacity, dissolve it in 10 cc. of chloroform, add 25 cc. of iodobromide solution, accurately measured from a burette, stopper the vessel securely, and allow it to stand for 30 min.[129] protected from light. Then add in the order named 30 cc. of a 1 N potassium iodide solution and 100 cc. of distilled water, and titrate the liberated iodine with tenth-normal sodium thiosulfate, shaking thoroughly after each addition of thiosulfate. When the iodine color becomes quite pale, add 1 cc. of a 1% starch indicator solution and continue the titration with thiosulfate until the blue color is discharged. Carry out a blank test at the same time with the same quantities of chloroform and iodobromide solution, allowing it to stand for the same length of time and titrating as directed. The difference between the number of cc. of thiosulfate consumed by the blank test and the actual test, multiplied by 1.269 and divided by the weight of sample taken, gives the iodine number. If more than half of the iodobromide solution is absorbed by the sample of the substance taken, the determination must be repeated, using a smaller sample of the substance under examination.

The iodobromide solution may be prepared by the following method:

Dissolve 13.2 g. of reagent iodine in 1,000 cc. of glacial acetic acid with the aid of gentle heat if necessary. Cool the solution to 25° and determine the iodine content in 20 cc. by titration with tenth-normal sodium thiosulfate. Add to the remainder of the solution a quantity of bromine equivalent to that of the iodine present. Preserve in glass-stoppered bottles, protected from light.

V. SPECIAL TESTS AND PROCEDURES

1. FLAVOR TESTS

A study of the odor and flavor of an oil, isolate, or synthetic is essential in judging quality and aids in the detection of adulteration. Comparison should always be made with an oil of good quality and of known purity. Organoleptic tests are unquestionably the most sensitive and satisfactory method for detecting slight spoilage in oils such as the citrus oils, and in

[127] The procedure given is essentially the official method of "The United States Pharmacopoeia," Thirteenth Revision, 647 (Hanus Method).

[128] The weight of the oil used is best determined by weighing by difference. A small bottle containing a few cc. of the oil and also a medicine dropper is accurately weighed; then about 8 or 9 drops of the oil are introduced into the Erlenmeyer flask, and the bottle with the residual oil and medicine dropper is again accurately weighed: the difference represents the weight of sample used. Small "petit cups" of glass may also be used; these cups (containing the requisite amount of oil, accurately weighed) are dropped into the Erlenmeyer flasks and are not removed during the determination.

[129] In the case of castor oil, allow the mixture to stand for 60 min.

detecting burned, pyroligneous "by-notes" resulting from improper distillation.

> *Procedure—Water Flavor Test:* Place ½ oz. of alcohol in an 8 oz. glass. Add 1 drop of oil and then 7 oz. of cold water, the water being added slowly with vigorous stirring. This should yield a clear or opalescent mixture, which does not separate oily droplets on the surface. The odor and flavor of these two water flavor tests should be carefully studied and evaluated.

In the case of dill oils, it is well to add 3 drops of glacial acetic acid to approximate more accurately the conditions under which this oil is usually employed. In the case of peppermint oils, it is best to add hot water; the flavor tests should be of uniform temperature and tasted while still warm.

> *Procedure—Sugar Syrup Test:* An acidified sugar syrup is prepared by adding 1 dram of 85% syrupy phosphoric acid and 7 drams of 50% citric acid to 1 gal. of simple syrup (U.S.P. quality: approximately 65% wt./wt.). Dilute 2 oz. of this prepared syrup with 2 oz. of cold water. Add 1 to 3 drops of a 10% alcoholic solution of the oil and mix thoroughly. The odor and flavor of these two sugar syrup tests should be carefully studied and evaluated.

These syrup tests are best prepared in widemouthed, screw-top bottles which permit of thorough mixing of the alcoholic solution and the syrup by vigorous shaking.

Such syrup flavor tests are especially valuable in evaluating citrus oils citrus concentrates, and oils and synthetics which duplicate the flavor of highly acidic fruits.

In the case of sweetening agents (such as vanillin, coumarin, and heliotropin), it is well to dispense with the acidic medium and to use instead a mixture of equal parts of simple syrup and water.

It should be remembered that comparison with a product of good quality is essential.

2. TESTS FOR HALOGENS

The presence of chlorine in a synthetic is usually indicative of insufficient purification. The detection of halogen in a reputedly natural essential oil or fraction is indicative of adulteration with a chlorine-containing synthetic. For example, cassia oils showing the presence of chlorine have probably been adulterated with impure synthetic cinnamic aldehyde.

Of the numerous procedures which have been suggested, the classical test with copper oxide (the so-called Beilstein test) proves by far the most convenient and rapid. Should this test prove inconclusive, the presence o absence of halogen should be confirmed by the combustion method, a more sensitive test.

Procedure—I. Beilstein Method: Wind the end of a No. 16 gage[130] copper wire into a tight spiral about 6 mm. long and 6 mm. in diameter. Fasten the other end of this wire to a wooden handle. Heat the wire in the nonluminous flame of a Bunsen burner until it glows without coloring the flame green.[131] Permit the wire to cool and reheat several times until a good coat of oxide has formed on the coil.[132] Add to the cooled spiral 2 drops of the material to be tested. Ignite and permit it to burn freely in air. The wire is again cooled and 2 more drops of the material are added and burned. This process is continued until a total of 6 drops has been added and ignited. Then hold the spiral in the oxidizing portion of a Bunsen flame, adjusted to about 1½ in. high. If the material is free from halogen, the flame will show no green color. The degree of persistence of green color is a rough indication of the amount of halogen present. A highly purified synthetic, free from halogen, will not show even a transient green color or flash of green.

Instead of the wire spiral described above, a piece of 30 mesh copper screening (1.5 cm. × 5.0 cm.) may be used. The screening should be rolled tightly around a copper wire and held in position by bending back the wire and twisting securely. Such a roll of copper screening will hold about 1 cc. of oil because of surface tension.

Certain nitrogen-containing compounds may give a positive test although no halogen is present. Also, the presence of free organic acids may cause a green colored flame since the copper salt may be sufficiently volatile; e.g., phenylacetic acid. Therefore, if a positive test is obtained, it is best to confirm such findings by the combustion method.

Procedure—II. Combustion Method:[133] A piece of filter paper about 5 × 6 cm. is folded and saturated with the oil to be tested. The paper is placed in a small porcelain evaporating dish which rests in a larger watch glass. The paper is ignited and covered immediately with a 2 liter beaker, the inner surface of which has been previously moistened with water. (The watch glass should be sufficiently large to extend beyond the rim of the beaker.) After the flame has died out, the beaker is permitted to remain in position for 5 min. The porcelain evaporating dish is removed, and the products of combustion, which have condensed on the inner surface of the beaker, are washed into the watch

[130] A No. 16 gage wire has a diameter of 0.065 in.

[131] Too intense a heat is to be avoided since the copper spiral will then fuse and will offer less surface in the subsequent test.

[132] These wires may be used repeatedly. After many determinations, the wire becomes somewhat porous and well coated with oxide. Such wires prove very satisfactory.

[133] *Ber. Schimmel & Co.* April (1890), 29; October (1904), 57.

glass with about 10 cc. of distilled water and then poured into a filter.[134] Add to the filtrate 1 drop of nitric acid and 1 cc. of 0.1 N silver nitrate solution. If the oil is free from halogen, no turbidity should result. Since this method will detect even minute traces of chlorine, it is absolutely necessary to run a blank.[135]

Several other tests have been described, such as the lime test,[136] the test employing sodium peroxide,[137] and the classical test employing molten metallic sodium.[138] The last named test is perhaps the most sensitive, but it suffers from the inherent disadvantages of working with metallic sodium.

A special test for the detection of side chain chlorine in cinnamic aldehyde has been accepted as official in the "National Formulary," Eighth Edition, Monograph on cinnamic aldehyde. It indicates the presence of chlorine only when it appears in the side chain. This is not intended as a general test for the detection of side chain halogen in all compounds. However, it has proven satisfactory for such synthetics as cinnamic aldehyde.

Procedure:[139] To a 1 cc. sample of cinnamic aldehyde add 10 cc. of commercial isopropanol, 1 cc. of nitric acid (1:1) and 1 cc. of 10% silver nitrate solution. Shake the mixture after the addition of each reagent. Heat to incipient boiling and permit the test tube to stand for 5 min. If chlorine is present *in the side chain*, opalescence or turbidity will result. Carry out simultaneously a blank in order to assure absence of chlorine in the reagents.

When recording the presence of halogen always designate the method employed. Also, an estimate of the relative amount of halogen present should be given; use may be made of such relative phrases as "strongly positive," "moderately positive," "slightly positive," "positive: traces," and "negative."

3. TESTS FOR HEAVY METALS

Heavy metals are often present as impurities in essential oils. It is especially important that oils be free from such impurities if they are to be used for medicinal purposes or in foodstuffs. Furthermore, the presence of

[134] The filter is best prepared by thoroughly washing a small filter paper in a glass funnel with distilled water until the water passing through fails to show any turbidity or opalescence when treated with a drop of nitric acid and 1 cc. of the silver nitrate solution.

[135] Some filter papers contain sufficient amounts of chlorine to give a positive test.

[136] Gildemeister and Hoffmann, "Die ätherischen Öle," 3d Ed., Vol. I, 779.

[137] "The United States Pharmacopoeia," Thirteenth Revision, 104.

[138] Directions for the ignition with metallic sodium have been admirably described by Mulliken in his classical work, "A Method for the Identification of Pure Organic Compounds," John Wiley & Sons, Inc., New York Vol. I (1904), 10.

[139] "The National Formulary," Eighth Edition, 153.

heavy metals in perfume oils will often cause discoloration in such products as soaps and cosmetic creams.

A very sensitive test for heavy metals has long been official in "The United States Pharmacopoeia"[140] to insure the absence of lead and copper. The test is based upon the fact that hydrogen sulfide will react with the chlorides of these metals to give dark colored sulfides.

The sulfides of most metals are black or brownish black. The following represent the exceptions: the sulfides of cadmium, arsenic and tin (stannic form) are yellow; of antimony, orange; and of zinc, white. This test is especially satisfactory for the determination of small amounts of copper or lead.

> *Procedure:* Shake 10 cc. of the oil with an equal volume of distilled water to which 1 drop of concentrated hydrochloric acid has been added, and pass hydrogen sulfide through the mixture until it is saturated. No darkening in color in either the oil or the water is produced in the absence of heavy metals. In order to discern any darkening if only traces of heavy metals are present, it is necessary to carry out simultaneously a blank determination to which no hydrogen sulfide is added: a comparison of the blank and the run will clearly indicate traces of heavy metals if present. Test tubes may be conveniently used for these determinations.

Often a scum will form at the surface between the oil and water layers. The formation of the scum is no indication of the presence of heavy metals, unless the scum is dark in color.

Oils manufactured in primitive stills or oils improperly stored in metal containers (especially if the oils are not thoroughly dried) or oils containing large amounts of free acids will often contain heavy metals. Anise, bay, sweet birch, cajuput, clove, geranium, and sassafras oils usually contain heavy metals when distilled commercially. Therefore, it is well to test those oils to ascertain whether or not they have been properly treated to remove such impurities.

A metallic impurity frequently encountered in essential oils is iron. Oils distilled using iron condensers and oils stored in imperfectly lined drums frequently show the presence of this impurity. Oils rich in phenols, or containing a phenol group, such as the salicylates, are often contaminated. Iron will not be precipitated by hydrogen sulfide in an acidic medium and, therefore, will not give a positive heavy metals test. Ammonium sulfide or sodium polysulfide will precipitate black ferrous sulfide.

The test as described above shows a high degree of sensitivity: ten parts per million of metallic lead in oil of cloves gives a positive test; the threshold value is approximately five parts per million.

[140] Thirteenth Revision, 658.

Removal of Heavy Metals.—For the removal of metallic impurities from essential oils, citric or tartaric acid is frequently employed, giving rise to complex citrates and tartrates which are insoluble and which may be filtered off.

> *Procedure:* Add to the oil a small amount of dry tartaric acid (usually ½ to 1 per cent will prove sufficient) and shake thoroughly. Permit the acid to settle and filter the supernatant liquid. If this method should fail to remove all the metallic impurities, agitate the oil with ½ to 1% of a saturated aqueous solution of tartaric acid, separate the oil, shake thoroughly with salt, and filter.

The removal of heavy metals from clove, bay, pimenta, and geranium oils frequently requires several treatments.

When reporting the presence of heavy metals in an oil, it is well to indicate the relative amount found by terms such as: strongly positive; positive; positive—small amounts; positive—traces.

4. TEST FOR DIMETHYL SULFIDE IN PEPPERMINT OILS

Dimethyl sulfide occurs as a normal constituent in peppermint oils. Upon rectification of the oil obtained by steam distillation from the plant, most of this volatile compound is lost because of its low boiling point (boiling point = 37.5°–38°). Hence, the presence of dimethyl sulfide in a peppermint oil is an indication that such an oil has not been rectified. "The United States Pharmacopoeia"[141] has made use of the following procedure to assure the absence of dimethyl sulfide found in nonrectified oils:

> *Procedure:* Distill 1 cc. from 25 cc. of peppermint oil and carefully superimpose the distillate on 5 cc. of a 6.5% mercuric chloride solution in a test tube. A white film does not form at the zone of contact within 1 min.

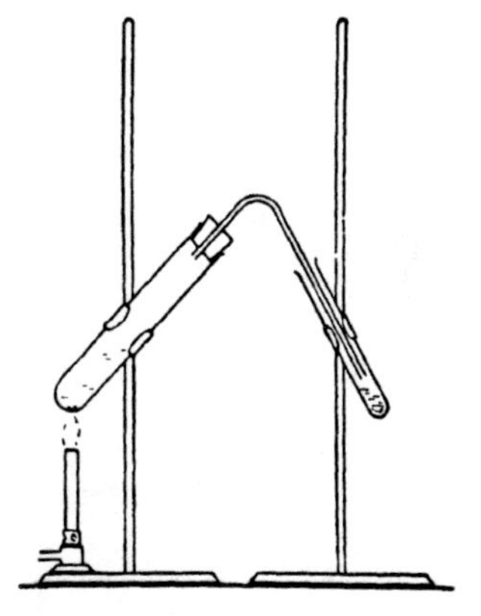

DIAGRAM 4.12. Apparatus for the determination of Dimethyl sulfide.

This test is based upon the reaction of dimethyl sulfide with mercuric chloride, giving a white sulfonium compound which is insoluble in saturated mercuric chloride solution. The following modification[142] of this official test is more sensitive and somewhat more reliable:

[141] Thirteenth Revision, 390 (Peppermint Oil).

[142] This modification has proved satisfactory during the last five years in the laboratories of Fritzsche Brothers, Inc.

Dry the oil by shaking thoroughly with a small amount of anhydrous sodium sulfate in a stoppered bottle and filter. Place 25 cc. of this dried and filtered oil in a large Pyrex test tube (diameter = 22 mm., length = 200 mm.) which is clamped to a ring stand at an angle of approximately 45°. Add a small piece of clay chip. Insert a tight-fitting cork equipped with a bent glass tube which extends 2 cm. through the cork. The other leg of this bent tube is inserted into a second test tube which contains 5 cc. of a 6.5% aqueous solution of mercuric chloride (see Diagram 4.12). The tube should not dip into the solution, but should extend to within 1 cm. of the surface. Apply gentle heat until the oil begins to boil. Heating is continued until the ring of condensing vapor rises to within 1 cm. of the end of the glass tube. If the oil has been carefully dried and the heating has been carried out slowly, no oil will distill over into the second test tube. The formation of a white scum on the surface of the mercuric chloride solution or on the sides of this second test tube indicates the presence of dimethyl sulfide in the oil.

5. TESTS FOR IMPURITIES IN NITROBENZENE

***a.* Test for Thiophene.**—Nitrobenzene which has been manufactured from an impure grade of benzene will give a positive thiophene test, if insufficiently purified. This is due to the fact that inferior grades of benzene contain thiophene, SCH:CHCH:CH, and that all thiophene compounds give an intense blue coloration when mixed with isatin, $C_6H_4NHCOCO$, and concentrated sulfuric acid, because of the formation of indophenin, $(C_{12}H_7NOS)x$.

Procedure:[143] Shake thoroughly 5 cc. of nitrobenzene and 0.5 cc. of concentrated sulfuric acid, in a test tube, and add a pinch of isatin, and again shake the mixture thoroughly. Permit the test tube to stand for 2 hr. No blue coloration should appear during this interval.

***b.* Soap Test.**—The soap test[144] is an empirical method of testing the purity of nitrobenzene. Since a large quantity of this synthetic is used to perfume soaps, it is necessary to carry out a soap test to determine whether or not the nitrobenzene in question will cause a discoloration of the soap.

Procedure: Into a large, wide Pyrex test tube of approximately 75 cc. capacity introduce 5 cc. of the nitrobenzene and 10 cc. of a 15% aqueous solution of potassium hydroxide. Heat the mixture to boiling over an open flame. It is important to

[143] This is a modification of the test described in "A. C. S. Analytical Reagents," American Chemical Society, Washington, D. C., March (1941), 38.

[144] Gildemeister and Hoffmann, "Die ätherischen Öle," 3d Ed., Vol. I, 673.

shake thoroughly the test tube while the mixture is heated and boiled, in order to prevent the formation of two layers. (The nitrobenzene and potassium hydroxide solution will then be thrown out of the test tube with explosive violence.) After boiling for 2 min., permit the test tube to stand, at room temperature, for one-half hour, and then filter the mixture through filter paper previously wetted with water. The potassium hydroxide solution passes through the wetted paper; the nitrobenzene is retained. The alkaline filtrate should be colorless, or at most show only a light yellow color. A full deep yellow indicates that the nitrobenzene has been insufficiently purified or is old. Such a product will require rectification before it is satisfactory for use in soaps.

6. TEST FOR PHELLANDRENE

Phellandrene readily yields a solid nitrite which occurs as a voluminous, flocculent precipitate in the following test.

Procedure:[145] Into a test tube introduce a solution of 5 g. of sodium nitrite in 8 cc. of water. Superimpose a solution of 5 cc. of the oil in 10 cc. of petroleum ether. Add slowly 5 cc. of glacial acetic acid, shaking the tube gently with a rotatory motion. A flocculent precipitate at the junction of the two layers indicates the presence of phellandrene.

If large amounts of phellandrene are present, the petroleum ether layer will solidify to a gel-like mass.

The crystals may be separated with a Büchner funnel, and purified by filtering, washing with water and methyl alcohol, and finally by dissolving the crystals in chloroform and then precipitating with methyl alcohol. Since there are eight possible isomers, the melting point has little meaning unless the physical isomers are separated.

7. TEST FOR FURFURAL

To detect the presence of furfural in an oil, the following procedure proves satisfactory. It is based on the water solubility of furfural and on the well-known color reaction of furfural with aniline in the presence of glacial acetic acid.

Procedure: Shake thoroughly 100 cc. of the oil with 25 cc. of distilled water in a separatory funnel. Permit the mixture to stand until a good separation is obtained. Filter the aqueous layer through wetted paper to give a clear solution. Add 1 cc. of the filtered aqueous layer to 5 cc. of a 2% solution of *freshly distilled* aniline in glacial acetic acid. The presence of small amounts of furfural will result in an intense deep red color within 5 min. If a negative test results, extract the filtered aqueous

[145] Wallach and Gildemeister, *Liebigs Ann.* **246** (1888), 282.

layer with 25 cc. of ether. Cautiously evaporate the ether on a steam bath. Add 1 cc. of distilled water and then 5 cc. of the acetic acid aniline solution. The appearance of an intense deep red color within 5 min. indicates the presence of traces of furfural.

If a red color is obtained, the furfural may be separated from a fresh sample of the oil (following the above procedure), and identified by the formation of a suitable derivative.

If only a small sample of the oil is available, the test may be carried out directly on the oil itself. Garratt[146] has suggested the following method for the determination of furfural.

Procedure: To 0.1 cc. of the oil in a test tube add from a burette 5 cc. of a 2% solution of freshly distilled aniline in glacial acetic acid. Protect from bright light, and allow to stand for 10 min. Examine in a Lovibond tintometer, and measure the "red value."

According to this authority, the test may have value for the detection of adulteration, if the adulterant has a relatively much higher furfural content than the oil to which it has been added, and if the adulterant has not been treated to remove the furfural. Garratt tentatively suggests its use for the detection of light camphor oil in rosemary oils; of clove oil in bay or pimenta berry oils; of Japanese mint oil in American peppermint oil.[147] It may also have value for the detection of added synthetic methyl salicylate to wintergreen and sweet birch oils.[148]

The "ten minute red values" obtained by Garratt[149] for reputedly genuine samples of these oils are given in Table 4.15.

TABLE 4.15. LOVIBOND TINTOMETER VALUES FOR FURFURAL-CONTAINING OILS

Oil	*"Ten Minute Red Value"*
Light Camphor	1.8 to 9.2
Rosemary	0.4 to 0.8
Clove	23.0
Pimenta Berry	1.1
Bay	1.4
Japanese Mint	4.5 to 7.4
American Peppermint	ca. 0.7
Methyl Salicylate	0.0

[146] *Analyst* 60 (1935), 369, 371.

[147] This is the basis of the color reaction for the detection of oil of *Mentha arvensis* (Japanese mint), in oil of *Mentha piperita* L. (peppermint oil); this test is official in "The United States Pharmacopoeia," Thirteenth Rev., 390. It has been the experience of the laboratories of Fritzsche Brothers, Inc., that this test is far from being satisfactory and that the test which was official in "The United States Pharmacopoeia," Eleventh Revision, is to be preferred. (See also "Detection of *Mentha arvensis* Oil," p. 343.)

[148] LaWall (*Am. J. Pharm.* 92 (1920), 891) reports the presence of furfural in wintergreen and sweet birch oils, and its absence in synthetic methyl salicylate.

[149] *Analyst* 60 (1935), 595.

The reader is referred to the original literature for further details; a list of "ten minute red values" is given for more than 50 different oils by Garratt.[150]

8. TEST FOR PHENOL IN METHYL SALICYLATE

Methyl salicylate which has been insufficiently purified will frequently contain phenol. The presence of this impurity affects markedly the odor and flavor of the synthetic. Hence, it is well to test all samples of methyl salicylate for phenol. For routine analyses the following simple procedure has proved quite satisfactory:

> *Procedure I:* Dissolve 5 cc. of the oil in 50 cc. of a 1 N aqueous potassium hydroxide solution. Heat on a steam bath for 2 hr., cool to room temperature, and acidify with sulfuric acid (1:3). Cautiously smell the flask for the distinct characteristic odor of phenol. If no such odor is apparent the sample may be considered free of objectionable amounts of phenol. If a phenolic "by-note" is observed, the presence of phenol should be confirmed by the method of Dodge. (See "Procedure II.")

The Dodge method[151] for the detection of phenol in methyl salicylate has proved of value as a qualitative method. Attempts have been made to convert this method to a quantitative procedure. However, these have proved unsatisfactory, particularly when applied to the natural oils of sweet birch and wintergreen.

> *Procedure II:* Into a 100 cc. Pyrex saponification flask introduce 10 cc. of the oil in question, and add 35 cc. of a 10% aqueous solution of sodium hydroxide, measured from a graduated cylinder. Connect an air-cooled reflux condenser and heat the flask on a steam bath for 2½ hr. Remove the flask and allow it to cool for 15 min. Neutralize the saponified mixture with dilute hydrochloric acid (1:3) until the solution is distinctly acid to blue litmus paper; this requires from 3.5 to 5 cc. of acid. The hydrochloric acid should be added slowly from a burette so that no precipitation occurs. Then slowly add from a burette enough of a saturated, *freshly prepared* sodium bicarbonate solution to just neutralize the mixture, and then an additional 0.5 cc. of the sodium bicarbonate solution. Filter into a 500 cc. distillation flask and distill with steam, using an efficient trap to prevent the mechanical carrying over of any of the solution. A 500 cc. Erlenmeyer flask may conveniently be used for this trap; the delivery tube from the side arm of the distillation flask should extend to within ½ in. of the bottom of the Erlenmeyer flask, and the delivery tube from the trap to the con-

[150] *Ibid.*
[151] *Drug Markets* **24** (1928), 509.

denser should not extend more than 1 in. below the rubber stopper into the flask[152] (see Diagram 4.13). Collect three 5 cc. portions of distillate and filter each distillate. Test for the presence of phenol by the addition of enough bromine water to give a permanent light brown color. If phenol is present to the extent of 0.01% or more, a crystalline precipitate of tribromophenol (melting point, 95°) will form within an hour.

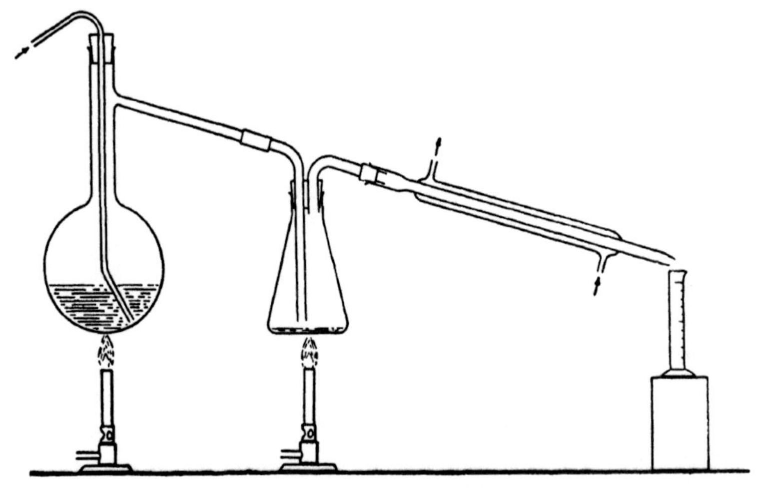

DIAGRAM 4.13. Apparatus for the detection of Phenol in Methyl Salicylate.

Procedure II is based upon the well-known fact that acids will react to form the corresponding sodium salts, but phenols will not form the corresponding phenolates when treated with sodium bicarbonate solutions. Thus the free phenol is steam distilled out of the solution in which is dissolved the nonvolatile sodium salicylate. Normal, pure wintergreen and sweet birch oils do not give a positive reaction with this procedure; certain constituents of the oils will distill which are capable of decolorizing the bromine water, but which do not form *crystalline* derivatives under the conditions outlined. Hence, a positive test for such oils indicates adulteration with phenol-containing synthetic methyl salicylate. *The test is to be considered positive only when definite crystal formation is observed within 1 hr. at room temperature in the bromine treated distillates.*

9. DETERMINATION OF ESSENTIAL OIL CONTENT OF PLANT MATERIAL AND OLEORESINS

A laboratory distillation of essential oil from plant material is often necessary in order to evaluate raw material to be used for large-scale com-

[152] The use of an efficient trap is mandatory to prevent any sodium salicylate from being carried over mechanically into the distillate.

mercial distillations. The determination of the essential oil content is also important in appraising the quality of spices and oleoresins.

Such determinations may be conveniently carried out in a special apparatus devised by Clevenger.[153] This apparatus offers the following advantages: compactness, cohobation of distillation waters, the actual distillation and separation of the essential oil (so that certain chemical and physical properties may be determined, and so that the odor and flavor of the oil may be studied) and an accurate determination of the essential oil content using only small quantities of plant material. Furthermore, this apparatus may be used to advantage for steam rectification of small amounts of essential oils.

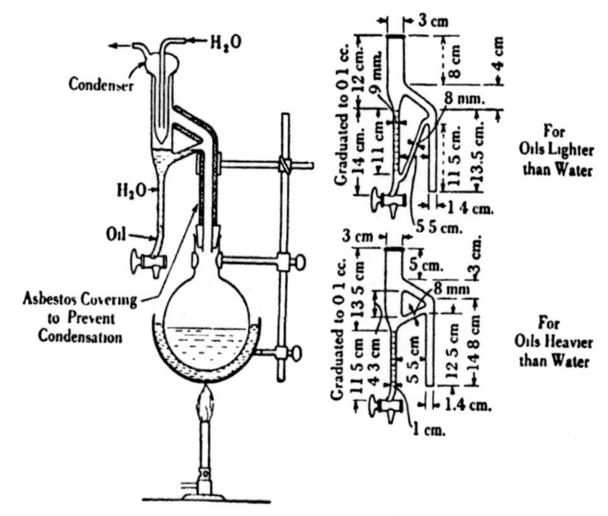

DIAGRAM 4.14. Apparatus for the determination of the volatile oil content of plant materials.

The apparatus consists of two specially designed oil traps and a small condenser of the "cold-finger" type[154] (see Diagram 4.14). Two traps are supplied; one, for oils lighter than water; the other, for oils heavier than water. The diagrams are self-explanatory.

Procedure: Place a sufficient quantity of the ground or chipped material[155] to yield 2 to 6 cc. of oil (preferably 4 cc.) in

[153] *Am. Perfumer* **23** (1928), 467.

[154] An Allihn type condenser having at least four bulbs is often substituted to prevent loss of oil and water vapor. However, if a moderate rate of distillation is maintained, the "cold-finger" condenser proves satisfactory, since the thin layer of condensed water forms an effective seal.

[155] It is very important that the material be ground or chipped into small pieces, especially in the case of woods, roots, and berries. The yield of oil is greatly increased, the time required for distillation is materially reduced, and the quality of the resulting oil is

a round bottom, short-necked flask of 2 liter capacity. Add sufficient water to the flask to correspond to 3–6 times the weight of the plant material; in general, 4 times the weight is sufficient. Attach the proper essential oil trap and the condenser to the flask, and add enough water to fill the trap. Place the flask in an oil bath, heated electrically or by a Bunsen burner to approximately 130°. Adjust the temperature of the bath so that a condensate of about 1 drop per sec. is obtained.

Continue the distillation until no further increase of oil is observed. Usually 5 to 6 hr. are sufficient, although in the case of the distillation of certain woods and roots a much longer period may be necessary. When the distillation has been completed, permit the oil to stand undisturbed so that a good separation is obtained, and so that the oil may cool to room temperature. Determine the number of cc. of oil obtained, and express the yield as a volume/weight percentage; i.e., number of cc. of oil per 100 g. of plant material.

In the event that a good separation is not obtained, the oil and water may be withdrawn from the trap into a graduated cylinder; the addition of sufficient salt to saturate the aqueous layer often aids in obtaining a sharp separation. Periodic withdrawal of the oil and water into such a cylinder is sometimes necessary when the oil being distilled has a specific gravity close to that of water, or when the oil consists of two main fractions—one lighter than water, and the other heavier than water.

The volume/weight may be converted into a weight/weight relationship by means of the following formula:

$$P = pD$$

where: P = wt./wt. percentage;
p = vol./wt. percentage at temperature $t°$;
D = density of oil at temperature $t°$.

This necessitates the determination of the specific gravity of the separated oil.

It is advisable to permit the separated oil to remain overnight in an uncorked bottle before evaluating the odor and flavor. Freshly distilled oils often have a peculiar weedy note, which soon disappears. The yield and the physicochemical properties of the resulting oil will agree closely with the results of a commercial distillation. However, the oil from a commercial or pilot still is generally superior to that obtained in the Clevenger apparatus in respect to odor and flavor.

improved. The distillation should be carried out immediately after the material has been ground to prevent loss of oil by evaporation. The sample to be examined should be representative of the lot or shipment in question.

When determining the essential oil content of oleoresins, it is best to bring the water to incipient boiling before adding the oleoresin[156] and to distill at a faster rate. The addition of clay chips and boiling tubes will prevent undue bumping.

10. DETERMINATION OF ETHYL ALCOHOL CONTENT OF TINCTURES AND ESSENCES

The determination of the content of ethyl alcohol in essences, tinctures and alcoholic extracts is frequently necessary. Because of the presence of volatile esters and essential oils, a determination by distillation alone is usually impossible. Consequently these interfering substances must be removed by washing with heptane or petroleum ether before such distillation is attempted. The two procedures given below give satisfactory results:

> *Procedure—Method I:* Pipette 25 cc. of the sample into a 500 cc. separatory funnel, noting the temperature. Add 100 cc. of a saturated salt solution and 100 cc. of petroleum ether. Shake thoroughly for 2 to 3 min., and permit the mixture to stand undisturbed until a good separation is obtained (usually within 5 to 60 min.).[157]
>
> Draw off the salt solution into a 1 liter distilling flask. Wash the petroleum ether layer with two successive 35 cc. portions of saturated salt solution, adding both washings to the solution in the distilling flask. Discard the petroleum ether layer. Add 100 cc. water to the contents of the distilling flask; also a small amount of solid phenolphthalein and enough 10% aqueous sodium hydroxide solution to make the contents alkaline to the indicator. Also add a few small clay chips and slowly distill until a distillate of about 70 cc. has been collected in a 100 cc. volumetric flask immersed in a beaker of cold water (use a straight tube water-cooled condenser). Add enough distilled water to make the volume up to about 90 cc. If the distillate remains water-white (or at most has a faint opalescence —not a turbidity), adjust the temperature to that originally observed and make up to 100 cc.[158]

[156] The weight of sample can be determined conveniently by placing approximately the amount of oleoresin required in a graduate and weighing before and after the sample has been introduced into the flask.

[157] If a sharp separation is not obtained pipette a 25 cc. sample into a 1 liter distilling flask, add 500 cc. of water and distill. Collect 100 cc. of distillate in a 250 cc. separatory funnel, saturate with salt, add 100 cc. petroleum ether, and shake thoroughly. After the layers have separated continue as directed in the second paragraph of the procedure.

The heptane distillation method may also be used (see "Procedure, Method II").

[158] If a turbid solution is obtained, transfer the distillate completely from the volumetric flask with the aid of distilled water into a 250 cc. separatory funnel. Add 50 cc. of petroleum ether and shake thoroughly. (It may be necessary to add enough salt to saturate the water-alcohol mixture in order to get a sharp separation.) When a sharp separation is obtained (this usually takes from 15 to 60 min.), draw off the lower layer into a 1 liter distilling flask. Wash the petroleum ether layer successively with a 50 cc. and a 25 cc. portion of saturated

Determine the specific gravity accurately and, from this, the alcohol percentage by volume (see Table 4.16). Multiply by 4 to obtain the alcohol content of the original material. If this value is above 25%,[159] determine the refractive index at 20° and compare with the value given in Table 4.17. The calculated index should not differ by more than 0.0002 from the experimentally determined value. A larger difference indicates the presence of some interfering substance in the alcoholic distillate. The determination should then be repeated, using the double distillation procedure previously described in footnote 158.

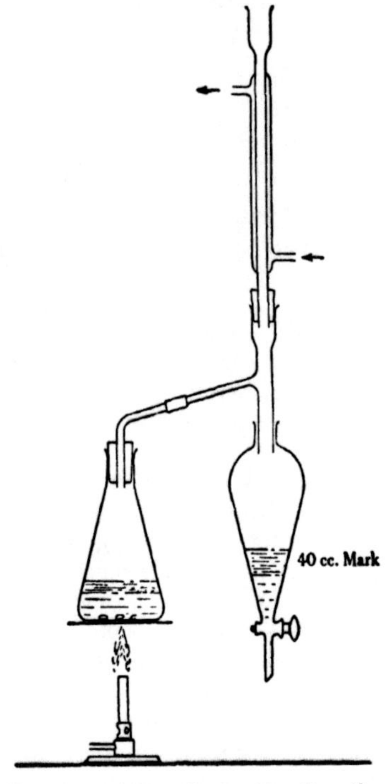

DIAGRAM 4.15. Apparatus for the determination of alcohol.

Procedure—Method II.—Into a 500 cc. Erlenmeyer flask pipette 25 cc. of the sample and add 50 cc. of distilled water and 25 cc. of *n*-heptane. To the 250 cc. separatory funnel add 40 cc. of *n*-heptane and connect the distilling tube and reflux condenser as shown in Diagram 4.15. Heat gently and distill slowly until 40 cc. of distillate have been collected under the heptane layer in the separatory funnel. Permit the contents of the funnel to stand undisturbed for 15 min. to attain room temperature and drain the distillate into a 50 cc. volumetric flask. Wash the residual heptane with two 4 cc. portions of distilled water, adding these washings to the volumetric flask. Fill the flask to the mark and determine the specific gravity of the mixture and calculate the alcoholic percentage by means of Table

salt solution and add these washings to the distilling flask. Add 100 cc. of water, a small amount of solid phenolphthalein, and make alkaline with 10 per cent sodium hydroxide solution. Continue as before, collecting a distillate of about 70 cc. If a turbid solution is again obtained resort to the heptane distillation method (see "Procedure, Method II").

[159] If the alcoholic percentage is less than 25 per cent, pipette 50 cc. of the distillate into a 500 cc. distilling flask, add 100 cc. distilled water, a few clay chips, and distill slowly until a distillate of 20 cc. has been collected in a 25 cc. volumetric flask. Adjust the temperature and make up to 25 cc. Continue as before, but multiply by 2 (instead of 4) to obtain the alcohol percentage in the original material.

TABLE 4.16. ALCOHOLOMETRIC TABLE*

Per Cent of C_2H_5OH by Volume, at 15.56°	Corresponding Per Cent of C_2H_5OH by Weight	Specific Gravity in Air at $\frac{25°}{25°}$	Specific Gravity in Air at $\frac{15.56°}{15.56°}$
0	0.00	1.0000	1.0000
1	0.80	0.9985	0.9985
2	1.59	0.9970	0.9970
3	2.39	0.9956	0.9956
4	3.19	0.9941	0.9942
5	4.00	0.9927	0.9928
6	4.80	0.9914	0.9915
7	5.61	0.9901	0.9902
8	6.42	0.9888	0.9890
9	7.23	0.9875	0.9878
10	8.05	0.9862	0.9866
11	8.86	0.9850	0.9854
12	9.68	0.9838	0.9843
13	10.50	0.9826	0.9832
14	11.32	0.9814	0.9821
15	12.14	0.9802	0.9810
16	12.96	0.9790	0.9800
17	13.79	0.9778	0.9789
18	14.61	0.9767	0.9779
19	15.44	0.9756	0.9769
20	16.27	0.9744	0.9759
21	17.10	0.9733	0.9749
22	17.93	0.9721	0.9739
23	18.77	0.9710	0.9729
24	19.60	0.9698	0.9719
25	20.44	0.9685	0.9708
26	21.29	0.9673	0.9697
27	22.13	0.9661	0.9687
28	22.97	0.9648	0.9676
29	23.82	0.9635	0.9664
30	24.67	0.9622	0.9653
31	25.52	0.9609	0.9641
32	26.38	0.9595	0.9629
33	27.24	0.9581	0.9617
34	28.10	0.9567	0.9604
35	28.97	0.9552	0.9590
36	29.84	0.9537	0.9576
37	30.72	0.9521	0.9562
38	31.60	0.9506	0.9548
39	32.48	0.9489	0.9533
40	33.36	0.9473	0.9517

TABLE 4.16.—*Continued*

Per Cent of C_2H_5OH by Volume, at 15.56°	Corresponding Per Cent of C_2H_5OH by Weight	Specific Gravity in Air at $\frac{25°}{25°}$	Specific Gravity in Air at $\frac{15.56°}{15.56°}$
41	34.25	0.9456	0.9501
42	35.15	0.9439	0.9485
43	36.05	0.9421	0.9469
44	36.96	0.9403	0.9452
45	37.87	0.9385	0.9434
46	38.78	0.9366	0.9417
47	39.70	0.9348	0.9399
48	40.62	0.9328	0.9380
49	41.55	0.9309	0.9361
50	42.49	0.9289	0.9342
51	43.43	0.9269	0.9322
52	44.37	0.9248	0.9302
53	45.33	0.9228	0.9282
54	46.28	0.9207	0.9262
55	47.25	0.9185	0.9241
56	48.21	0.9164	0.9220
57	49.19	0.9142	0.9199
58	50.17	0.9120	0.9177
59	51.15	0.9098	0.9155
60	52.15	0.9076	0.9133
61	53.15	0.9053	0.9111
62	54.15	0.9030	0.9088
63	55.17	0.9006	0.9065
64	56.18	0.8983	0.9042
65	57.21	0.8959	0.9019
66	58.24	0.8936	0.8995
67	59.28	0.8911	0.8972
68	60.33	0.8887	0.8948
69	61.38	0.8862	0.8923
70	62.44	0.8837	0.8899
71	63.51	0.8812	0.8874
72	64.59	0.8787	0.8848
73	65.67	0.8761	0.8823
74	66.77	0.8735	0.8797
75	67.87	0.8709	0.8771
76	68.98	0.8682	0.8745
77	70.10	0.8655	0.8718
78	71.23	0.8628	0.8691
79	72.38	0.8600	0.8664
80	73.53	0.8572	0.8636

TABLE 4.16.—*Continued*

Per Cent of C_2H_5OH by Volume, at 15.56°	Corresponding Per Cent of C_2H_5OH by Weight	Specific Gravity in Air at $\frac{25°}{25°}$	Specific Gravity in Air at $\frac{15.56°}{15.56°}$
81	74.69	0.8544	0.8608
82	75.86	0.8516	0.8580
83	77.04	0.8487	0.8551
84	78.23	0.8458	0.8522
85	79.44	0.8428	0.8493
86	80.66	0.8397	0.8462
87	81.90	0.8367	0.8432
88	83.14	0.8335	0.8401
89	84.41	0.8303	0.8369
90	85.69	0.8271	0.8336
91	86.99	0.8237	0.8303
92	88.31	0.8202	0.8268
93	89.65	0.8167	0.8233
94	91.03	0.8130	0.8196
95	92.42	0.8092	0.8158
96	93.85	0.8053	0.8118
97	95.32	0.8011	0.8077
98	96.82	0.7968	0.8033
99	98.38	0.7921	0.7986
100	100.00	0.7871	0.7936

4.16. Multiply by 2 in order to obtain the alcohol content of the original material.

It should be noted that certain low boiling, water-soluble constituents, such as acetic acid and acetone, will interfere in such a determination. However, most low boiling esters are readily absorbed by the heptane layer.

11. DETERMINATION OF WATER CONTENT

a. Determination by the Bidwell-Sterling Method.—The most convenient method for the determination of water in essential oils, oleoresins, and drugs is by the water determination apparatus of Bidwell and Sterling.[160] The sample to be tested is distilled in this apparatus with a liquid immiscible with water, such as toluene. The special trap collects and measures the condensed water, the excess solvent overflowing and returning to the still.

Procedure: Connect the apparatus as shown in Diagram 4.16. Introduce into the 500 cc. flask, sufficient material, accurately weighed, to yield from 2 to 4 cc. of water. Add about 200 cc. of toluene to the flask and also fill the receiving trap with

[160] *J. Ind. Eng. Chem.* **17** (1925), 147.

Table 4.17. Refractive Indices* of Ethyl Alcohol-Water Mixtures from 1–25%

Percentage of Alcohol (by Volume) at 15.56°	*Refractive Index at 20°*
1	1.33342
2	1.33391
3	1.33443
4	1.33495
5	1.33549
6	1.33602
7	1.33657
8	1.33711
9	1.33768
10	1.33824
11	1.33882
12	1.33940
13	1.33997
14	1.34057
15	1.34116
16	1.34176
17	1.34236
18	1.34297
19	1.34359
20	1.34420
21	1.34482
22	1.34544
23	1.34606
24	1.34666
25	1.34726

* These values are based on the Reference Tables of "Methods of Analysis of the A.O.A.C.," 4th Ed. (1935), 663–670.

toluene, poured through the top of the condenser. Heat the flask gently by means of a Bunsen burner or electric hot-plate until the toluene begins to boil. Distill at a rate of about 2 drops per sec. until most of the water has passed over. Then increase the rate of distillation to about 4 drops per sec. When no further increase in collected water is observed, continue the distillation for an additional 15 min. Permit the apparatus to cool. When the water and toluene have separated completely, read the volume of water, and calculate the percentage present in the substance.

If the condenser and moisture trap have been thoroughly cleaned with chromic acid cleaning solution, the tendency of droplets to adhere is greatly minimized. Should such droplets of water be observed on the sides of the condenser, they may be forced down by brushing the inner tube of the condenser with a small brush previously saturated with toluene.

A convenient method for detecting the presence of dissolved water in essential oils, such as rose and bay, has been described under "Solubility," see p. 252.

b. Determination by Karl Fischer Method.—For the determination of mere *traces* of water, the method employing the Karl Fischer water titration reagent will prove exceptionally sensitive.

The Karl Fischer water titration reagent[161] is a solution of iodine, sulfur dioxide, and pyridine in methyl alcohol. The method depends on the oxidation of sulfur dioxide by iodine in the presence of water to form sulfuric and hydriodic acid.

$$SO_2 + 2I + 2H_2O \rightarrow H_2SO_4 + 2HI$$

The reaction is conducted in the presence of pyridine which acts as an acid acceptor, thus enabling the reaction to go to completion. The end point is indicated by a color change from yellow to reddish-brown, the latter being produced by the free iodine in the reagent when an excess of the reagent is added.

The method is applicable to a large number of organic and inorganic compounds, both liquid and solid. The exact limitations of the method have not been determined, but it can probably be used on all organic and inorganic compounds that do not react with the reagent and that are not naturally colored red or brown. It is known to be applicable to organic compounds such as hydrocarbons, alcohols, esters, carboxylic acids (except formic), halogen derivatives of hydrocarbons, phenols, nitro compounds, amines, and heterocyclic compounds. It is not applicable to aldehydes and ketones, nor to reducing compounds which react readily with iodine in the cold. The active hydrogen in primary and secondary amines must be blocked by solution in glacial acetic acid before titrating with the reagent.

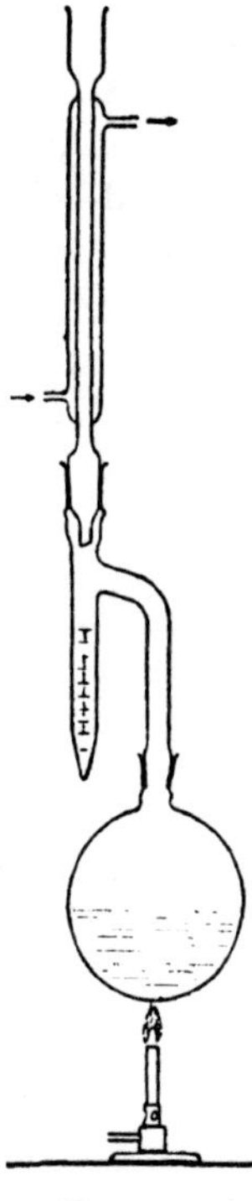

DIAGRAM 4.16. Apparatus for the determination of water.

Liquids are dissolved in a mutual solvent for both the sample and the reagent before titrating. Solids may be analyzed by pulverizing and dissolving or suspending in dry methyl alcohol. It is not essential that the material be soluble in methyl alcohol, as the hygroscopic nature of both methyl alcohol and the reagent will act to extract the water from the sample.

The solvent used in preparing the sample for analysis will contain some moisture, hence a blank titration must be made using the same volume of solvent and the *same size flask*, as the moisture in the air space is an integral part of the blank. To check the end point, breathe into the flask and the

[161] *Angew. Chem.* **48** (1935), 394.

end point will disappear, but an additional drop or two of the reagent should bring back the reddish-brown color. The choice of solvents is wide: methyl alcohol, dioxane, glacial acetic acid, chloroform, etc.

When attempting new applications of the method, i.e., with new or unknown compounds, the reactivity of the compound with the Fischer reagent must first be determined. If the compound is inert toward the reagent, the method is applicable. Also, the reagent is so avidly hygroscopic that it will dehydrate hydrated compounds. The degree of such dehydration (number of mols of water reacting with the reagent) must be determined beforehand.

All apparatus must be thoroughly dried and every precaution must be made to exclude atmospheric moisture during the titration. The titration is carried out in a small flask (125 cc. Erlenmeyer) and taken to completion rapidly. This method will detect, in general, 0.0005 g. of water, equivalent to 0.005 per cent when using a 10 g. sample.

Procedure: Pipette 10 cc. of methyl alcohol into each of three 125 cc. glass stoppered Erlenmeyer flasks, which should be kept stoppered as much as possible. Weigh accurately from a weighing pipette about 0.1 g. of distilled water into each flask. Titrate with the Karl Fischer reagent to the color change (the color should change from a straw yellow to a reddish-brown when the end point is reached). At the same time run a blank on the methyl alcohol. Calculate the water equivalent of the reagent by means of the following formula:

$$E = \frac{w}{A - B}$$

where: E = water equivalent of the reagent (in grams of water per cc.);
w = weight in grams of water used;
A = cc. of reagent used for the determination;
B = cc. of reagent used for the blank.

Into a 125 cc. Erlenmeyer flask weigh a sufficiently large sample of the material to be tested to yield approximately 0.1 g. of water. Add 10 cc. of methyl alcohol and titrate. Run a blank at the same time on the alcohol. The water content may be calculated from the following formula:

$$\text{Percentage of water} = \frac{100(A - B)E}{w}$$

where: A = cc. of reagent used for the determination;
B = cc. of reagent used for the blank;
E = water equivalent of the reagent;
w = weight of sample in grams.

It is necessary to standardize the reagent daily.

The Karl Fischer water titration reagent may be purchased from chemical supply houses, or may be prepared in the following way:

> Place 1 liter of dry methyl alcohol and 400 cc. of pyridine in a 2 liter reservoir of an automatic burette. Add 127 g. of iodine, stopper the bottle and swirl until the iodine is completely dissolved. Cool the bottle in a salt-ice mixture for one-half hour and then add 100 g. of sulfur dioxide,[162] weighing by difference on a balance. The resulting solution is very hygroscopic and must be kept stoppered as much as possible. Then remove the bottle from the ice bath and insert the siphon and burette unit. Thoroughly grease the ground glass joint between the bottle and burette to give an airtight seal. Fit a calcium chloride drying tube to the opening at the top of the burette and between the bottle and hand aspirator, which is used to fill the burette. The tip of the burette is fitted with a 2-hole rubber stopper which fits the neck of the 125 cc. Erlenmeyer flask. Protect the tip of the burette when not in use.

It is best to age the solution for two to four days before using so that the variation in standardizing from day to day will be minimized.

12. DETERMINATION OF STEAROPTENE CONTENT OF ROSE OILS

Oil of rose contains as a natural constituent a mixture of solid paraffinic hydrocarbons known collectively as "stearoptene." The highly purified stearoptene is odorless and hence contributes little to the odor value of the oil. However, for many years the quality of rose oils was judged superficially by the "melting point" of the oil; oils with high "melting points" were assumed to be unadulterated. As a consequence there arose the practice of adding spermaceti, tristearin, high melting paraffins, and guaiac wood oil as adulterants.

"The United States Pharmacopoeia"[163] requires a certain minimum content of stearoptene and describes a limiting test for its determination.

> *Procedure I:* Introduce 1 cc. of oil of rose into a 25 cc. glass-stoppered, graduated cylinder and add 1 cc. of chloroform: a clear solution should result. Then add 19 cc. of 90% alcohol (by volume): crystals of stearoptene should crystallize out of the solution within 24 hr., the temperature being maintained at 25°.

A rough indication of the amount of stearoptene present in the oil can be obtained by this modified official test. Oils with high stearoptene contents will deposit an abundant amount of crystalline material immediately; oils with low stearoptene contents will sometimes separate only one or two

[162] Dry the sulfur dioxide by bubbling through concentrated sulfuric acid. Commercial sulfur dioxide of refrigeration grade is sufficiently pure for the preparation of the reagent.
[163] Thirteenth Revision, 456.

well-formed crystals after standing 24 hr.; some oils will show no separation of crystals whatsoever. The appearance of the crystals is also important; only through experience will an essential oil chemist be able to draw conclusions as to possible adulteration from the appearance of the separated material.

This test will also indicate whether or not the oil has been properly dried; a cloudy solution in one volume of chloroform is usually indicative of the presence of water in the oil.

For the determination of the amount of stearoptene, the oil is usually dissolved in dilute alcohol and chilled; the relatively insoluble paraffins separate out and can be filtered off and weighed. It is customary to use 75 per cent alcohol[164] for this determination, although certain investigators have recommended the use of 85 per cent alcohol[165] or acetone.[166]

Procedure II: Dissolve 5 g. of the oil in 50 cc. of 75% alcohol (by volume) with the aid of gentle heat if necessary. Cool the solution in an ice bath at 0° for 2 hr. and filter off the separated stearoptene with suction, using a well cooled Büchner funnel. Wash the stearoptene with a 50 cc. portion of 75% alcohol cooled to 5°. Remove as much of the alcohol as possible by suction, and then transfer the cake of stearoptene to a tared evaporating dish. Break up the cake with a spatula and dry in a desiccator for 24 hr. Weigh, and calculate the percentage of stearoptene present in the original oil.

To be assured of the absence of adulterants, it is necessary to examine the separated stearoptene.

The naturally occurring paraffinic hydrocarbons in rose stearoptene consist of at least two components[167] having melting points of 22° and 41°.[168] The mixture separated from rose oil should melt between 32° and 37°; usually at about 33–34°.[169] Additions of spermaceti, guaiac wood oil, and many readily available solid paraffins will raise the melting point.

Spermaceti, tristearin, or other fatty acid esters may be detected by an abnormally high ester number of the separated stearoptene. Occasionally it is possible to isolate the fatty acids from the saponified material.

Guaiac wood oil consists mainly of the alcohol, guaiol; its presence will be revealed by a high ester number after acetylation of the separated stearoptene.

[164] *Ber. Schimmel & Co.*, April (1889), 37.

[165] Burgess, "Chemistry of Essential Oils and Artificial Perfumes" (Parry), D. Van Nostrand Co., Inc., New York, 1921, 402.

[166] Jeancard and Satie, *Bull. soc. chim.* [3] 31, (1904), 934.

[167] Flückiger, "Pharmakognosie," 3d Ed., 170.

[168] Gildemeister and Hoffmann, "Die ätherischen Öle," 3d Ed., Vol. I, 302.

[169] Parry, "Chemistry of Essential Oils and Artificial Perfumes," D. Van Nostrand Co., New York (1921), 402.

High melting paraffins are very difficult to detect when used as adulterants for rose oils. The appearance of the stearoptene may reveal their presence; a peculiar granular structure is frequently indicative of such additions. The appearance of the crystals which separate in the test described under Procedure I is sometimes helpful in this connection.

The congealing point of the rose oil itself is also indicative of the amount of stearoptene present in the oil. The congealing point of rose oil[170] has been defined as that temperature at which the first crystals appear when the oil is subjected to slow cooling. (This is quite different from the true congealing point of oils such as anise. See "Congealing Point," p. 253.) Determine the "congealing point" of the oil by the following technique:

> *Procedure III:* Place 10 cc. of the oil in a test tube having a diameter of 15 mm.; suspend a thermometer in the oil in such a way that it touches neither the sides nor the bottom; warm the contents of the tube to about 5° above the point of saturation; stir well; then permit the oil to cool slowly until the first crystals appear; read the temperature. Repeat the determination.

As a general rule, good Bulgarian oils produced by the usual methods[171] show a congealing point of 18° to 23°.[172]

13. DETERMINATION OF SAFROLE CONTENT OF SASSAFRAS OILS

The congealing point of sassafras oils gives a good estimate of the safrole content.

> *Procedure:* Determine the congealing point of the sassafras oil (see p. 253 for details), and estimate the safrole content from Table 4.18.

TABLE 4.18. CONGEALING POINT AND SAFROLE CONTENT

Per Cent Safrole	*Congealing Point*
100	11.0°
90	7.5
80	4.6
70	1.7
60	−1.3

Table 4.18[173] will give values of the safrole content with an accuracy of about 2 per cent if the congealing point is above 2°.

[170] Raikow, *Chem. Ztg.* 22 (1898), 149.

[171] Pure Bulgarian oils of exceptionally fine quality produced by rotary distillation showed congealing points as low as +13°. (See also Guenther and Garnier, *Am. Perfumer* 25 (June–December 1930).)

[172] Gildemeister and Hoffmann, "Die ätherischen Öle," 3d Ed., Vol. II, 837.

[173] This table was prepared by the laboratories of Fritzsche Brothers, Inc.; it is based on the congealing points of known mixtures of safrole and pinene and safrole and eugenol.

14. DETERMINATION OF CEDROL CONTENT OF CEDARWOOD OILS

For the determination of cedrol in cedarwood oils, Rabak[174] has suggested the following method:

> One hundred parts of oil are agitated vigorously with 6 parts of 65% alcohol (by volume) for one to two minutes in a widemouthed, stoppered flask. Sudden and complete solidifications of the emulsion thus formed usually result if the oil contains a sufficient quantity of cedrol. If it fails to solidify, add a small quantity of crystalline cedrol to the emulsion, and cool in a refrigerator for several hours. Filter the solidified mass with the aid of a well cooled Büchner funnel and wash the fine silky crystals with a few drops of cold 98% alcohol. Weigh the dry crystals. The cedrol may be purified by dissolving it in hot alcohol, then cooling and filtering the mass.

In general, an analytical method based upon the actual separation of a constituent by physical means will not give completely accurate results. However, *comparative data* may be obtained provided all experimental conditions are carefully controlled.

This method appears to be of value only for obtaining comparative data when two oils are examined simultaneously; all experimental conditions should be maintained as identical as possible.

15. DETERMINATION OF THE COLOR VALUE OF OLEORESIN CAPSICUM

In order to standardize the color of oleoresin capsicum it has been found that a very close match to the natural color can be attained with the proper mixture of solutions of potassium dichromate and cobalt chloride. The color standard is prepared as follows:

> Into a 50 cc. Nessler tube pipette 5 cc. of a 0.1 N potassium dichromate solution[175] (4.904 g. $K_2Cr_2O_7$ per liter) and 0.5 cc. of a 0.5 N cobaltous chloride solution (5.948 g. $CoCl_2 \cdot 6H_2O$ per 100 cc.) and make up to 50 cc. with distilled water.

The color value of the oleoresin is defined as the number of cc.'s of acetone, multiplied by 100, which are necessary to add to 1 cc. of a 1 per cent solution of the oleoresin capsicum in acetone, in order to match the color standard as outlined above. The height of the liquid in the Nessler tube should be about 8 in. and the color should be matched by looking down into the column, and *not laterally*.

> *Procedure:* Weigh accurately 1.00 g. of oleoresin and make up to 100 cc. with acetone. Pipette 1 cc. of this 1% solution

[174] *Am. Perfumer* **23** (1929), **727.**

[175] The potassium dichromate and the cobalt chloride used for these solutions should be of the grade known as "analytical reagent."

into a 50 cc. volumetric flask and make up to 50 cc. with acetone.[176] Pour this dilute solution (0.02%) into a burette. Introduce sufficient of this solution into an empty 50 cc. Nessler tube to approximate the color of the standard (viewed through the length of the tube). Then add sufficient acetone to bring the volume up to about 45–47 cc. and make the final adjustment of color by addition of small amounts of the dilute solution (0.02%) from the burette. Finally add sufficient acetone to bring the volume to exactly 50 cc. and check the color match. Color value

$$= 100 \left[\frac{50 - (0.02)\ (\text{no. of cc. of dilute solution required})}{(0.02)\ (\text{no. of cc. of dilute solution required})} \right]$$

Using this procedure, an accuracy of about ±1,000 units can be obtained. The color values will vary between 5,000 and 25,000 for commercial oleoresins; a value of 14,000 is generally considered very satisfactory.

The procedure may be modified to permit the use of 100 cc. Nessler tubes and the colors of the standard, and the solution of oleoresin may be accurately matched with a Nesslerimeter.

VI. DETECTION OF ADULTERANTS

1. DETECTION OF FOREIGN OILS IN SWEET BIRCH AND WINTERGREEN OILS

For a rapid evaluation of the quality of sweet birch and wintergreen oils, the alkali solubility test often proves of value. (See also "Solubility," p. 252.)

Procedure:[177] Introduce 2 cc. of the oil in a 25 cc. glass-stoppered, graduated cylinder and add 23 cc. of an aqueous solution of potassium hydroxide prepared by dissolving 6.5 g. of potassium hydroxide (analytical grade) in sufficient distilled water to yield 100 cc. of solution. Shake thoroughly and permit the cylinder to stand undisturbed for 24 hr.: no oily separation should result, although a separation of a *solid* waxy material is indicative of a normal oil.

Since the natural waxy separation melts at a relatively low temperature, care should be exercised in interpreting the results of this test in warm weather.

It is well to study the odor of the solution or any insoluble portion. Since the potassium phenolate of methyl salicylate is practically odorless, additions of foreign, odor-bearing substances may be detected.

[176] This procedure is satisfactory for oleoresins with a color value of 4,900 or higher; if the color value is lower, a stronger solution should be used.

[177] This is a slight modification of the test described in "The United States Pharmacopoeia," Tenth Revision, 239.

2. DETECTION OF PETROLEUM AND MINERAL OIL

a. Oleum Test.—The saturated paraffinic hydrocarbons, found in petroleum oils, are chemically very inert; they are not destroyed by fuming sulfuric acid. Other compounds are attacked, giving rise to reaction products which are soluble in sulfuric acid.

> *Procedure:*[178] Place 20 cc. of fuming sulfuric acid in a *dry* cassia flask[179] of 150 cc. capacity, and cool thoroughly in an ice-salt mixture. Add *slowly* 5 cc. of the oil in question from a small burette. The oil should be added drop by drop, with frequent shaking and cooling in the ice-salt mixture, since too rapid addition of the oil is apt to cause the liberated sulfur dioxide to carry part of the acid and oil out of the flask. After the oil has been added, the flask is again shaken and permitted to stand at room temperature for 10 min. It is then warmed on a steam bath for 5 min. with frequent agitation. The flask is permitted to cool to room temperature and is then filled with 95% sulfuric acid. After standing overnight, the mineral oil will rise into the neck and separate as a colorless, or straw-colored liquid. As a confirmatory test, a small amount of the separated mineral oil may be removed from the cassia flask (by means of capillary action, using a glass tube drawn out to a small tip). The refractive index of this separated oil should be less than 1.4400.

A flavor test often will prove of value for the detection of kerosene. In this connection, see the discussion of adulteration of "Orange Oils of French Guinea," Vol. III.

Since petroleum fractions often contain aromatic and unsaturated compounds as well as paraffins, the separation of the paraffinic portion described above does not usually represent the total amount of added petroleum. In general, such actual separation usually is a small percentage of the adulterant.

The test may be rendered more sensitive by preliminary fractionation of the oil.

The addition of petroleum fractions to an oil causes a lowering of the specific gravity, index and optical rotation. The solubility of the oil usually is affected: this is the basis of the well-known Schimmel Test for citronella oils described below.

b. Schimmel Tests.

The "Old Schimmel Test."[180]—In order to limit the amount of adulteration of citronella oils with petroleum fractions, the chemists of Schimmel

[178] This procedure is essentially the Oleum Test of "The National Formulary," Eighth Edition, 543 (Turpentine Oil).

[179] A narrow necked Babcock bottle may be used in place of the cassia flask; this offers the further advantage of permitting the bottle and contents to be centrifuged for better separation.

[180] *Ber. Schimmel & Co.*, October (1889), 22; (1917), 14.

nd Company introduced the well-known Schimmel Test. Several modifi-ations of this test have been proposed, but the trade accepts the following rocedure in writing contracts for oils.

> *Procedure:* Into a glass-stoppered, graduated cylinder introduce exactly 1 cc. of the oil. Add dropwise 80% alcohol until a clear solution results. This should occur at 1 to 2 volumes. Add sufficient 80% alcohol to bring the amount of added alcohol to 10 volumes. The solution may show a slight opalescence, but should not separate oily droplets even after standing for several hours. When adding the alcohol, violent shaking should be avoided to prevent an emulsion that will separate only after very prolonged standing.

A citronella oil meets the Schimmel Test if it yields a clear solution in to 2 volumes of 80 per cent alcohol and *does not separate oily droplets* when he amount of alcohol added is increased to 10 volumes. This test limits he amount of added petroleum fractions to about 10 per cent. If more than his amount has been added, oily droplets will form *on the surface* of the lcoholic solution. Additions of fatty oils will result in the formation of ily droplets which settle to the bottom.

The "New Schimmel Test."[181]—At a later date the description of the riginal test was modified resulting in the so-called "New Schimmel Test." This test is somewhat more stringent than the "Old Schimmel Test" de-cribed above. However, the trade has not accepted the new version. A lescription of this test follows:

> Oil of citronella Ceylon must be clearly soluble in from 1 to 2 volumes of 80% alcohol by volume at 20°. Upon the further addition of alcohol of the same strength, the solution should show an opalescence at the most, but no turbidity or direct cloudiness. The alcohol must be added slowly, drop by drop; the addition being at once interrupted if a cloudiness or turbidity appears. The alcohol is then added slowly, drop by drop, until the point of highest or maximum cloudiness or turbidity is obtained. The mixture is carefully set aside and maintained at 20° to observe if any oily constituents separate out. Ten volumes of 80% alcohol at the most are added. If oil separates out immediately or after prolonged standing, the oil does not pass the "New Schimmel Test." Strong or violent shaking must be avoided since any possible oily separation will become finely dispersed and will not separate out on standing.

Many oils will show an oily separation at the point of highest cloudiness r turbidity, but will show no oily separation if 10 volumes of 80 per cent lcohol are added.

[181] *Ber. Schimmel & Co.* (1923), 18.

The "Raised Schimmel Test."[182]—In order to limit adulteration with mineral spirits to 5 per cent, the "Raised Schimmel Test" was introduced. This test has never attained commercial importance.

> Oil of citronella Ceylon is mixed with 5% of kerosene and the "Old Schimmel Test" is applied, disregarding any intermediate stages of cloudiness or turbidity; i.e., simply add 80% alcohol up to 10 volumes. A fresh unadulterated citronella oil will show no oily separation. Oils containing small amounts of petroleum will show an oily separation either immediately or after prolonged standing at 20°.

This test is by far the most stringent of the three.

3. DETECTION OF ROSIN

In testing for rosin as an adulterant in oils, the following pertinent properties of this substance should be borne in mind. It is a nonvolatile material, and consequently may be concentrated in the residue by distillation of the oil under vacuum or at atmospheric pressure; it is found also in the evaporation residue. Rosin consists primarily of complex acids and, therefore, will increase the acid number of an oil or of the evaporation residue if such residue normally consists of solid esters or paraffins; this is specifically of importance in the case of citrus oils. Rosin is soluble in most organic solvents, including petroleum ether, benzene, and xylene; since cinnamic aldehyde (the main constituent of cassia oil) is practically insoluble in petroleum ether, this permits a convenient separation of added rosin for this oil, and is the basis of "The United States Pharmacopoeia" test described below. Rosin gives a dark green copper salt when treated with cupric acetate; this salt is sufficiently soluble in petroleum ether to impart to this solvent a green color. Rosin is a relatively high melting solid, normally a hard, noncrystalline material which fractures readily; hence the consistency of the evaporation residue is frequently altered if rosin is present.

a. Detection of Rosin in Balsams and Gums.

> *Procedure:*[183] Place in a small mortar 1 g. of the substance, powdered or crushed if necessary, and add 10 cc. of purified petroleum ether. Triturate well for 1 or 2 min. Filter into a test tube and add to the filtrate 10 cc. of a freshly prepared aqueous solution of cupric acetate (1 g. in 200 cc.). Shake well and allow the liquids to separate. The petroleum ether layer should not show a green color.

[182] *Ber. Schimmel & Co.*, April (1904), 29; April (1910), 32; April (1911), 47.
[183] "The United States Pharmacopoeia," Thirteenth Revision, 688.

b. Detection of Rosin in Cassia Oils.

> *Procedure I:*[184] Shake about 2 cc. of the oil in a test tube with 10 cc. of petroleum ether. Permit the liquids to separate and decant the benzene layer into a second test tube. Add an equal volume of cupric acetate solution (1 in 1000); a green color indicates the presence of rosin in the oil.

It is well to carry out simultaneously a test with an oil known to be free of rosin, to act as a blank. Unfortunately, tests based upon color reactions have not proved too reliable in mixtures as complex as essential oils; nevertheless, this test will give an indication of the presence or absence of rosin.

> *Procedure II:*[185] About 50 g. of the oil, accurately weighed, are distilled from a tared distilling flask over an open flame. Continue the distillation until decomposition is evidenced by the formation of white fumes within the flask; this usually occurs at a temperature of about 280°. Cool the flask and weigh; calculate the percentage of residue.

This test will reveal adulteration with nonvolatile material such as rosin, if large amounts have been added. Normal oils show a distillation residue of 6 to 8 per cent, or at most 10 per cent, according to Gildemeister and Hoffmann.[186] Furthermore, the residue should be tacky, but not hard and brittle. According to Allen,[187] formerly of Hongkong, the residue should not be higher than 5 per cent for a pure oil. Treff[188] has pointed out that distillation should be carried out rapidly, since the amount of residue obtained is greatly dependent upon the rate of distillation.

> *Procedure III:* Determine the acid number of the oil in the usual manner. If the oil is pure and has been properly stored, the acid number should not be greater than 15.

c. Detection of Rosin in Orange Oils.

> *Procedure:* Determine the evaporation residue in the usual manner. In the case of pure oils this residue upon cooling should be soft and waxy, not hard, brittle or tacky. The acid number of the residue should lie between 11 and 28, the ester number between 118 and 157.[189]

[184] "The United States Pharmacopoeia," Thirteenth Revision, 132.

[185] *Ber. Schimmel & Co.*, October (1889), 15. Gildemeister and Hoffmann, "Die ätherischen Öle," 3d Ed., Vol. II, 631.

[186] "Die ätherischen Öle," 3d Ed., Vol. II, 631.

[187] D. Allen, private communication.

[188] *Z. angew. Chem.* **39** (1926), 1308.

[189] Gildemeister and Hoffmann, "Die ätherischen Öle," 3d Ed., Vol. III, 79. These data apply to Italian orange oils. However, the values for oils from other origins do not appear to differ materially from these limits.

4. THE DETECTION OF TERPINYL ACETATE

It has been pointed out in the "Determination of Esters" that certain esters are not completely saponified under the standard analytical conditions if the time of reflux is limited to 1 hr. Terpinyl acetate is such an ester.

Additions of esters of this type to readily saponifiable esters (such as linalyl acetate) will be revealed by a difference in the ester numbers obtained by saponification for periods of 1 and 2 hr., respectively. Under standard conditions linalyl acetate is completely saponified in a period of 30 min.; terpinyl acetate requires about 2 hr. Hence, an appreciable difference between the ester numbers determined after heating for 30 min. and for 1 hr. (or 2 hr.) indicates the presence of certain foreign esters, such as terpinyl acetate, in oils containing only readily saponifiable esters (e.g., bergamot oil and lavender oil). If only small amounts of terpinyl acetate have been added, the difference will be too small to draw any definite conclusions. However, by modifying the experimental conditions, such small differences may be greatly magnified. The method outlined below is the classical method developed by the chemists of Schimmel and Company[190] for the detection of terpinyl acetate as an adulterant in bergamot oils; it is also applicable to lavender oils and to synthetic linalyl acetate. With further modification, it can be used for the detection of terpinyl acetate and terpineol in numerous oils; *such applications, however, should be made with discretion.*

> *Procedure:* Pipette 2 cc. of the oil into each of three tared saponification flasks and weigh accurately. To flask I add 10 cc. of 0.5 N alcoholic sodium hydroxide solution and 25 cc. of alcohol. To flask II add 20 cc. of the alkali solution, but no alcohol. To flask III add 10 cc. of the alkali solution and 5 cc. of alcohol. (The alkali solution should be measured accurately from a burette or pipette.) The contents of flask I and flask III are refluxed on a steam bath for a period of 1 hr.; the contents of flask II, for 2 hr. Calculate the ester numbers for the three determinations.

In the case of pure bergamot oils, the difference between ester number I and ester number II will not be greater than 5; the usual value lies below 3. In the case of an oil adulterated with 4 per cent terpinyl acetate, the difference amounts to about 10.0; with 10 per cent terpinyl acetate, about 19.0.[191] Furthermore, in the case of pure oils, ester number III will be approximately the arithmetical mean of ester number I and ester number II.

For oils containing larger amounts of ester, the size of the sample must be reduced; 1 cc. will often prove sufficient. In the case of synthetic linalyl

[190] *Ber. Schimmel & Co.*, October (1911), 115.
[191] *Ber. Schimmel & Co.*, October (1911), 115.

acetate, a 1 cc. sample should be used and the quantities of alkali should be doubled.

Fractional saponification may also be used to detect the presence of terpineol by carrying out the determination on an acetylized oil; great discretion must be used, however, since terpineol and certain difficultly saponifiable esters may be present as natural constituents, or the process of acetylation may result in the formation of such esters. Recourse to fractionation of the oil or of the acetylized oil with subsequent fractional saponification of the proper fraction may frequently prove of value. Table 4.19 gives the boiling points of terpineol and terpinyl acetate at various pressures:

TABLE 4.19

Pressure in Mm. of Hg.	*Boiling Point*	
	*α-Terpineol**	*Terpinyl Acetate*
5	92.4°	90–94°†
10	104.0°	110–115°‡
760	217.5°	220°‡

* von Rechenberg, "Einfache und fraktionierte Destillation in Theorie und Praxis," Schimmel & Co., Miltitz, Leipzig (1923), 257.
† Gildemeister and Hoffmann, "Die ätherischen Öle," 3d Ed., Vol. I, 647.
‡ Bouchardat and Lafont, *Ann. chim. phys.* [6] 9 (1886), 515.

5. DETECTION OF TURPENTINE OIL

The addition of turpentine oil as an adulterant generally reduces the specific gravity and affects the solubility and optical rotation of most essential oils. Its presence may be proved in oils which contain no pinene as a natural constituent by the separation and identification of α-pinene, the main constituent of turpentine oils.

Highly purified *d*-α-pinene has the following properties:

Boiling Point	= 155–156°
Specific Gravity at 15°	= 0.864
Refractive Index at 20°	= 1.4656
Specific Rotation	= + 48°24′
Solubility at 20°	4 vol. of 90% alcohol and more.

The boiling point of α-pinene lies below that of most of the terpenes and oxygenated constituents found in essential oils. Consequently, in testing for the presence of pinene it is customary to fractionate the oil, collecting the first 10 per cent, or better the distillate coming over below 160° at atmospheric pressure.

Procedure:[192] Distill a 50 cc. sample of the oil from a three bulb, 125 cc. Ladenburg flask, collecting only the first 5 cc.

[192] The procedure as given is essentially the official method of the Association of Official Agricultural Chemists, 6th Ed., 374, for the detection of pinene in orange and lemon oils.

Mix this distillate with 5 cc. of glacial acetic acid and cool to 0° in a freezing bath. Add 10 cc. of amyl nitrite and then add dropwise, with constant stirring, 2 cc. of dilute hydrochloric acid (2:1). Permit the mixture to stand in the freezing bath for 15 min. and collect the crystals which form on a Büchner funnel. Wash thoroughly with alcohol. Permit the crystals to dry at room temperature and dissolve in a small amount of chloroform. Add methyl alcohol to the chloroform solution dropwise until the nitrosochlorides precipitate out. Separate the crystals by filtration and dry at room temperature. Mount in a fixed oil (olive oil) and examine microscopically. Pinene nitrosochloride[193] crystals have irregular pyramidal ends (melting point, 103°).

6. DETECTION OF ACETINS

The acetic acid esters of glycerin are occasionally employed as adulterants in order to increase the apparent ester content. Since all three acetins are relatively soluble in water they may easily be washed out and tested for by the procedure described below. The least soluble of the three is triacetin; even this, however, is soluble in water to the extent of about 7 per cent. In order to insure the removal of most of the triacetin, a 5 per cent alcoholic solution is employed.

Procedure:[194] Shake 20 cc. of the oil with 40 cc. of 5% alcohol in a 125 cc. glass-stoppered, separatory funnel. When the mixture has separated completely withdraw 30 cc. of the alcoholic solution by means of a pipette and place it in a 125 cc. Erlenmeyer flask. Neutralize the solution with 0.5 N sodium hydroxide, using a 1% phenolphthalein solution as indicator. Then add exactly 5 cc. of 0.5 N alcoholic sodium hydroxide and heat the mixture on a steam bath for 1 hr. Remove the flask and allow the mixture to cool. Titrate the excess of alkali with 0.5 N hydrochloric acid. At least 4.7 cc. of the acid should be used for this neutralization.

This test is not specific for acetins; if large amounts of other water-soluble esters are present, these will appear in the dilute alcoholic layer.

7. DETECTION OF ETHYL ALCOHOL

Alcohol has been used frequently as an adulterant, since it is a cheap and available diluent for essential oils. The presence of ethyl alcohol as an adulterant may be readily detected by several simple tests.

[193] Limonene nitrosochloride, which may also be present, crystallizes in needles.

[194] The procedure as given is essentially that of "The United States Pharmacopoeia," Thirteenth Revision, 285, described under Oil of Lavender.

> *Procedure I:* Determine accurately the refractive index and specific gravity of the oil. Then shake thoroughly an equal volume of oil and saturated salt solution in a separatory funnel. Permit the oil to separate completely and determine the refractive index and specific gravity of this washed oil. These should not differ materially from those of the original oil. An approximation of the amount of added alcohol may be obtained from a consideration of these values.

This procedure is not specific for alcohol and will detect other water-soluble adulterants.

> *Procedure II:* Place 50 cc. of the oil (previously dried with anhydrous sodium sulfate) in a 100 cc. Ladenburg flask and distill slowly over an open flame. Collect and measure the distillate below 100°. Since most constituents of essential oils boil much above 100°, unadulterated oils generally show no distillate at this temperature. However, if a distillate is obtained dilute to 10 cc. with distilled water. Test a 5 cc. portion for ethyl alcohol by the iodoform test and the residual 5 cc. portion by the ethyl benzoate test.
>
> *Iodoform Test:* To 5 cc. of the diluted distillate add 10 drops of a 10% sodium hydroxide solution and sufficient iodine-potassium iodide solution drop by drop until a faint, permanent yellow color is obtained, indicating an excess of iodine. Allow the test tube to stand undisturbed for 5 min. The formation of yellow, flat, hexagonal crystals with the peculiar odor of iodoform indicates a positive reaction. If no positive result is obtained, heat the test tube to 60° for 1 min. in a beaker of water and permit the mixture to stand for 1 hr.
>
> The iodine-potassium iodide solution is prepared by dissolving 2 g. of potassium iodide in 8 cc. of distilled water and adding 1 g. of iodine; stir until solution is complete.
>
> *Ethyl Benzoate Test:* To 5 cc. of the dilute distillate add 5 drops of benzoyl chloride and 2 cc. of a 10% sodium hydroxide solution. Warm on a steam bath. The fruity odor of ethyl benzoate indicates the presence of ethyl alcohol.

The iodoform test will give a positive reaction with any compound containing a $CH_3\overset{\overset{O}{/\!/}}{C}$—group united to either a carbon or a hydrogen atom, or to any chemical which is oxidized under the conditions of the test to a compound having such a structure. In particular, acetone will give a positive iodoform test. In the ethyl benzoate test, all low boiling aliphatic alcohols will give fruity odors. However, only ethyl alcohol will give positive results with both the iodoform and ethyl benzoate tests.

The presence of ethyl alcohol materially lowers the flash point of most essential oils. There exist insufficient published data on the normal limits of the flash points of the unadulterated oils to draw valid conclusions from the results of flash-point determinations.

Oils containing relatively large amounts of alcohol will form milky emulsions with water. Use of this fact may be made for a quick test.

8. DETECTION OF METHYL ALCOHOL

The following procedure is based upon the fact that methyl alcohol may readily be oxidized to formaldehyde by potassium permanganate in the presence of dilute phosphoric acid. The resulting formaldehyde can then be detected by means of the reaction with chromotropic acid (1,8-dihydroxynaphthalene-3,6-disulfonic acid) which gives a violet color in the presence of sulfuric acid. The chemistry of this color reaction is unknown.

The following compounds give no reaction with chromotropic acid: acetaldehyde, aromatic aldehydes, butyraldehyde, chloralhydrate, crotonaldehyde, glyoxal, isobutyraldehyde, isovaleraldehyde, oenanthal, propionaldehyde. Fructose, furfural, glyceraldehyde, robinose and sucrose all give yellow colors. Other sugars, acetones and carboxylic acids do not react. High concentrations of furfural give red color.

This test is satisfactory for the detection of methyl alcohol in the presence of ethyl alcohol.

> *Procedure:*[195] Mix 2 drops of the alcohol in question in a test tube with 2 drops of 5% phosphoric acid and 2 drops of 5% potassium permanganate solution. After 1 min., add a little solid sodium bisulfite with shaking until the mixture is decolorized. If any brown precipitate of the oxide of manganese remains undissolved, add a further drop or two of phosphoric acid and a little more sodium bisulfite. When the solution is entirely colorless, add 8 cc. of 72% sulfuric acid and a small amount of finely powdered chromotropic acid. Shake the mixture well and then heat to 60° for 10 min. A violet color which deepens on cooling, indicates the presence of methyl alcohol.

According to Feigl the identification limit is 3.5 γ methyl alcohol; the concentration limit, 1:13600.

9. DETECTION OF HIGH BOILING ESTERS

***a.* Detection of Various Esters.**—Relatively odorless esters frequently are added to essential oils to increase the apparent ester content. Fortunately, most such esters are high boiling and permit of easy separation. The best general method for the detection of such added esters is to separate the acids and identify them. Detection of added esters of acetic and formic

[195] Feigl, "Laboratory Manual of Spot Tests," 193, published by Academic Press Inc., New York (1943).

acid (by isolation and identification of the acids) is not practical since these acids usually occur as natural constituents of essential oils.

Procedure:[196] Saponify 10 cc. of the oil for 2 hr. with 20 cc. of 0.5 N alcoholic potassium hydroxide.[197] Add 25 cc. of water and evaporate off most of the alcohol.[198] Wash out the unsaponified oil by shaking with 3 equal portions of ether. The aqueous solution is then made distinctly acid with hydrochloric acid (1:3) and again shaken out with ether. The ethereal solution will now contain the relatively insoluble acids, such as benzoic, cinnamic, oleic, phthalic, and lauric acid. Upon evaporation of the ether these may be recovered. The aqueous solution will contain the readily water-soluble acids, such as citric, oxalic, and tartaric acid. This solution should, therefore, be made just alkaline to phenolphthalein and an excess of saturated barium chloride solution added. After warming for about 10 min., a crystalline precipitate of the insoluble barium salts will be obtained from which the acids can be liberated and identified.

The chemists of Schimmel and Company[199] devised a method for the detection of esters of acids which are not readily volatile with steam—e.g., succinates, citrates, oxalates, and the esters of the higher fatty acids.

Procedure: Determine the saponification number of the oil in the usual manner. Then add a few drops of 0.5 N alcoholic sodium hydroxide to the contents of the saponification flask and evaporate to dryness on a steam bath. Dissolve the residue in 5 cc. of water and add 2 cc. of dilute sulfuric acid (1:3). Distill off the volatile acids with steam, using the apparatus shown in Diagram 4.17. The distillation should be carried out at such a rate that a distillate of 250 cc. is collected in the receiver at the end of 30 min.; the volume of the liquid in the saponification flask should be kept at about 10 cc. with the aid of the small flame. Collect a further 100 cc. of distillate in a second receiver. Add a few drops of a 1% alcoholic phenolphthalein solution to each receiver and titrate the free acids with 0.5 N potassium hydroxide solution. The first 250 cc. contain most of the volatile acids; the next 100 cc. should require only 1 or 2 drops of the alkali. From the total amount of alkali required to neutralize the acids, acid number II is calculated. A large difference between the saponification number and acid number

[196] Parry, "The Chemistry of Essential Oils," D. Van Nostrand Co., Inc., New York (1922), Vol. II, 321.

[197] If the oil has a high ester number, a larger amount of alkali will be required.

[198] Some chemists prefer to evaporate to dryness and then take up the residue in a small amount of water.

[199] *Ber. Schimmel & Co.*, October (1910), 43.

II indicates the presence of esters of acids only slightly volatile with steam.[200]

The presence of the high boiling glyceryl acetates is not revealed by either of the procedures described above, since the acid liberated is acetic acid, which is volatile with steam, and which occurs naturally in many oils (see "Detection of Acetins," p. 338).

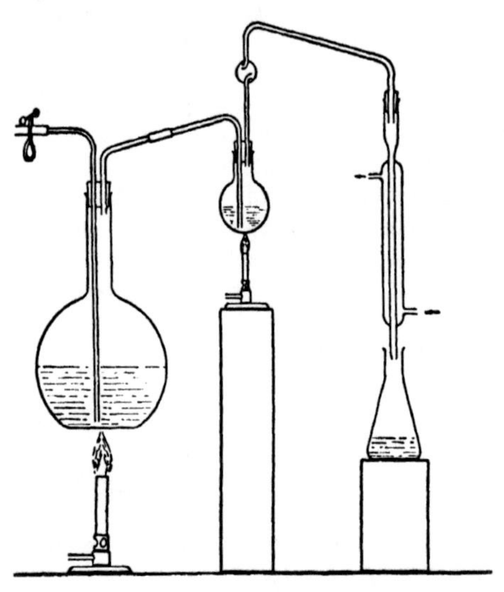

DIAGRAM 4.17. Apparatus for the detection of high boiling esters.

b. **Detection of Phthalates.**—This method is based upon a preliminary saponification of the oil, followed by a separation of phthalic acid as the lead salt. The separation is not specific since certain acids other than phthalic (e.g., oxalic, citric, and phosphoric) give rise to insoluble lead salts. Therefore, it is important to regenerate the acid and determine its melting point.

Procedure:[201] Introduce 2 g. of the oil in a 100 cc. saponification flask. Add 25 cc. of an alcoholic sodium hydroxide solution prepared by dissolving 1.25 g. of metallic sodium in 100 cc. of 95% alcohol.[202] Saponify for 1 hr. Remove and permit the flask to cool to room temperature and then immerse it in an ice-salt mixture. After standing for 30 min. filter off the precipitated sodium salts, using a well-cooled Büchner funnel. Wash

[200] This procedure was originally proposed for the examination of bergamot oils; pure oils showed a difference between the saponification number and acid number II of not more than 7.

[201] See Naves and Sabetay, "Phthalic Esters," *Perfumery Essential Oil Record* 29 (1938), 25.

[202] If the oil has a very high ester number, a larger amount of alkali will be required.

these crystals with ice cold anhydrous alcohol. A precipitate at this point may be indicative of any number of organic acids (phthalic, salicylic, citric or tartaric). Transfer the salt to a 250 cc. beaker and dry in an oven at 105° for 2 hr. Cool and add 40 to 50 cc. of distilled water and 2 or 3 cc. of glacial acetic acid. Heat this solution to the boiling point and add 30 cc. of a 10% lead acetate solution. Upon thoroughly cooling in an ice bath, the lead salt of phthalic acid will precipitate out almost quantitatively. The lead salts of benzoic acid, cinnamic acid, and salicylic acid are soluble and remain in the filtrate. Separate the lead salt of phthalic acid by filtration. Regenerate the phthalic acid with acid, recrystallize and determine the melting point. Phthalic acid melts at about 206°.[203]

10. DETECTION OF *MENTHA ARVENSIS* OIL

Several color reactions have been proposed to distinguish between the oil distilled from *Mentha piperita* L. and the oil from *Mentha arvensis* L. In common with most color reactions, these tests are not always reliable with mixtures as complex as essential oils.

The test described below is the official test of "The United States Pharmacopocia."[204]

> *Procedure:* Mix in a dry test tube 3 drops of oil of peppermint with 5 cc. of a solution of 1 volume of nitric acid in 300 volumes of glacial acetic acid, and place the tube in a beaker of boiling water. In from 1 to 5 min. the liquid develops a blue color which on continued heating deepens and shows a copper colored fluorescence and then fades leaving a golden yellow solution.

The characteristic color changes described in this procedure do not occur if an oil distilled from *Mentha arvensis* L. is examined: the acid solution then attains a light yellow color which shows no appreciable change during the 5 min. of heating.

It should be remembered that the color changes described are characteristic of the oil from *Mentha piperita* L.; mixtures of this oil and *Mentha arvensis* L. give the color changes described. Therefore, the test cannot be used to detect adulteration with *Mentha arvensis* L.

Several other color reactions have been described for these oils in the literature.

11. DETECTION OF VARIOUS ADULTERANTS

The physical and chemical properties of several common adulterants (which have not been thoroughly discussed previously) are briefly noted here to aid the essential oil chemist.

[203] Phthalic anhydride may be formed; the anhydride melts at 131°.
[204] Eleventh Revision, 259.

I. Cedarwood Oil.—This is usually found in the last fractions owing to the high boiling points of its constituents.

d_{15}.............0.951 to 0.960
α_D.............$-28°28'$ to $-35°39'$
n_D^{20}...........1.5030 to 1.5059
Sol. 20°.........Often insoluble in 10 vol. 90% alc.

II. Copaiba Oil.—This also is found in the last fractions.

d_{15}...........0.901 to 0.905
α_D.............$-11°18'$ to $-14°22'$
n_D^{20}..........1.4972 to 1.4990
Sol. 20°.........Insoluble in 10 vol. 90% alc.

III. Gurjun Balsam Oil.—This is a high boiling oil.

d_{15}...........0.918 to 0.930
α_D.............$-35°0'$ to $-130°0'$
n_D^{20}...........1.5010 to 1.5050
Sol. 20°.........Insoluble in 10 vol. 90% alc.

The following color reaction for this oil has been recommended:

> To a mixture of 10 cc. of glacial acetic acid and 5 drops of concentrated nitric acid, add 5 drops of the oil: gurjun oil gives a purple-violet color within 2 min.

A rather elaborate test has been described by Deussen and Philipp[205] involving the preparation and isolation of gurjun-ketone semicarbazone—melting point, 234°.

IV. Fatty Oils.—Such oils greatly increase the ester number and evaporation residue of an oil. They are not volatile with steam, and cannot be distilled without decomposition except at exceptionally low pressures. In general, they are very insoluble in 90 per cent alcohol and frequently insoluble in 95 per cent alcohol; castor oil proves an exception, being readily soluble in 95 per cent alcohol. The saponified oil frequently shows much foaming, owing to the formation of soaps.

VII. A PROCEDURE FOR THE INVESTIGATION OF THE CHEMICAL CONSTITUENTS OF AN ESSENTIAL OIL

Assurance of the purity of the essential oil is of primary importance in an investigation of its chemical constituents. If there is the slightest doubt as to whether or not the oil may have been contaminated or adulter-

[205] *Liebigs Ann.* 369 (1909), 57.

ated, then such an oil is worthless for the examination, because the results obtained after much labor will be open to question. Therefore, it is best for the investigator to distill the oil from the botanical, or to supervise the distillation in the producing region or factory. Such distillations should be carried out on a commercial scale in the manner in which the oil of commerce is produced; otherwise, misleading results may be obtained. If this is impossible, the oil should be obtained directly from a prime source of unquestionable repute.

A representative sample of the oil to be investigated should be analyzed carefully. All physical and chemical properties should be determined, including specific gravity, optical rotation, refractive index, solubility and the percentages of esters, aldehydes, ketones, phenols, acids and alcohols. These physicochemical properties should be compared with values given in the literature for normal pure oils. Further examination should not be attempted if these properties show any suspicious deviation from normal values. Such deviation might indicate accidental contamination, adulteration, or the production of an abnormal oil.

Although an oil may have been distilled from the proper botanical material, nevertheless, it may not represent the normal article of commerce. Such factors as the degree of maturity of the botanical frequently exert an important influence on the composition of the oil. Consider, for example, oil of coriander. If an oil is distilled from the immature and green coriander seed it will show a high decyl aldehyde content, sometimes attaining a value as high as 70 per cent. As the seed matures, the aldehyde content of the oil decreases and the linaloöl content increases, until finally an oil is obtained from mature seed which shows an aldehyde content of about 1 per cent. Needless to say, the oil having this low aldehyde content is the oil accepted in commerce as normal oil of coriander.

A further difficulty exists in the proper selection of the botanical. Sometimes there are many species within a plant family but only one or more yields the desired oil or oils; the eucalypts are a good example. Occasionally there are found several varieties of the same species which may yield different oils upon distillation. The production of juniper berry oil from *Juniperus communis* L. growing in America gives rise to an oil which differs from the normal commercial product formerly obtained from *Juniperus communis* L. grown in Central Europe. This has been explained by the fact that the American oil is distilled from a variety of the true *Juniperus communis* L.; viz., *Juniperus communis* L. var. *depressa* Pursh. Physiological varieties of the same species of certain plants are also known (e.g., *Eucalyptus dives*).

The geographical location of the growing section may exert an effect upon the composition and quality of the oil. This probably results from the nature of the soil, the altitude at which the plant grows, as well as factors such as intensity of sunlight, rainfall and temperature.

Consideration should be given to the methods of distillation and production of the commercial oil and to the handling of the botanical before distillation. Some plants should be distilled as soon as cut, some after sun drying for a day, some after thorough drying in the shade, some after drying and storage for several years. For details, the reader is referred to the section in Chapter III on "Practice of Distillation."

All of the above factors should be carefully considered, and as much information as possible concerning the botany, geographical source, maturity, preliminary treatment of the plant material and method of production of the oil should be included in the report on the chemical constituents of the oil.

The amount of oil used for the examination is a limiting factor. The availability and the cost of the oil enter in most commercial and academic investigations. For oils that are available in relatively unlimited quantity, the difficulty of handling large amounts in a research laboratory must be considered. Such difficulty may be overcome if the manufacturing plant or factory cooperates in the investigation. It then becomes possible to fractionate large quantities of the oil, even hundreds of pounds, and to investigate the individual fractions or aliquot parts of such fractions. Constituents occurring in minute amounts have been identified by such a procedure. Without benefit of this preliminary fractionation, it is difficult to handle much more than 15 liters of an oil in the laboratory.

For an oil which has not been investigated previously, the first step is a general examination, followed by an investigation which endeavors to discover as many of the constituents as possible. This usually reveals those constituents which occur in substantial amounts. Frequently, indications of the occurrence of other constituents are thereby obtained, whose presence, however, cannot be established conclusively. A subsequent investigation directed solely to the isolation and identification of such individual constituents often will prove successful.

It is obvious that no comprehensive procedure can be given which will prove applicable to all essential oils. The following notes are intended merely as an aid to the chemist embarked upon such an investigation. From a study of the physicochemical properties of the oil, a general plan for the investigation is formulated.

If the oil shows a large percentage of free acids, phenols or carbonyl compounds it is usually advisable to remove these components before fractionation. Any free acids should always be removed before further treatment of small amounts, it may be better to fractionate the oil and then separate these components from the enriched fraction or fractions.

Occasionally solid constituents (such as camphor, menthol, safrole, or anethole) may be separated from the whole oil by freezing, followed by filtration or centrifuging. Since such separations are never quantitative it may be advisable to freeze out these components from the enriched fraction rather than from the whole oil. If the solid constituents occur in large amounts, one may resort to a preliminary freezing, followed by fractionation of the filtrate so obtained. The enriched fractions should then be frozen and the material thus further separated added to that obtained from the original oil. The difficulty of maintaining sufficiently low temperatures during the filtration, especially for large amounts of oil, may make a separation from the whole oil impractical. In general, for the isolation and purification of the various constituents it is necessary to resort to chemical methods in addition to purely physical means.

After such preliminary treatment as indicated above, the oil or residual oil should be fractionated. This will result in a separation of the oil into a low boiling terpene fraction, and intermediate fraction, a fraction rich in oxygenated constituents, a second intermediate fraction, a fraction containing the sesquiterpene constituents, and a distillation residue. The residue usually contains polymerization products and high boiling constituents, such as azulenic compounds, and the naturally occurring waxes in the case of citrus oils obtained by expression. These waxes show a tendency to "fix" part of the volatile components. If present to any appreciable extent these waxes should be freed from the more volatile components by steam distillation or by the addition of a water-soluble glycol (e.g., diethylene glycol), followed by vacuum distillation.[206] The latter procedure will remove most of the volatile material from the waxes, leaving a relatively inodorous residue. The glycol may then be removed from the natural constituents by washing out with water or sodium chloride solution.

Should the original analysis show a high ester content it is usually best to fractionate the oil before saponification so that the ester may be obtained in a state of relative purity for a determination of physical properties. Its components may then be identified after saponification. Since the corresponding free alcohol usually is present with the ester, saponification of the whole oil (followed by fractionation) may be preferable, especially if only small amounts of ester are present.

The treatment of an oil or fraction with reagents for the purpose of separating and purifying various constituents may cause drastic changes to occur. This may give rise to new chemical compounds not originally present as such in the oil. Intra- and intermolecular rearrangements may occur as well as degradations and dehydrations. Such possibilities must be considered in the evaluation of the final results of the investigation.

For the identification of individual constituents which have been separated and purified from the oil, two general procedures are employed: (1) The determination of physical properties including melting point (or congealing point), boiling point, specific gravity, optical rotation, refractive index and solubility in alcohol of varying strengths. (2) The preparation of suitable derivatives, preferably solid compounds of definite melting point capable of purification by recrystallization. In general, the identification may be considered established if no depression is observed in the melting point when a derivative of the constituent is mixed with the corresponding derivative of a sample of known purity and constitution. The reader is referred to Volume II on the "Constituents of Essential Oils" for the properties of the individual compounds and for data on the melting points of certain frequently employed derivatives. In many cases compounds obtained by oxidation, reduction, and condensations may be used for identification.

Other methods are often employed in establishing the identity of a constituent or derivative: combustion to determine the percentage of carbon and hydrogen and to establish the empirical formula; molecular weight determinations, especially by cryoscopic methods; molecular refraction; ignition of metallic salts, especially the silver salts of organic acids; determinations of the percentage of halogen in chlorides and bromides; and other procedures.

Detailed procedures for the separation of chemical groups and for the isolation and purification of individual constituents are given in Volume II dealing with the "Constituents of Essential Oils."

VIII. SUGGESTED ADDITIONAL LITERATURE

The references given in the following pages do not represent a complete survey of the literature; they are intended merely as a guide for those interested in pursuing further the study of analytical procedures applied to essential oils and related products.

Frequently the citation does not refer to the initial article by an author, but to a later publication which includes modifications and improvements.

In this list, there have been included the following:

1. Established methods which have been superceded by other procedures.
2. New methods as yet not thoroughly established or investigated.
3. Important methods which do not have general value for routine analysis.
4. Suggested modifications of well-known methods.

5. Criticisms of suggested methods.

6. Evaluations of methods, often with comparative data.

7. Several reviews, which have added value because of their references to the literature.

The author assumes no responsibility as to the reliability of these methods.

Sampling and Storage.

Analysis of oils of sweet orange and lemon. Mechanism and evaluation of their deterioration or alteration. Y. R. Naves, *Parfums France* **10** (1932), 225. *Chem. Abstracts* **26** (1932), 5704.

Analysis of oil of lemon and oil of sweet orange. Measuring the deterioration of the oils by oxidation. Y. R. Naves, *Parfums France* **12** (1934), 314. *Chem. Abstracts* **29** (1935), 2307.

Standard methods of sampling. W. W. Scott, "Standard Methods of Chemical Analysis," D. Van Nostrand Co., Inc., (1939), 5th Ed., 1301.

Carbonyl and peroxide number in the analysis of ethereal oils. Y. R. Naves, *Fette u. Seifen* **48** (1941), 677. *Chem. Zentr.* II (1942), 109. *Chem. Abstracts* **37** (1943), 6409.

Determination of stability of oils and fats. N. D. Sylvester, L. H. Lampitt and A. N. Ainsworth, *J. Soc. Chem. Ind.* **61** (1942), 165. *Chem. Abstracts* **37** (1943), 2599.

Specific Gravity.

Change in the specific weight of ethereal oils by heat. K. Irk, *Pharm. Zentralhalle* **55** (1914), 831. *Chem. Abstracts* **9** (1915), 509.

Optical Rotation.

Optical rotation and chemical constitution. Werner Kuhn, *Ber.* **63B** (1930), 190. *Chem. Abstracts* **24** (1930), 1554.

Effect of solvents on the optical rotation of menthene, bornylene and borneol. Irene M. McAlpine, *J. Chem. Soc.* (1932), 543. *Chem. Abstracts* **26** (1932), 2905.

Molecular dispersion, refractive index, and rotatory power. Value in the study and chemical analysis of essential oils. A review. Y. R. Naves, *Parfums France* **10** (1932), 253. *Chem. Abstracts* **27** (1933), 372.

Rules of optical rotation and their application to the investigation of constitution and configuration. Karl Freudenberg, *Ber.* **66B** (1933), 177. *Chem. Abstracts* **27** (1933), 2418.

The simplest principles and laws of optical rotation. Werner Kuhn, *Ber.* **66B** (1933), 166. *Chem. Abstracts* **27** (1933), 2418.

Measuring rotatory dispersion in the ultraviolet range by photoelectric polarimetry. Y. R. Naves, *Parfums France* **11** (1933), 185. *Chem. Ab-*

stracts **28** (1934), 3181. Use of Brulat and Chatelain's photoelectric polarimeter. Y. R. Naves, *Chem. Abstracts* **26** (1932), 5455, 5847.

The accurate determination of the rotatory power of essential oils. Y. R. Naves and M. G. I., *Parfums France* **13** (1935), 253. *Chem. Abstracts* **30** (1936), 572.

Identification of essential oils by the effect of solvents on their rotatory powers. Y. R. Naves and B. Angla, *Compt. rend.* **213** (1941), 570. *Chem. Abstracts* **37** (1943), 3877.

Determination of ethereal oils by changes in optical rotation. Y. R. Naves, *Fette u. Seifen* **49** (1942), 183. *Chem. Abstracts* **37** (1943), 6405.

Optical activity of terpenes. Influence of solvent on rotation of bornyl and isobornyl methyl ethers. W. Hückel, H. Kaluba, *Liebigs Ann.* **550** (1942), 269. *Chem. Abstracts* **37** (1943), 3421.

Refractive Index.

Change in the refractive index of ethereal oils by heat. K. Irk, *Pharm. Zentralhalle* **55** (1914), 789. *Chem. Abstracts* **9** (1915), 509.

Immersion liquids for determining refractivity of solid substances by the embedding method. A. Mayrhofer, *Mikrochemie* **3** (1931), 52. *Chem. Abstracts* **25** (1931), 1417.

Evaporation Residue.

Determination of residue on evaporation of essential oils. Établissements Antoine Chiris, *Parfums France* **10** (1932), 114. *Chem. Abstracts* **26** (1932), 4413.

Flash Point.

W. W. Scott, "Standard Methods of Chemical Analysis," D. Van Nostrand Co., Inc., New York (1939), Vol. II, 1744, 1752, 1732.

Determination of Acids.

Identification of organic acids by partition between ethyl ether and water. O. C. Dermer and V. H. Dermer, *J. Am. Chem. Soc.* **65** (1943), 1653. *Chem. Abstracts* **37** (1943), 5697.

Determination of Esters.

Procedure for the examination of the ethers (esters) of essential oils. B. Angla, *Ann. chim. anal. chim. appl.* **18** (1936), 145. *Chem. Abstracts* **30** (1936), 5723.

Determination of Alcohols.

Quantitative determination of the active hydrogen in organic compounds. T. Zerevitinov, *Ber.* **41** (1908), 2233. *Chem. Abstracts* **2** (1908), 2810.

Quantitative estimation of hydroxy, amino and imino derivatives of organic compounds by means of the Grignard reagent, and the nature of the changes taking place in solution. H. Hibbert, *Proc. Chem. Soc.* **28** (1912), 15. *J. Chem. Soc.* **101** (1912), 328. *Chem. Abstracts* **6** (1912), 1744.

Notes on the acetification (acetylation) of Java citronella oil. T. H. Durrans, *Perfumery Essential Oil Record* **3** (1912), 123. *Chem. Abstracts* **6** (1912), 2975.

New color reaction for alcohols and alcoholic hydroxy groups. L. Rosenthaler, *Chem. Ztg.* **36** (1912), 830. *Chem. Abstracts* **6** (1912), 3251.

Use of pyridine as a solvent in the estimation of hydroxyl groups by means of alkyl magnesium halides. A. P. Tanberg, *J. Am. Chem. Soc.* **36** (1914), 335. *Chem. Abstracts* **8** (1914), 1277.

Magnesium organic method for the determination of hydroxyl groups. T. Zerevitinov, *Ber.* **47** (1914), 1659. *Chem. Abstracts* **8** (1914), 2728.

Pyridine as solvent in the determination of active hydrogen in organic compounds by means of methyl magnesium iodide. T. Zerevitinov, *Ber.* **47** (1914), 2417. *Chem. Abstracts* **9** (1915), 75.

Formic acid as a reagent in essential oil analysis. W. H. Simmons, *Analyst* **40** (1915), 491. *Chem. Abstracts* **10** (1916), 664.

Determination of alcohols by acetylation. Anon., *Perfumery Essential Oil Record* **7** (1916), 374. *Chem. Abstracts* **11** (1917), 866.

Determination of alcohols in essential oils. T. T. Cocking, *Perfumery Essential Oil Record* **9** (1918), 37. *Chemist Druggist* **74** (1913), 87. *Chem. Abstracts* **12** (1918), 1102.

Determination of volatile alcohols. A. Grün and T. Wirth, *Z. deut. Öl- Fett-Ind.* **41** (1921), 145. *Chem. Abstracts* **15** (1921), 1869.

Pharmacopoeial assay for alcohols in santal oil to include the true acetyl value. C. Harrison, *J. Assocn. Official Agr. Chem.* **4** (1921), 425. *Chem. Abstracts* **15** (1921), 2333.

Determination of alcohols and phenols in ethereal oils by acetylation with pyridine (especially santalol, menthol, eugenol). H. W. van Urk, *Pharm. Weekblad* **58** (1921), 1265. *Chem. Abstracts* **15** (1921), 3891.

Determination of alcohols by acetylizing. Hans Wolff, *Chem. Umschau* **29** (1922), 2. *Chem. Abstracts* **16** (1922), 1374.

Estimation of easily dehydrated alcohols in essential oils. L. Glichitch, *Compt. rend.* **177** (1923), 268. *Chem. Abstracts* **17** (1923), 3226. *Bull. soc. chim.* **33** (1923), 1284. *Chem. Abstracts* **17** (1923), 3904.

A color reaction of the alcoholic hydroxyl. Walter Parri, *Giorn. farm. chim.* **73** (1924), 109. *Chem. Abstracts* **18** (1924), 2667.

A reagent for multivalent alcohols. The R-acid or β-naphthol-3,6-disulfonic acid. P. Thomas and A. Misca, *Bull. soc. stiinte Cluj.* **2** (1924), 224. *Chem. Abstracts* **19** (1925), 3074.

Determination of alcohols and phenols in ethereal oils by means of magnesium methyl iodide. T. Zerevitinov, *Z. anal. Chem.* **68** (1926), 321. *Chem. Abstracts* **21** (1927), 153.

Notes on the determination of total alcohols in oil of citronella. Justin Dupont and Louis Labaune, *Chim. ind.* **17** (1927), 905. *Chem. Abstracts* **21** (1927), 2960.

Determination of alcohols. A. Verley, *Bull. soc. chim.* **43** (1928), 469. *Chem. Abstracts* **22** (1928), 3113.

Determination of phenols and alcohols. A. Verley, *Am. Perfumer* **24** (1929), 233. *Chem. Abstracts* **23** (1929), 4422.

Detection and identification of primary phenylethyl alcohols in essential oils and in mixtures of perfumes. S. Sabetay, *Ann. chim. anal. chim. appl.* **11** (1929), 193. *Chem. Abstracts* **23** (1929), 4770.

The determination of citronellol and rhodinol in presence of geraniol and nerol. L. S. Glichitch and Y. R. Naves, *Parfums France* **8** (1930), 326. *Chem. Abstracts* **25** (1931), 1033.

The nitro chromic acid reaction for the detection of primary and secondary alcohols with special reference to saccharides. Wm. R. Fearson and David M. Mitchell, *Analyst* **57** (1932), 372. *Chem. Abstracts* **26** (1932), 4011.

The determination of alcohols in essential oils. Établissements Antoine Chiris, *Parfums France* **10** (1932), 114. *Chem. Abstracts* **26** (1932), 4414.

Citronella oil, critical survey on the analytical methods. J. Zimmermann, *Perfumery Essential Oil Record* **23** (1932), 128. *Chem. Abstracts* **26** (1932), 5701.

Determination of alcohols by Zerevitinov's method. Établissement Antoine Chiris, *Parfums France* **10** (1932), 247. *Chem. Abstracts* **26** (1932), 5704.

Application of the method of Franchimont to the estimation of the components of essential oils. Marinoa de Mingo, *Rev. acad. cienc. Madrid* **29** (1932), 150. *Chem. Abstracts* **27** (1933), 2251.

Bulgarian otto of rose. E. J. Parry and J. H. Seager, *Perfumery Essential Oil Record* **24** (1933), 149. *Chem. Abstracts* **27** (1933), 3776.

Bulgarian otto of roses and its rhodinol content. L. S. Glichitch and Y. R. Naves, *Parfums France* **11** (1933), 154. *Chem. Abstracts* **27** (1933), 5481.

Estimation of the primary alcohol content of essential oils by phthalization. L. S. Glichitch and Y. R. Naves, *Chimie & industrie*, Special No., June (1933), 1024. *Chem. Abstracts* **28** (1934), 575.

Estimation of primary alcohol content of essential oils L. S. Glichitch and Y. R. Naves, *Parfums France* **11** (1933), 235. *Chem. Abstracts* **28** (1934), 3182.

Lavender oil from the province of Savona. Andrea Gandini and Terenzio Vignola, *Ann. chim. applicata* **24** (1934), 431. *Chem. Abstracts* **29** (1935), 4129.

Rapid determination of primary and secondary alcohols in essential oils. S. Sabetay, *Compt. rend.* **199** (1934), 1419. *Chem. Abstracts* **29** (1935), 6359.

Analytical results with citronella oils. D. R. Koolhaas, *Indische Mercuur* **58** (1935), 429. *Chem. Abstracts* **29** (1935), 5991.

The determination of the alcohols in sandalwood oil. R. Delaby and Y. Breugnot, *Bull. sci. pharmacol.* **42** (1935), 385. *Chem. Abstracts* **29** (1935), 6699.

Determination in the presence of tertiary alcohols of the free primary and secondary alcohol contents of essential oils by acetylation in pyridine. S. Sabetay, *Bull. soc. chim.* [5] **2** (1935), 1716. *Chem. Abstracts* **30** (1936), 567.

Determination of tertiary alcohols by cold formylation. S. Sabetay, *Ann. fals.* **39** (1936), 225. *Chem. Abstracts* **30** (1936), 4787.

A method for the rapid detection and approximate determination of primary alcohols in the presence of secondary and tertiary alcohols by the formation of trityl ethers. Sébastien Sabetay, *Compt. rend.* **203** (1936), 1164. *Chem. Abstracts* **31** (1937), 1728.

Investigation of monohydric primary, secondary and tertiary alcohols. The micro method of determining the rate of esterification. Shunsuke Murahashi, *Sci. Papers Inst. Phys. Chem. Research (Tokyo)* **30** (1936), 272. *Chem. Abstracts* **31** (1937), 3001.

Simplified procedure for determining primary alcohols by phthalization in benzene solution. Sébastien Sabetay and Y. R. Naves, *Ann. chim. anal. chim. appl.* **19** (1937), 35. *Chem. Abstracts* **31** (1937), 2550.

Simple and rapid procedure for the determination of primary alcohols and certain secondary alcohols by hot, pyridinic phthalization and a practical technic for identifying esterifiable alcohols in the form of acid phthalates. Sébastien Sabetay and Y. R. Naves, *Ann. chim. anal. chim. appl.* **19** (1937), 285. *Chem. Abstracts* **32** (1938), 456.

Determination of alcohol and phenol functions. Elie Raymond and Emile Bouvetier, *Compt. rend.* **209** (1939), 439. *Chem. Abtsracts* **33** (1939), 9198.

Phthalization in hot pyridine. S. Sabetay, *Ann. chim. anal. chim. appl.* **2** (1939), 289. *Chem. Abstracts* **34** (1940), 692.

Determining alcohols in essential oils. John E. S. Han, *Am. Perfumer* **41**, No. 2 (1940), 35. *Chem. Abstracts* **34** (1940), 7066. *Am. Perfumer* **42**, No. 6 (1941), 41. *Chem. Abstracts* **36** (1942), 219.

Iodoform micro test for higher alcohols and ketones. F. Stodola, *Ind. Eng. Chem., Anal. Ed.* **15** (1943), 72. *Chem. Abstracts* **37** (1943), 1673.

Viscometric method for determining free menthol in peppermint oil. L. J. Swift and M. H. Thornton, *Ind. Eng. Chem., Anal. Ed.* **15** (1943), 422. *Chem. Abstracts* **37** (1943), 5550.

Periodate reaction applied to cosmetic ingredients. Determination of glycerol, ethylene glycol and propylene glycol. Erwin S. Shupe, *J. Assocn. Official Agr. Chem.* **26** (1943), 249. *Chem. Abstracts* **37** (1943), 5551.

Volatile plant substances. Application of selective formylation of borneol, 3-octanol and benzyl alcohol in the presence of linaloöl and its esters in the analysis of essential oils. Y. R. Naves, *Helv. Chim. Acta* **27** (1944), 942. *Chem. Abstracts* **39** (1945), 1017.

Estimation of alcohols in essential oils. Y. R. Naves, *Perfumery Essential Oil Record* **36** (1945), 92. *Chem. Abstracts* **39** (1945), 3119.

Determination of Aldehydes and Ketones.

Bisulfite addition compounds. Reinking, Dehnel and Labhardt, *Ber.* **38**, (1905), 1069.

The assay of benzaldehyde and oil of bitter almond. F. D. Dodge, *Orig. Com. 8th Intern. Congr. Appl. Chem.* **17**, 15. *Chem. Abstracts* **6** (1912), 3162.

Study of some methods for the determination of aldehydes. B. G. Feinberg, *Orig. Com. 8th Intern. Congr. Appl. Chem.* **1**, 187. *Chem. Abstracts* **6** (1912), 3251.

Constitution of the aldehyde and ketone bisulfite compounds. F. Raschig, *Ber.* **59B** (1926), 859. *Chem. Abstracts* **20** (1926), 2816.

Constitution of aldehyde and ketone bisulfites. F. Raschig and W. Prahl, *Liebigs Ann.* **448** (1926), 265. *Chem. Abstracts* **20** (1926), 3156.

The alleged potassium hydroxymethane sulfonate of Max Müller. F. Raschig and W. Prahl, *Ber.* **59B** (1926), 2025. *Chem. Abstracts* **21** (1927), 223.

I. Constitution of the aldehyde and ketone bisulfites. G. Schroeter, *Ber.* **59B** (1926), 2341. *Chem. Abstracts* **21** (1927), 386.

II. Chemical constitution of aldehyde and ketone bisulfites. G. Schroeter and M. Sulzbacher, *Ber.* **61B** (1928), 1616. *Chem. Abstracts* **23** (1929), 94.

Microchemical method for determining semicarbazones and its application to the analysis of ketones. Ralph P. Hobson, *J. Chem. Soc.* (1929), 1384. *Chem. Abstracts* **23** (1929), 4648.

Assay of citronella. F. D. Dodge, *Am. Perfumer* **24** (1929), 11. *Chem. Abstracts* **23** (1929), 5004.

A colorimetric method for determination of camphor. Angelo Castiglioni, *Ann. chim. applicata* **26** (1930), 53.

Identification of flavoring constituents of commercial flavors. Optical properties of the semicarbazones of certain aldehydes and ketones. John B. Wilson and George L. Keenan, *J. Assocn. Official Agr. Chem.* **13** (1930), 389. *Chem. Abstracts* **24** (1930), 5431.

A method for determining aldehydes based on the Cannizzaro and Claisen reactions. L. Palfray, S. Sabetay and D. Sontag, *Compt. rend.* **194** (1932), 1502. *Chem. Abstracts* **26** (1932), 4011.

Determination of aldehydes and ketones in essential oils. H. Schmalfuss, H. Werner and R. Kraul, *Z. anal. Chem.* **87** (1932), 161. *Chem. Abstracts* **26** (1932), 4681.

Analysis of oils of sweet orange and lemon: I. Observations on the determination of aldehydes. Y. R. Naves, *Parfums France* **10** (1932), 198. *Chem. Abstracts* **26** (1932), 5175.

Hydroxylamine method for determinations of aldehydes and ketones in essential oils. R. C. Stillman and R. M. Reed, *Perfumery Essential Oil Record* **23** (1932), 278. *Chem. Abstracts* **26** (1932), 5702.

Semimicro method for the determination of cinnamaldehyde in cinnamon bark. S. Rivas Goday, *Bol. farm. militar* **10** (1932), 18. *Anales farm. bioquim suplemento* **3**, 17. *Chem. Abstracts* **26** (1932), 5703.

Determination of compounds containing the carbonyl group by means of 2,4-dinitrophenylhydrazine. O. Fernández and L. Socías, *Rev. acad. cienc. Madrid* **28** (1932), 330. *Chem. Abstracts* **26** (1932), 4556.

2,4-Dinitrophenylhydrazine in quantitative determination of carbonyl compounds. O. Fernández, L. Socías and C. Torres, *Anales soc. españ. fís. quím.* **30** (1932), 37. *Chem. Abstracts* **26** (1932), 2395.

Description of the Cannizzaro condensation method for determination of aldehydes. S. Sabetay, L. Palfray and D. Sontag, *Ann. chim. anal. chim. appl.* **15** (1933), 251. *Chem. Abstracts* **27** (1933), 3777.

New method for determining aldehydes by quantitative Cannizzarization. L. Palfray, S. Sabetay and D. Sontag, *Chimie & industrie,* Special No., June (1933), 1037. *Chem. Abstracts* **28** (1934), 434.

The determination of aldehydes and ketones with hydroxylamine salt. Herrmann Schultes, *Angew. Chem.* **47** (1934), 258. *Chem. Abstracts* **28** (1934), 4336.

Extension of the Cannizzaro reaction to aliphatic and arylaliphatic aldehydes. Sébastien Sabetay and Leon Palfray, *Compt. rend.* **198** (1934), 1513. *Chem. Abstracts* **28** (1934), 4718.

Detection of aldehydes and ketones in essential oils and drugs. R. Fischer and A. Moor, *Arch. Pharm.* **272** (1934), 691. *Chem. Abstracts* **28** (1934), 5597.

A new number applicable to aldehydes: Cannizzaro number. Application to the assay of bitter almonds. L. Palfray, S. Sabetay and D. Sontag,

Chimie & industrie, Special No., April (1934), 863. *Chem. Abstracts* **28** (1934), 5927. See also *Chem. Abstracts* **29** (1935), 2939.

Application of several methods for the determination of aldehydes in essential oils. Maria Anna Schwartz, *Ann. chim. applicata* **24** (1934), 352. *Chem. Abstracts* **28** (1934), 7419.

Estimation of aldehydes by the bisulfite method. A. Eric Parkinson and E. C. Wagner, *Ind. Eng. Chem., Anal. Ed.* **6** (1934), 433. *Chem. Abstracts* **29** (1935), 427.

Determination of camphor in galenicals by means of 2,4-dinitrophenylhydrazine. C. H. Hampshire and G. R. Page, *Quart. J. Pharm. Pharmacol.* **7** (1934), 558. *Chem. Abstracts* **29** (1935), 550.

Influence of free acid on determination of aldehydes and ketones by hydroxylamine hydrochloride. L. Palfray and S. Tallard, *Compt. rend.* **199** (1934), 296. *Chem. Abstracts* **29** (1935), 2478.

Improved hydroxylamine method for the determination of aldehydes and ketones. Displacement of oxime equilibria by means of pyridine. W. M. D. Bryant and Donald M. Smith, *J. Am. Chem. Soc.* **57** (1935), 57. *Chem. Abstracts* **29** (1935), 1749.

The determination of menthone in peppermint essence with hydroxylamine. G. Parraud, *Bull. sci. pharmacol.* **42** (1935), 337. *Chem. Abstracts* **29** (1935), 5990.

2,4-Dinitrophenylhydrazine as a quantitative reagent for carbonyl compounds. Benzophenone and acetone. G. W. Perkins and Myles W. Edwards, *Am. J. Pharm.* **107** (1935), 208. *Chem. Abstracts* **29** (1935), 6166.

A highly sensitive reaction for the characterization and determination of citral. J. Bougault and E. Cattelain, *J. pharm. chim.* **21** (1935), 437. *Chem. Abstracts* **29** (1935), 7578.

Volumetric determination of camphor by the hydroxylamine method. Robert Vandoni and Gérard Desseigne, *Bull. soc. chim.* [5] **2** (1935), 1685. *Chem. Abstracts* **30** (1936), 56.

Analysis of essence of cumin. S. Sabetay and L. Palfray, *Ann. chim. anal. chim. appl.* **17** (1935), 289. *Chem. Abstracts* **30** (1936), 240.

Determination of the carbonyl group in camphor, menthone, pulegone, citral and furfural with 2,4-dinitrophenylhydrazine. L. Socias Viñals, *Anal. acad. nac. farm. Madrid* (1935), 1. *Anales farm. bioquim. suppl.* **6** (1935), 84. *Chem. Abstracts* **30** (1936), 239.

Citral and its sulfonates. F. D. Dodge, *Am. Perfumer* **32**, No. 3 (1936), 67. *Chem. Abstracts* **30** (1936), 3403.

Use of 2,4-dinitrophenylhydrazine as a reagent for carbonyl compounds. N. R. Campbell, *Analyst* **61** (1936), 391. *Chem. Abstracts* **30** (1936), 5534.

Semicarbazides. *m*-Tolylsemicarbazide as a reagent for the identification of aldehydes and ketones. Peter P. T. Sah, Si-Min Wang and Chung-

Hsi Kao, *J. Chinese Chem. Soc.* **4** (1936), 187. *Chem. Abstracts* **31** (1937), 655.

New series of reagents for the carbonyl group, their use for extraction of ketonic substances and for microchemical characterization of aldehydes and ketones. André Girard and Georges Sandulesco, *Helv. Chim. Acta* **19** (1936), 1095. *Chem. Abstracts* **31** (1937), 1006.

Determination of camphor in the form of 2,4-dinitrophenylhydrazine in concentrated and in weak tinctures of camphor. Maurice-Marie Janot and Marcel Mouton, *J. pharm. chim.* **23** (1936), 547. *Chem. Abstracts* **31** (1937), 2750.

Use of the reagent of Girard and Sandulesco for the isolation of ketones from volatile and animal drugs. G. Sandulesco and S. Sabetay, *Riechstoff Ind. Kosmetik* **12** (1937), 161. *Chem. Abstracts* **31** (1937), 8822.

Determination of camphor in alcoholic solutions by the dinitrophenylhydrazine method. Elmer M. Plein and Charles F. Poe, *Ind. Eng. Chem., Anal. Ed.* **10** (1938), 78. *Chem. Abstracts* **32** (1938), 2686.

Determination of carbonyl compounds by means of hydroxylamine hydrochloride. A. Reclaire and R. Frank, *Perfumery Essential Oil Record* **29** (1938), 212. *Chem. Abstracts* **32** (1938), 8302.

Simplified procedure for the analytical oximation of aldehydes and ketones. S. Sabetay, *Bull. soc. chim.* [5] **5** (1938), 1419. *Chem. Abstracts* **33** (1939), 1268.

Notes on analysis of essential oils. Francis D. Dodge, *Am. Perfumer* **40**, No. 5 (1940), 41. *Chem. Abstracts* **34** (1940), 4862.

Benzylideneaminomorpholine compounds (for identification of aromatic aldehydes). L. Dugan, Jr., and H. Haendler, *J. Am. Chem. Soc.* **64** (1942), 2502. *Chem. Abstracts* **37** (1943), 130.

Vanillin determination. D. T. Englis and D. J. Hanahan, *Ind. Eng. Chem., Anal. Ed.* **16** (1944), 505.

Detection of α-dicarbonyl compounds. C. A. Tarnutzer, L. A. Rittschof and C. S. Boruff, *Ind. Eng. Chem., Anal. Ed.* **16** (1944), 621.

A study of the determination of some official aldehydes. M. E. Martin, K. Kelly and M. Green, *J. Am. Pharm. Assocn. (Sci. Ed.)* **35** (1946), 220.

Determination of Phenols.

Quantitative determination of carvacrol and thymol in volatile oils. Kremers and Schreiner, *Pharm. Rev.* **14** (1896), 221.

New bromine method for the determination of aromatic phenols. Its special application to thymol. A. Seidell, *J. Wash. Acad. Sci.* **1**, 196. *Chem. Abstracts* **6** (1912), 203.

Identification of phenols in essential oils. R. M. Reed, *Perfumery Essential Oil Record* **24** (1933), 190. *Chem. Abstracts* **27** (1933), 4346.

Identification of phenols with 2,4-dinitrochlorobenzene. R. W. Bost and Frank Nicholson, *J. Am. Chem. Soc.* **57** (1935), 2368. *Chem. Abstracts* **30** (1936), 1761.

Detection and determination of thymol and carvacrol in essential oils. Y. Mayor, *Parfumerie moderne* **31** (1937), 5. *Chem. Abstracts* **31** (1937), 2749.

Determination of Cineole.

Determination of cineole in eucalyptus and cajuput oils. C. T. Bennett, *Chemist Druggist* **72** (1908), 55. *Chem. Abstracts* **2** (1908), 1325.

The determination of cineole (eucalyptol) in eucalyptus oils. O. Wiegand and M. Lehmann, *Chem. Ztg.* **32** (1908), 109. *Chem. Abstracts* **2** (1908), 1855.

Estimation of cineole by the resorcinol method. C. T. Bennett, *Perfumery Essential Oil Record* **3** (1912), 269. *Chem. Abstracts* **7** (1913), 863.

Notes on the determination of the cineole content of eucalyptus oils. G. A. Harding, *Analyst* **39** (1914), 475. *Chem. Abstracts* **9** (1915), 510.

Estimation of cineole in oil of eucalyptus. J. L. Turner and R. C. Holmes, *Am. J. Pharm.* **87** (1915), 101. *Chem. Abstracts* **9** (1915), 1093.

Determination of eucalyptol in eucalyptus oils. C. T. Bennett and M. S. Salamon, *Perfumery Essential Oil Record* **10** (1919), 211. *Chem. Abstracts* **14** (1920), 1001.

A new method for the estimation of cineole in eucalyptus oil. T. T. Cocking, *Pharm. J.* **105** (1920), 81. *Chemist Druggist* **93** (1920), 1032. *Chem. Abstracts* **14** (1920), 2967.

Eucalyptol determination. C. T. Bennett and M. S. Salamon, *Perfumery Essential Oil Record* **11** (1920), 302. *Chem. Abstracts* **15** (1921), 146.

o-Cresol method for eucalyptol determination. C. T. Bennett and M. S. Salamon, *Perfumery Essential Oil Record* **12** (1921), 11. *Chem. Abstracts* **15** (1921), 1188.

Cresineol method for the determination of cineole. T. T. Cocking, *Perfumery Essential Oil Record* **12** (1921), 339. *Chem. Abstracts* **16** (1922), 1292.

Cresineol method for the determination of cineole. C. E. Sage and J. D. Kettle, *Perfumery Essential Oil Record* **12** (1921), 44. *Chem. Abstracts* **15** (1921), 1597.

Determination of cineole in essential oils. G. Walker, *J. Soc. Chem. Ind.* **42** (1923), 497T. *Chem. Abstracts* **18** (1924), 566.

Estimation of cineole in essential oils by the Cocking process. L. S. Cash and C. E. Fawsitt, *J. Proc. Roy. Soc. N. S. Wales* **57** (1923), 157. *Chem. Abstracts* **18** (1924), 1730.

Determination of cineole in essential oils. α-Naphthol method. T. T. Cocking, *Perfumery Essential Oil Record* **15** (1924), 10. *Chem. Abstracts* **18** (1924), 1030.

Phosphoric acid method for cineole in eucalyptus oils. R. E. Shapter, *Perfumery Essential Oil Record* **15** (1924), 423. *Chem. Abstracts* **19** (1925), 701.

Color test for cineole. E. J. Schorn, *Perfumery Essential Oil Record* **16** (1925), 83. *Chem. Abstracts* **19** (1925), 1754.

Estimation of cineole. Determination of cineole in camphor oils. T. T. Cocking, *Perfumery Essential Oil Record* **18** (1927), 254. *Chem. Abstracts* **21** (1927), 3252.

Cineole determination. J. Allan, *Chemist Druggist* **107** (1927), 615. *Chem. Abstracts* **22** (1928), 2638.

Occurrence of a number of varieties of *Eucalyptus dives* as determined by chemical analysis of the essential oil. II. With remarks on the o-cresol method for estimation of cineole. A. R. Penfold and F. R. Morrison, *J. Proc. Roy. Soc. N. S. Wales* **62** (1928), 72. *Perfumery Essential Oil Record* **19** (1928), 468. *Chem. Abstracts* **23** (1929), 475.

Estimation of cineole in eucalyptus oil. P. A. Berry, *Australasian J. Pharm.* (1929), 203. *Chem. Abstracts* **23** (1929), 3302.

The determination of cineole in cajuput oil. A. Reclaire and D. B. Spoelstra, *Ber. Afdeel. Handelsmuseum Ver. Koloniaal Inst.* No. 54 (1930), 8. *Chem. Abstracts* **25** (1931), 2521.

Determination of cineole in essential oils. Second report. John Allan, et al., *Analyst* **56** (1931), 738. *Chem. Abstracts* **26** (1932), 255.

Determination of cineole in eucalyptus oils. A. T. S. Sissons, *Soc. Chem. Ind. Victoria, Proc.* **32** (1932), 681. *Chem. Abstracts* **27** (1933), 565.

Determination of cineole in eucalyptus oil. P. A. Berry and T. B. Swanson, *Australian New Zealand Assocn. Advancement Sci. Sydney Meeting*, August (1932), 15. *Chem. Abstracts* **27** (1933), 1991.

New molecular compounds of eucalyptol. F. D. Dodge, *J. Am. Pharm. Assocn.* **22** (1933), 20. *Chem. Abstracts* **27** (1933), 2142.

The determination of the freezing point of a mixture of pure o-cresol and pure cineole in molecular proportions. P. A. Berry and T. B. Swanson, *Australasian J. Pharm.* **14**, 550. *Perfumery Essential Oil Record* **24** (1933), 224. *Chem. Abstracts* **27** (1933), 4975.

Macro-, micro-, and histochemical detection of cineole. R. Wasicky and E. Gmach, *Scientia Pharm.* **5** (1934), 113. *Chem. Abstracts* **29** (1935), 1577.

Allyl Isothiocyanate.

Assay of mustard oil and mustard essence. Gadamer, *Arch. Pharm.* **237**, 110. *Chem. Zentr.* II (1899), 457.

Assay of mustard oil and mustard essence. Grützner, *Arch. Pharm.* **237**, 110. *Chem. Zentr.* I (1899), 1227.

Assay of volatile mustard oil. P. Roeser, *J. pharm. chim.* [6] **15** (1902), 361. *Chem. Zentr.* I (1902), 1254.

The volumetric assay of allyl mustard oil. M. Kuntze, *Arch. Pharm.* **246** (1908), 58. *Chem. Abstracts* **2** (1908), 1857.

New process of the determination of allyl mustard oil in powdered black mustard. R. Meesemaecker and J. Boivin, *J. pharm. chim.* [8] **11** (1930) 478. *Chem. Abstracts* **25** (1931), 771.

Estimation of mustard oil in semen sinapis via the D. A. B. 6. Hans Kaiser and Otto Leeb, *Süddeut. Apoth. Ztg.* **73** (1933), 612. *Chem. Abstracts* **28** (1934), 256.

Determination of allyl mustard oil in mustard flour. R. Gros and G. Pichon, *J. pharm. chim.* [8] **19** (1934), 249. *Chem. Abstracts* **28** (1934), 5179.

Simple volumetric estimation of mustard oil in spiritus sinapis. H. Kaiser and E. Fürst, *Apoth. Ztg.* **50** (1935), 1734. *Chem. Abstracts* **30** (1936), 1177.

Determination of Ascaridole.

Oil of chenopodium and ascaridole. Y. R. Naves, *Parfums France* **13** (1935), 4. *Chem. Abstracts* **29** (1935), 2663.

Determination of Methyl Anthranilate.

Rapid identification of methyl anthranilate. S. Sabetay, *Ann. fals.* **28** (1935), 478. *Chem. Abstracts* **30** (1936), 702.

Determination of Essential Oil Content of Plant Material.

Apparatus for the determination of volatile oil. J. F. Clevenger, *J. Am. Pharm. Assocn.* **17** (1928), 346. *Chem. Abstracts* **22** (1928), 2439.

Rapid determination of ethereal oils in alcoholic solutions. G. Rosenberger, *Parfümeur* **3** (1929), 78. *Chem. Abstracts* **23** (1929), 5542.

Assay of drugs yielding volatile oils. G. R. A. Short, *Perfumery Essential Oil Record* **22** (1931), 208. *Chem. Abstracts* **25** (1931), 5245.

The estimation of volatile oil in cloves. L. G. Mitchell and S. Alfend, *J. Assocn. Official Agr. Chem.* 15 (1932), 293. *Chem. Abstracts* **26** (1932), 4414.

Simplified determination of essential oils in plants. A. S. Ginzberg, *Khim. Farm. Prom.* **8–9** (1932), 326. *Chem. Abstracts* **27** (1933), 372.

Estimation of essential oils in drugs and plant material. R. Wasicky, G. Rotter and T. Alber, *Pharm. Presse, Wiss.-prakt. Heft* (1933), 57. *Chem. Abstracts* **27** (1933), 3557.

Estimation of essential oils in drugs. Gerhard Scholz, *Apoth. Ztg.* **49** (1934), 1690. *Chem. Abstracts* **29** (1935), 1577.

Estimation of essential oil in drugs. Horkheimer, *Pharm. Ztg.* **80** (1935), 148. *Chem. Abstracts* **29** (1935), 2660.

Determination of the essential oil content and yield of plants and drugs. Y. R. Naves and M. G. I., *Parfums France* **13** (1935), 197. *Chem. Abstracts* **29** (1935), 7583.

Improved method for estimation of essential oil content of drugs. T. Tusting Cocking and G. Middleton, *Perfumery Essential Oil Record* **26** (1935), 207. *Chem. Abstracts* **29** (1935), 7014.

Estimation of essential oils in drugs, and the oil content of peppermint, sage, fennel and caraway. L. Kofler and G. V. Herrensschwand, *Arch. Pharm.* **273** (1935), 388. *Chem. Abstracts* **30** (1936), 568.

Estimation of essential oils in drugs and plant material. R. Wasicky, F. Graf and S. Bayer, *Scientia Pharm.* **6** (1935), 101. *Chem. Abstracts* **30** (1936), 570.

Essential oil industry of foreign lands. C. A. Browne, *J. Chem. Education* **11** (1934), 131. *Chem. Abstracts* **28** (1934), 2125.

Improved method for the estimation of essential oil content of drugs. T. Tusting Cocking and G. Middleton, *Quart J. Pharm. Pharmacol.* **8** (1935), 435. *Chem. Abstracts* **30** (1936), 813.

New method for the determination of essential oils in drugs and fluid extracts. Hans Kaiser and Elizabeth Fürst, *Süddeut. Apoth.-Ztg.* **76** (1936), 265. *Chem. Abstracts* **30** (1936), 4269.

Evaluation of crude drugs containing ethereal oils. H. Theo Mijnhardt, *Pharm. Weekblad* **73** (1936), 791. *Chem. Abstracts* **30** (1936), 6131.

The determination of volatile oils in spices. E. L. Krugers Dagneaux, *Chem. Weekblad* **33** (1936), 544. *Chem. Abstracts* **30** (1936), 8421.

Estimation of volatile oil in plant material. W. A. N. Markwell, *Perfumery Essential Oil Record* **27** (1936), 325. *Chem. Abstracts* **31** (1937), 504.

Determination of volatile oil in drugs. H. O. Meek and F. G. Salvin, *Perfumery Essential Oil Record* **28** (1937), 274. *Chem. Abstracts* **31** (1937), 8112.

Estimation of ethereal oils in spices. P. A. Rowaan and A. J. van Duuren, *Chem. Weekblad* **34** (1937), 534. *Chem. Abstracts* **32** (1938), 4683.

Determination of volatile oil in drugs. H. O. Meek and F. G. Salvin, *Quart. J. Pharm. Pharmacol.* **10** (1937), 471. *Chem. Abstracts* **32** (1938), 6001.

Determination of volatile oils in vegetable drugs. Louis Goldberg, R. K. Snyder, E. H. Wirth and E. N. Gathercoal, *J. Am. Pharm. Assocn.* **27** (1938), 385. *Chem. Abstracts* **32** (1938), 7668.

Estimation of essential oil in drugs. O. Moritz, *Arch. Pharm.* **276** (1938), 368. *Chem. Abstracts* **32** (1938), 8690.

A simple apparatus for the estimation of the essential oil content of vegetable, drugs and spices. Yu Hsieh and Ying Hung, *J. Chinese Chem. Soc.* **8** (1941), 32. *Chem. Abstracts* **37** (1943), 221.

Determination of volatile oils in drugs with the aethometer. Zd. Zachystal, *Chem. Listy* **35** (1941), 231. *Chem. Abstracts* **37** (1943), 3224.

The elucidation of essential oil compositions. M. M. Rama Rao, *Soap, Perfumery and Cosmetics* **15** (1942), 214. *Chem. Abstracts* **37** (1943), 500.

Comparative study of the determination of volatile oils in medicinal and spice plants. K. H. Bauer and L. R. Pohloudek, *Pharm. Ind.* **9** (1942), 181. *Chem. Zentr.* II (1942), 1066. *Chem. Abstracts* **37** (1943), 5194.

A comparison of the methods for determining the percentages of volatile oils in drugs. R. Holdermann and H. Pfäffle, *Deut. Apoth.-Ztg.* **57** (1942), 142. *Chem. Zentr.* II (1942), 810. *Chem. Abstracts* **37** (1943), 5555.

Determination of essential oils in plant material. K. H. Bauer and R. Pohloudek, *Pharm. Zentralhalle* **84** (1943), 223. *Chem. Abstracts* **39** (1945), 387.

Determination of Water Content.

The solubility of water in essential oils. J. C. Umney and S. W. Bunker *Perfumery Essential Oil Record* **3** (1912), 101, 197. *Chem. Abstracts* **6** (1912), 2487, 2976.

The quantitative determination of water in substances by means of alkyl magnesium halogen compounds. Th. Zerewitinoff, *Z. anal. Chem.* **50**, 680. *Chem. Abstracts* **6** (1912), 203.

Preliminary notes on the direct determination of moisture. G. L. Bidwell and W. F. Sterling, *Ind. Eng. Chem.* **17** (1925), 147. *Chem. Abstracts* **19** (1925), 620.

The normal moisture content of essential oils. L. S. Glichitch, *Parfums France* **34** (1925), 351. *Chem. Abstracts* **20** (1926), 798.

Determination of moisture in essential oils. R. M. Reed, *Oil and Soap* **9**, No. 3 (1932), 66. *Chem. Abstracts* **26** (1932), 2555.

Determination of moisture in aromatic hydrocarbons. Yu. L. Khmel'-nitskiĭ, A. I. Doladugin and A. V. Guseva, *Zavodskaya Lab.* **11** (1945), 534. *Chem. Abstracts* **40** (1946), 2417.

Test for Heavy Metals.

Detection and determination of traces of metals in essential oils, concretes, and *enfleurage* products. Y. R. Naves, *Parfums France* **12** (1934), 89, 116, 139. *Chem. Abstracts* **28** (1934), 5928.

The action of essential oils in alcoholic solution on various metals. G. A. Rosenberger, *Seifensieder-Ztg.* **64** (1937), 967. *Chem. Abstracts* **32** (1938), 2073.

The heavy metals test for volatile oils. Frederick K. Bell and John C. Krantz, Jr., *J. Am. Pharm. Assocn.* **31** (1942), 533. *Chem. Abstracts* **37** (1943), 2136.

Investigation of the heavy metals test in "The United States Pharmacopoeia," Twelfth Revision. W. W. Edman and C. H. Bundy, *Proc. Sci. Sect. Toilet Goods Assocn.* No. 4 (1945), 28–32, Discussion, 32. *Chem. Abstracts* **40** (1946), 2589.

Test for Halogens.

Benzyl alcoholic solutions of potassium hydroxide and their applications. Determination of halogens. Sébastien Sabetay and Jean Bléger, *Bull. soc. chim.* **47** (1930), 114. *Chem. Abstracts* **24** (1930), 2082.

Safrole.

Determination of safrole in essential oils. Y. Huzita and K. Nakahara, *J. Chem. Soc. Japan* **62** (1941), 5. *Chem. Abstracts* **37** (1943), 3882.

Detection of Petroleum.

The detection of petroleum in ethereal oils. J. Zimmermann, *Chem. Weekblad* **31** (1934), 132. *Chem. Abstracts* **28** (1934), 2467.

Methyl and Ethyl Alcohol.

Determination of ethyl and methyl alcohol in natural essential oils. R. Garnier and L. Palfray, *Perfumery Essential Oil Record* **26** (1935), 259. *Chem. Abstracts* **29** (1935), 6698.

Presence of methanol and formaldehyde in essential oils and in ethanol solutions of essential oils. Y. R. Naves, *Parfums France* **13** (1935), 60, 91. *Chem. Abstracts* **29** (1935), 5597.

Detection of High Boiling Esters.

Detection of diethyl phthalate in spirit. Henryk Szancer, *Pharm. Zentralhalle* **70** (1929), 502. *Chem. Abstracts* **23** (1929), 5006.

Analysis of essential oils: determination of ethyl phthalate by K phthalate method. S. Sabetay, *Ann. fals.* **28** (1935), 100. *Chem. Abstracts* **29** (1935), 3778.

Phthalic esters—detection, identification and determination in essential oils, natural perfume substances and synthetic perfumes. Y. R. Naves and S. Sabetay, *Perfumery Essential Oil Record* **29** (1938), 22. *Chem. Abstracts* **32** (1938), 2687.

Determination of ethyl phthalate in presence of essential oils, natural perfumes and synthetic perfumes. Y. R. Naves and S. Sabetay, *Bull. soc. chim.* [5] **5**, (1938), 102. *Chem. Abstracts* **32** (1938), 2688.

Detection of *Mentha arvensis* Oil.

Detection of Japanese mint oil in peppermint oils. D. C. Garratt, *Analyst* **60** (1935), 369. *Chem. Abstracts* **29** (1935), 5596.

The application of the furfural test for mint oils to other essential oils D. C. Garratt, *Analyst* **60** (1935), 595. *Chem. Abstracts* **29** (1935), 7580.

Miscellaneous.

The hydrogen number of some essential oils and essential oil products. Oils of sassafras, anise, fennel, clove, and pimenta. A. R. Albright, *J. Am. Chem. Soc.* **36** (1914), 2188. *Chem. Abstracts* **8** (1914), 3838.

Catalytic hydrogenation of essential oils. L. Palfray and S. Sabetay, *15th Congr. chim. ind.* [*Bruxelles,* September (1935)] (1936), 762. *Chem. Abstracts* **30** (1936), 5725.

High-pressure catalytic hydrogenations. Essential oils and esters. L. Palfray and S. Sabetay, *Bull. soc. chim.* [5] **3** (1936), 682. *Chem. Abstracts* **30** (1936), 4461.

Special equipment for catalytic hydrogenation at high pressures. L. Palfray, *Bull. soc. chim.* [5] **3** (1936), 508. *Chem. Abstracts* **30** (1936), 3282.

Methods of diene syntheses and their interest in the chemistry of perfumes. Y. R. Naves, *Parfums France* **12** (1934), 255. *Chem. Abstracts* **29** (1935), 1581.

The detection and estimation of α-phellandrene in essential oils. A. J. Birch, *J. Proc. Roy. Soc. N. S. Wales* **71** (1937), 54. *Chem. Abstracts* **31** (1937), 8109.

Determination of coumarin in vanilla extract by a modification of the steam distillation method. Ira J. Duncan and R. B. Dustman, *Ind. Eng. Chem., Anal. ed.* **9** (1937), 416. *Chem. Abstracts* **31** (1937), 8057.

Precise method for the determination of coumarin, meliolotic acid and coumaric acid in plant tissue. Willard L. Roberts and Karl P. Link, *J. Biol. Chem.* **119** (1937), 269. *Chem. Abstracts* **31** (1937), 6286.

Determination of coumarin in sweet clover. Comparison of the steam distillation and alcohol extraction methods. Ira J. Duncan and R. B. Dustman, *Ind. Eng. Chem., Anal. Ed.* **9** (1937), 471. *Chem. Abstracts* **31** (1937), 8826.

Detection of coumarin in drugs. P. Casparis and E. Manella, *Pharm. Acta Helv.* **19** (1944), 158. *Chem. Abstracts* **39** (1945), 153.

The determination of coumarin in plant substances. Milos Cerny, *Chem. Obzor.* **18** (1943), 149. *Chem. Abstracts* **39** (1945), 3328. *Chem. Zentr.* I (1944), 40.

Coumarin determination. D. T. Englis and D. J. Hanahan, *Ind. Eng. Chem., Anal. Ed.* **16** (1944), 505. *Drug & Cosmetic Ind.*, October (1944), 465.

A color reaction for coumarin. R. B. Gelchinskaya, *Selektsiya i Semenovodstvo* No. 6 (1940), 27. *Khim. Referat. Zhur.* **4,** No. 1 (1941), 89. *Chem. Abstracts* **37** (1943), 1673.

Oil of birch and methyl salicylate: some new color reactions for the differentiation of oil of wintergreen. G. N. Watson and L. E. Sayre, *J. Am. Pharm. Assocn.* **3** (1914), 1658. *Chem. Abstracts* **9** (1915), 512.

Color reaction of geranium oil and certain commercial rhodinols. S. Sabetay, *Riechstoff Ind.* **8** (1933), 26. *Chem. Abstracts* **27** (1933), 2530.

Blue hydrocarbon occurring in some essential oils. Method of separation of "azulenes" with mineral acids. A. E. Sherndal, *J. Am. Chem. Soc.* **37** (1915), 167. *Chem. Abstracts* **9** (1915), 596.

A color reaction for azulene sesquiterpenes. S. Sabetay and H. Sabetay, *Compt. rend.* **199** (1934), 313. *Chem. Abstracts* **28** (1934), 6721.

Determination of methoxyl and ethoxyl groups. C. L. Palfray, *Documental sci.* **4** (1935), 1. *Chem. Abstracts* **30** (1936), 5150. *Chem. Zentr.* I (1935), 3821.

The methyl index of some balsams, rosins and drugs of animal origin. M. M. Janot and S. Sabetay, *Bull. sci. pharmacol.* **42** (1935), 529. *Chem. Abstracts* **30** (1936), 1944.

The determination of methoxyl-ethoxyl. W. W. Scott, "Standard Methods of Chemical Analysis," D. Van Nostrand Co., Inc., New York (1939), Vol. II, 2527.

New method of determining relative surface tension (capillary activity). Its application to the measurement of essential oils and related substances. Arno Müller, *J. prakt. Chem.* **134** (1932), 158. *Chem. Abstracts* **26** (1932), 4514.

Determination of viscosity. W. W. Scott, "Standard Methods of Chemical Analysis," D. Van Nostrand Co., Inc., New York (1939), Vol. II, 1718, 1719, 1724.

Antimony trichloride, a new reagent for the double bond. S. Sabetay, *Compt. rend.* **197** (1933), 557.

Reactions of terpenes with antimony trichloride. Victor E. Levine and Eudice Richman, *Biochem. J.* **27** (1933), 2051. *Chem. Abstracts* **28** (1934), 3392.

The characterization of double bonds by antimony trichloride. R. Delaby, S. Sabetay and M. Janot, *Compt. rend.* **198** (1934), 276. *Chem. Abstracts* **28** (1934), 2321.

Ethereal oils containing sulfur and their examination. S. L. Malowan, *Der Parfümeur* **4** (1930), 21. *Chem. Abstracts* **24** (1930), 1703.

Estimation of the Tillmans chloramine number in essential oils. P. W. Danckwortt and J. Hotzel, *Arch. Pharm.* **275** (1937), 468. *Chem. Abstracts* **31** (1937), 8111.

Raman effect of the terpenes. Some monocyclic terpenes. G. Dupont, P. Daure and J. Levy, *Bull. soc. chim.* **51** (1932), 921. *Chem. Abstracts* **27** (1933), 26.

Citronellol—rhodinol isomerism and Raman spectra. Y. R. Naves, G. Brus and J. Allard, *Compt. rend.* **200** (1935), 1112. *Chem. Abstracts* **29** (1935), 3465.

Light scattering, Raman spectra and allied physical properties of some essential and vegetable oils. C. Dakshinamurti, *Proc. Indian Acad. Sci.* **5A** (1937), 385. *Chem. Abstracts* **31** (1937), 6070.

Examination of essential oils by measurement of absorption in the ultraviolet. D. van Os and K. Dykstra. *J. Pharm. Chim.* **25** (1937), 437, 485. *Chem. Abstracts* **31** (1937), 8118.

Raman spectra of some terpene aldehydes. R. Manzoni-Ansidei, *Atti accad. Ital., Rend. classe sci. fis., mat. nat.* [7], **1** (1940), 558. *Chem. Abstracts* **37** (1943), 831.

Determination of absolute oil (irone) in concrete oil of iris. L. S. Glichitch and Y. R. Naves, *Parfums France* **9** (1931), 371. *Chem. Abstracts* **26** (1932), 2553.

General Analysis.

The oxidation assay of essential oils. F. D. Dodge, *Orig. Com. 8th Intern. Congr. Appl. Chem.* **6**, 86. *Chem. Abstracts* **6** (1912), 2976.

Notes on the anlysis of some essential oils. F. D. Dodge, *J. Am. Pharm. Assocn.* **3** (1914), 1664. *Chem. Abstracts* **9** (1915), 512.

Microchemical characterization of essential oils. L. Rosenthaler, *Pharm. Acta Helv.* **1** (1926), 117. *Chem. Abstracts* **21** (1927), 1870.

The analysis of perfume ingredients and essential oils with attention to their use in toilet preparations. H. P. Kaufmann, J. Baltes and F. Josephs, *Fette u. Seifen* **44** (1937), 506. *Chem. Abstracts* **32** (1938), 4724.

Recent progress in chemical methods applied to the functional analysis of essential oils. S. Sabetay and Y. R. Naves, *Compt. rend. 17th Congr. chim. ind., Paris,* September–October (1937), 777. *Chem. Abstracts* **32** (1938), 6805.

Specification and analytical evaluation of essential oils and natural perfumes. Y. R. Naves, A review and general discussion—(1) The value of analytical characteristics—*Perfumery Essential Oil Record* **31** (1940), 61. (2) Analytical interpretation—*Ibid.*, 86.

Constitutents of essential oils and natural perfumes. Notes on the investigation of their nature. Y. R. Naves, A review and discussion. *Perfumery Essential Oil Record* **31** (1940), 161.

Analytical assay of perfumery raw material. M. M. Rama Rao, *Soap, Perfumery and Cosmetics* **14** (1941), 757, 760, 770. *Chem. Abstracts* **37** (1943), 226; also **37** (1943), 500.

The analytic assay of perfumery raw materials. Chemical tests. M. M. Rama Rao, *Soap, Perfumery and Cosmetics* **15** (1942), 37, 52, 99, 114. *Chem. Abstracts* **37** (1943), 500.

The elucidation of essential oil compositions. M. M. Rama Rao, *Soap, Perfumery and Cosmetics* **15** (1942), 214. *Chem. Abstracts* **37** (1943), 500.

Structure of certain acyclic isolates. M. F. Carroll. *Perfumery Essential Oil Record* **38**, No. 7 (1947), 226.

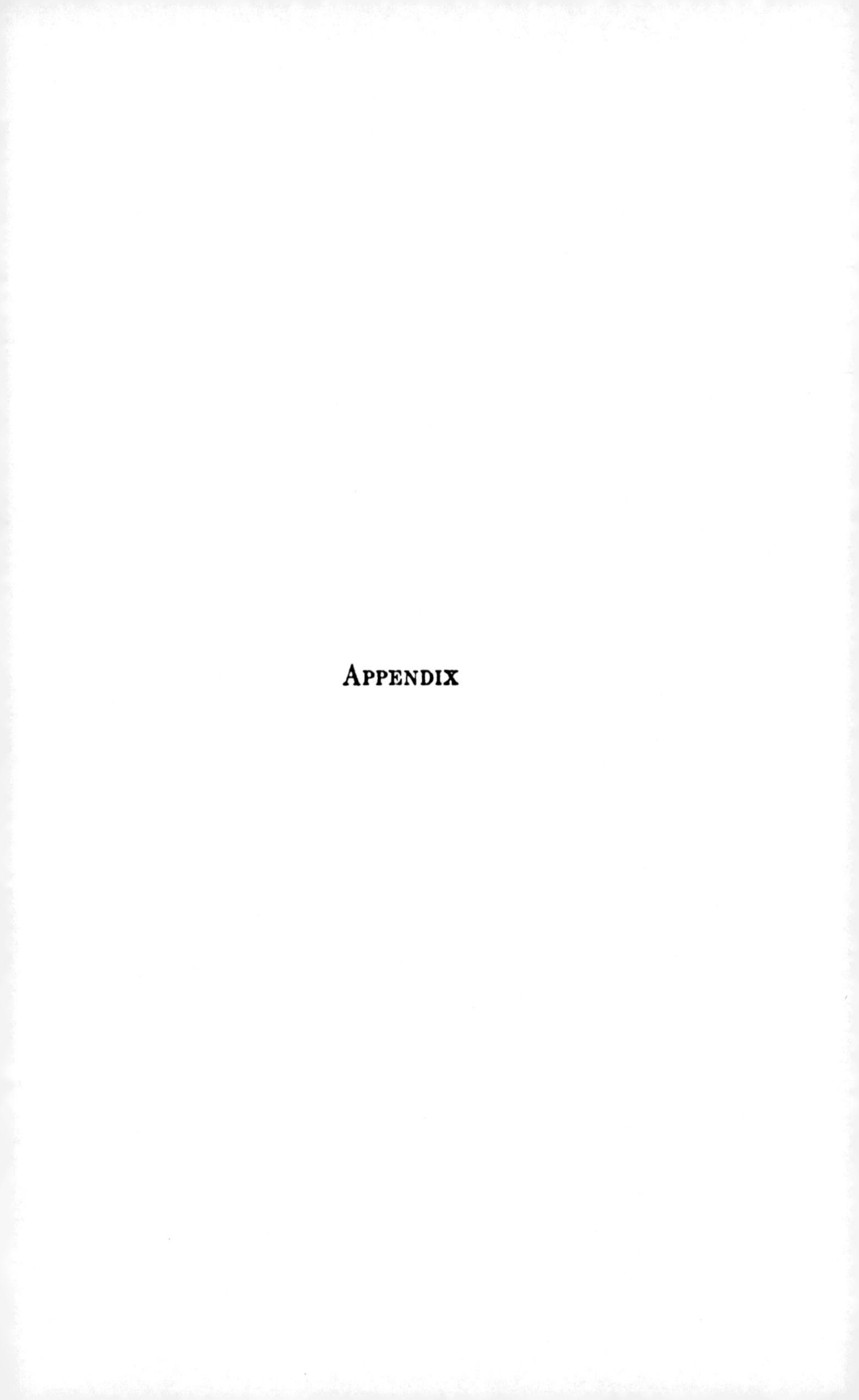

Appendix

Note. All temperatures in this book are given in degrees centigrade unless otherwise noted.

Appendix

I. USE OF ESSENTIAL OILS

Essential or, as they are also called, volatile or ethereal oils, find an amazingly wide and varied application in many industries for the scenting and flavoring of all kinds of consumers' finished products, some of them luxuries, most of them necessities in our advanced civilization. Many of these products contribute directly to our health, happiness and general well being. To underestimate their importance is to disregard entirely the physiological advantage of continuing to have available these accustomed necessities of our daily life.

Some volatile oils are more or less powerful external or internal antiseptics, others possess an analgesic, haemolytic, or antizymatic action, still others act as sedatives, stimulants and stomachics. The anthelmintic properties of certain volatile oils, especially wormseed oil, are well known. A great deal has been published on this subject, in books and papers on pharmacology, pathology and physiology, especially on the antiseptic and bactericidal activities of volatile oils, but many of the findings remain confusing, contradictory and require further elucidation. Much work will still have to be done on this fascinating and promising topic which cannot be discussed here as it would exceed by far the scope of this treatise.[1]

Spices with their flavor principles, volatile oils, have been used as flavoring materials since time immemorial. Yet, not always is it sufficiently realized that they are actually indispensable to man in order to bring about proper digestion of food. The digestive juices containing digestive enzymes such as pepsin, trypsin, lipase, amylase, etc., are secreted into the stomach and intestines only when stimulated by the smell and taste of pleasantly flavored food. The mouth "becomes watery" and so does the stomach. As the individual digests more food with a pleasant taste, more digestive juices will be secreted, a fact equally true in the reverse.

The wide use of volatile oils in perfumes, cosmetics and the scenting of soaps hardly needs to be mentioned.

Increasingly, volatile oils and their aromatic isolates serve also for the covering of somewhat objectionable odors, as, for instance, in the case of artificial leathers. Acceptable and useful articles can now be made from raw materials that were formerly discarded or overlooked because of dis-

[1] The reader is referred to the paper "Physiological Aspects of the Essential Oils," by G. Malcolm Dyson.—*Perfumery Essential Oil Record*, Special Number, 21 (1930), 287.

agreeable odors. In most instances the incorporation of aromatics into products such as synthetic rubbers and latices has opened new and profitable fields for manufacturers.

Few people realize that in the course of a single day, from morning to night, we use or consume a great variety of volatile oils which originate from many corners of the world. All of us thereby contribute to the employment of innumerable workers and their families, often primitive peoples in distant lands. Frequently these small producers depend for their income upon our continued use of these oils which have thus become really "essential" not only to these growers for their livelihood, but also to our industries so that they may be able to manufacture their specialties, many of them marketed internationally. The essential oils industry, as such, is a small one, apt to be overlooked in the economy of a country. Its total yearly turnover may be estimated as amounting to only a few scores of millions of dollars, but the turnover of the consumers' finished goods, which require small additions of essential oils, reaches into many billions per year. Countless is the number of people who are involved in the developing, manufacturing, controlling, advertising, marketing and selling of these products.

The following list will enumerate some of the various industries employing volatile oils, aromatic isolates, or combinations. For convenience sake, they are listed alphabetically, not according to importance. While in the case of the toilet goods industry, it is possible to group the products as belonging to this one industry, such a fine distinction cannot always be made with other products. Therefore, the terms "manufacturer" and "industry" will have to be applied interchangeably as the groupings may require. Neither can a clear line be drawn between the products manufactured by these various industries.

Adhesives

Glues
Paper and industrial tapes
Pastes
Porcelain cements
Rubber cements
Scotch tapes, etc.

Animal Feed Industry

Cat foods
Cattle feeds
Dog foods, etc.

Automobile Industry

Automobile finishing supplies
Cleaners
Polishes
Soaps, etc.

Baked Goods Industry

Biscuits
Cakes
Crackers
Doughnuts
Fruit cakes
Icings
Mince meat
Pies
Pretzels
Puddings
Sandwich fillings, etc.

Canning Industry

Fish
Meats
Sauces
Soups, etc.

Chewing Gum Industry

Chewing gums
Coated gums, etc.

Condiment Industry

Catsups
Celery and other salts
Chili sauces
Mayonnaises
Mustards
Pickled fish
Relishes
Salad dressings
Table sauces
Vinegars, etc.

Confectionery Industry

Chocolates
Fondants
Gum drops
Hard candies
Jellies
Mints
Panned goods
Soft center candies, etc.

Dental Preparations

Dentists' preparations
Mouth washes
Tooth pastes
Tooth powders, etc.

Exterminators and Insecticide Supplies

Bedbug sprays
Cattle sprays
Cockroach powders
Fly sprays
Japanese beetle attractants
Mosquito repellents
Naphthalene blocks
Paradichlorobenzene blocks
Plant sprays
Rat baits
Rodent odor eliminators, etc.

Extract Industry

Commercial extracts
Home extracts, etc.

FOOD INDUSTRY (GENERAL)

Cheeses
Cornstarch puddings
Dehydrated soups, meats and vegetables
Gelatin desserts
Mince meats
Pie fillers
Prepared cake mixes
Rennet desserts
Sauerkraut
Vegetable oils and fats, etc.

HOUSEHOLD PRODUCTS

Bluings
Deodorants
Furniture polishes
Laundry soaps
Room sprays
Starches
Vacuum cleaner pads, etc.

ICE CREAM INDUSTRY

Ice creams
Ices
Prepared ice cream mixes
Sherbets, etc.

INSECTICIDE INDUSTRY

Attractants
Disinfectants
Insecticides
Repellents
Sprays, etc.

JANITOR'S SUPPLIES

Detergents
Disinfectants
Floor polishes
Floor waxes
Scrub soaps
Sink cleaners
Sweeping compounds, etc.

MEAT PACKING INDUSTRY

Bolognas
Frankfurters
Prepared meats
Sausages, etc

PAINT INDUSTRY

Bituminous paints
Casein paints
Enamels
Lacquers
Paint and varnish removers
Paint diluents
Paints
Rubber paints
Synthetic coatings
Varnishes, etc.

Paper and Printing Industry

Carbon papers
Crayons
Drinking cups
Industrial tapes
Inking pads
Labels
Paper bags and food wrappers
Printing and writing inks
Printing paper
Typewriter ribbons
Writing paper, etc.

Perfume and Toilet Industry

Baby preparations
Bath preparations
Body deodorants
Colognes
Creams
Depilatories
Eye shadows
Facial masks
Hair preparations
Handkerchief extracts
Incense
Lipsticks
Lotions
Manicure preparations
Powders
Room and theatre sprays
Rouges
Sachets
Shaving preparations
Suntan preparations
Toilet waters, etc.

Petroleum and Chemical Industry

Bluing oils
Fuel oils
Grease deodorants
Greases
Lubricating oils
Naphtha solvents
Neoprene
Organic solvents
Petroleum distillates
Polishes
Sulfonated oils
Tar products
Waxes, etc

Pharmaceutical Industry

Antiacid tablets and powders
Cough drops
Elixirs
Germicides
Hospital sprays
Hospital supplies
Inhalants
Laxatives
Liniments
Medicinal preparations
Ointments
Patent medicines
Tonics
Vitamin flavor preparations
Wholesale druggists' supplies, etc.

Pickle Packing Industry

Dill pickles
Fancy cut pickles
Sour pickles
Sweet pickles, etc.

Preserve Industry

Fruit butters
Jams
Jellies

Rectifying and Alcoholic Beverage Industry

Bitters
Cordials
Rums
Vermouths
Whiskies
Wines, etc.

Rubber Industry

Baby pants
Gloves
Natural and synthetic latices
Shower curtains
Surgical supplies
Synthetic rubber products of all kinds
Toys
Water proofing compounds, etc.

Soap Industry

Cleaning powders
Detergents
Household soaps
Laundry soaps
Liquid hand soaps
Scrub soaps
Shampoos
Sweeping compounds
Technical soaps
Toilet soaps, etc.

Soft Drink Industry

Carbonated beverages
Cola drinks
Fountain syrups
Ginger ales
Root beers
Soda fountain supplies
Soft drink powders
Sundae toppings, etc.

Textile Processing Products

Artificial leather and fabric coatings
Dyes
Hosiery sizing
Linoleum
Oil cloths
Sisal deodorants
Textile chemicals
Textile oils
Upholstery materials
Water proofing materials, etc.

Tobacco Industry

Chewing tobaccos
Cigarettes
Cigars
Smoking tobaccos
Snuffs

VETERINARY SUPPLIES

Cattle sprays
Deodorants
Dog and cat soaps
Insect powders
Mange medicines and ointments, etc.

DIVERSIFIED INDUSTRIES

Alcohol denaturing compounds
Candles
Ceramics
Cleaners' products
Embalming fluid deodorants
Optical lenses
War gas simulants, etc.

II. THE STORAGE OF ESSENTIAL OILS

From the outset it should be stated that little indeed is known about the actual processes which cause the spoilage of an essential oil. Usually it is attributed to such general reactions as oxidation, resinification, polymerization, hydrolysis of esters, and to interreaction of functional groups. These processes seem to be activated by heat, by the presence of air (oxygen), of moisture, and catalyzed by exposure to light and in some cases, possibly by metals. There is no doubt that oils with a high content of terpenes (all citrus oils, pine needle oils, oil of turpentine, juniper berry, etc.) are particularly prone to spoilage, due probably to oxidation, and especially resinification. Being unsaturated hydrocarbons, the terpenes absorb oxygen from the air. Light seems to be of lesser importance as a factor causing deterioration, than is moisture.

Essential oils containing a high percentage of esters (oil of bergamot, lavender, etc.) turn acid after improper storage, due to partial hydrolysis of esters. The aldehyde content of certain oils (lemongrass, for example) gradually diminishes, yet much more slowly than if the isolated aldehyde (citral, in this case) were stored as such. Quite probably the essential oil contains also some natural antioxidants, yet unknown, which to a certain extent protect the aldehyde while it is contained in the oil. Fatty oils, with a few exceptions, are very prone to oxidation, but such spoilage can be retarded or prevented altogether by the addition of suitable antioxidants, such as hydroquinone or its monomethyl ether. Certain types of essential oils, especially those containing alcohols (geranium oil, for example), are quite stable and stand prolonged storage. Still others, patchouly and vetiver, for instance, improve considerably on aging; in fact, they should be aged for a few years before being used in perfume compounds.

As a general rule, any essential oil should first be treated to remove metallic impurities, freed from moisture and clarified, and then be stored in well-filled, tightly closed containers, at low temperature and protected from light. Bottles of hard and dark colored glass are eminently suitable for small quantities of oil, but larger quantities will have to be stored in metal drums, heavily tin lined, if possible. A layer of carbon dioxide or nitrogen gas blown into the container before it is sealed will replace the layer of air above the oil and thereby assure added protection against oxidation.

Previous to storing, as pointed out, the oil should be carefully clarified and any moisture removed as the presence of moisture seems to be one of the worst factors in the spoilage of an essential oil. The small lots can be dehydrated quite readily by the addition of anhydrous sodium sulfate, by thoroughly shaking, standing and filtration. Calcium chloride must never be used for dehydration of an essential oil, as this chemical is apt to form complex salts with certain alcohols. Larger commercial lots of oil are not always easy to clarify. Some oils, such as vetiver, give a great deal of trouble. The simplest procedure is to add a sufficient amount of common salt to the lot, to stir the mixture for a while, and to let it stand until the supernatant oil has become clear and can be drawn off the tank. The lower layer will be cloudy and needs to be filtered clear. If filtration through plain filter paper does not give a clear oil, kieselguhr or specially prepared filtering clay should be placed into the filter. Care must be exercised in the selection of the filtering medium as some media, activated carbon for example, may react chemically with certain constituents of the oil and affect its quality. Large quantities of oil should be filtered through filter presses which are readily available through any supply house. Centrifuging in high-speed centrifuges is an excellent means of clarifying essential oils. Not only moisture but also waxy material depositing after a certain period of storage, if possible at low temperature in a freezing room, can thus be eliminated.

Some lots of essential oils, especially those with a high content of phenols (clove, bay, thyme, origanum, etc.) arrive from the producing fields often in a crude form and dark colored, due to the presence of metallic impurities. Such lots must be decolorized before they can be placed at the disposal of the consumer. In many cases the dark color may be removed by the formation of complex salts with certain organic acids. For this purpose sufficient powdered tartaric acid is added to the oil, the mixture stirred for some time and permitted to settle. The supernatant clear oil can finally be drawn off, while the lower layer has to be filtered until clear. If the treatment with solid tartaric acid does not give satisfactory results, a concentrated aqueous solution of the acid is added to the oil. After thoroughly stirring, the mixture is allowed to stand until the two liquid layers separate clearly. The upper part of the oil layer should then be sufficiently clear to be drawn off, while the lower layer and especially the intermediary layer, need further treatment by clarification and filtration. Here again high-speed centrifugings are of great help. In cases where the color cannot be eliminated by treatment with organic acids, the oil will have to be clarified by redistillation or rectification. (For details, see section on Distillation in Chapter 3.)

III. TABLES OF BOILING POINTS OF ISOLATES AND SYNTHETICS AT REDUCED PRESSURE

These tables of boiling points at reduced pressures are taken from the excellent work of von Rechenberg: "Einfache und Fraktionierte Destillation in Theorie und Praxis," published in 1923 by Schimmel and Company, Miltitz bei Leipzig.

Unfortunately, this outstanding effort of von Rechenberg is little known in the essential oil industry. Moreover, it never has been readily available to the American chemist, since few libraries in the United States have this important book on their shelves.

In the following pages will be found the boiling point (in degrees centigrade) at various pressures (expressed in millimeters of mercury) of some two hundred isolates and synthetics. Von Rechenberg's complete listing includes an additional two hundred items, but these have been purposely omitted since they do not deal directly with this industry.

The individual compounds are arranged in order of ascending boiling point at 760 mm. pressure.

INDEX TO TABLES

INDEX TO TABLES—(*Cont.*)

INDEX TO TABLES—(*Cont.*)

B.P. 760 mm. M.P.	1 Acetaldehyde 22.38° −123.3°	2 Methyl Formate 31.8° ..	3 Ethyl Ether 34.60° −116.2°	4 *n*-Pentane 36.06° −130.8°
800 mm.	23.76°	33.16°	36.01°	37.53°
700	20.14	29.59	32.30	33.67
600	16.17	25.68	28.24	29.43
500	11.51	21.10	23.46	24.46
400	+5.90	15.58	17.72	18.48
300	−0.88	8.91	10.78	11.25
200	−9.67	+0.25	+1.77	+1.87
100	−24.13	−13.13	−12.15	−12.63
90	−25.17	−15.00	−14.10	−14.66
80	−27.37	−17.16	−16.35	−17.01
70	−29.65	−19.41	−18.69	−19.45
60	−32.24	−21.71	−21.54	−22.21
50	−35.22	−24.89	−24.39	−25.38
40	−38.97	−28.59	−28.24	−29.39
30	−43.63	−33.17	−33.01	−34.36
20	−49.41	−38.86	−38.93	−40.53
15	−53.47	−42.85	−43.09	−44.85
10	−58.52	−47.82	−48.26	−50.24
8	−61.28	−50.54	−51.08	−53.18
6	−65.03	−53.93	−54.61	−56.86
5	−66.85	−56.03	−56.87	−59.13

B.P. 760 mm. M.P.	5 Methyl Propyl Ether 39.14° .	6 Ethyl Formate 54.46° −80.5°	7 Acetone 56.48° −94.3°	8 Methyl Acetate 57.15° −98.05°
800 mm.	40.58°	55.92°	59.95°	58.62°
700	36.80	52.08	54.09	54.76
600	32.67	47.88	49.86	50.50
500	27.81	42.94	44.90	45.58
400	21.97	36.99	38.93	39.62
300	14.91	30.81	31.72	32.41
200	+5.73	19.66	22.34	23.01
100	−8.43	6.08	7.87	8.59
90	−10.41	4.07	5.85	6.57
80	−12.70	+1.74	3.51	4.23
70	−15.08	−0.68	+1.07	+1.79
60	−17.78	−3.43	−1.69	−0.96
50	−20.88	−6.58	−4.86	−4.13
40	−24.79	−10.56	−8.87	−8.12
30	−29.65	−15.50	−13.82	−13.07
20	−35.67	−21.63	−20.07	−19.24
15	−39.90	−25.93	−24.29	−23.54
10	−45.16	−31.28	−29.66	−28.91
8	−48.04	−34.20	−32.61	−31.85
6	−51.63	−37.85	−36.18	−35.51
5	−53.93	−40.11	−38.54	−37.96
4	..	..	..	−40.54

B.P. 760 mm. M.P.	9 Chloroform 61.26° −63.7°	10 Methyl Alcohol 64.88° −94°	11 *n*-Hexane 69.00° ..	12 Ethyl Acetate 77.12° −83.4°
800 mm.	62.82°	66.19°	70.60°	78.66°
700	58.72	62.79	66.40	74.61
600	54.22	58.97	61.80	70.18
500	48.95	54.62	56.40	64.97
400	41.60	49.40	49.90	58.71
300	34.92	42.97	42.05	51.14
200	24.96	34.40	31.85	41.31
100	9.57	20.90	16.10	26.13
90	7.42	18.97	13.90	24.01
80	4.92	16.84	11.35	21.55
70	+2.33	14.46	8.70	19.00
60	−0.60	11.78	5.70	16.10
50	−3.97	8.60	+2.25	12.78
40	−8.22	4.99	−2.10	8.59
30	−13.50	+0.38	−7.50	+3.38
20	−20.04	−5.82	−14.20	−3.08
15	−24.64	−10.04	−18.90	−7.61
10	−30.35	−15.73	−24.75	−13.25
8	−33.48	−19.39	−27.95	−16.33
6	−37.38	−22.51	−31.94	−20.18
5	−39.79	−24.83	−34.41	−22.56
4	−42.73	−27.60	−37.42	−25.46

B.P. 760 mm M.P.	13 Ethyl Alcohol 78.30° −114.5°	14 Methyl Propionate 79.82°	15 Benzene 80.19° +5.4°	16 Cyclohexane 80.81° +6.4°
800 mm.	79.61°	81.38°	82.13°	82.47°
700	76.30	77.29	77.83	78.11
600	72.40	72.80	73.11	73.32
500	68.06	67.54	67.58	67.71
400	62.85	61.20	61.12	60.95
300	56.50	52.55	52.87	52.79
200	47.93	43.61	42.43	42.18
100	34.40	28.25	26.29	25.80
90	32.47	26.12	24.04	23.52
80	30.35	23.62	21.43	20.86
70	27.98	21.04	18.61	18.11
60	25.33	18.12	15.54	14.99
50	22.21	14.75	11.99	11.40
40	18.52	10.51	7.65	6.88
30	13.92	+5.25	+2.11	+1.26
20	7.72	−1.28	−4.75	−5.77
15	+3.52	−5.86	−9.45	−10.59
10	−2.16	−11.56	−15.56	−16.67
8	−5.16	−14.69	−18.85	−20.00
6	−8.95	−18.58	−22.92	−24.15
5	−11.24	−20.98	−25.45	−26.72
4	−14.00	−23.92	−28.54	−29.62

B.P. 760 mm. M.P.	17 Propyl Formate 81.25°	18 Isopropyl Alcohol 82.42° −85.8°	19 Trimethyl Carbinol 82.57° +25 to +25.5°	20 Dipropyl Ether 89.65° .
800 mm.	82.83°	83.71°	83.92°	91.32°
700	78.68	80.35	80.42	86.93
600	74.14	76.57	76.48	82.12
500	68.80	72.26	72.00	76.47
400	62.38	67.09	66.61	69.66
300	54.63	60.73	60.00	61.45
200	44.55	52.25	51.16	50.78
100	28.99	38.89	37.24	34.30
90	26.82	36.97	35.25	32.00
80	24.30	34.87	33.06	29.63
70	21.69	32.51	30.61	26.56
60	18.72	29.91	27.85	23.42
50	15.31	26.79	24.65	19.81
40	11.02	23.14	20.85	15.26
30	+5.68	18.57	16.09	9.61
20	−0.94	12.43	9.70	+2.60
15	−5.58	8.26	+5.36	−2.32
10	−11.36	+2.62	−0.51	−8.44
8	−14.52	−0.35	..	−11.79
6	−18.46		.	−15.96
5	−21.90		.	−18.53
4	−23.87			−21.70

B.P. 760 mm. M.P.	21 Methyl Isobutyrate 92.66°	22 *n*-Propyl Alcohol 97.52° −127°	23 Isobutyl Formate 97.68°	24 *n*-Heptane 98.61° −91.3°
800 mm.	94.29°	98.47°	99.34°	100.33°
700	90.02	95.37	94.99	95.82
600	85.34	91.47	90.22	90.89
500	79.86	86.97	84.62	85.10
400	73.25	81.60	77.88	78.13
300	65.28	74.99	69.75	69.72
200	54.91	66.19	59.18	58.78
100	38.91	52.32	42.85	41.90
90	36.68	50.32	40.57	39.54
80	34.09	48.13	37.93	36.81
70	31.39	45.69	35.18	33.97
60	28.35	42.93	32.08	30.75
50	24.84	39.74	28.50	27.05
40	20.42	35.95	23.99	22.39
30	14.94	31.21	18.40	16.60
20	8.13	24.83	11.45	9.42
15	+3.35	20.05	6.58	+4.38
10	−2.59	14.64	+0.52	−1.89
8	−6.46	11.55	−2.80	−5.32
6	−9.90	7.67	−6.93	−9.60
5	−12.40	5.27	−9.49	−12.25
4	−15.46	2.44	−12.41	−15.47

B.P. 760 mm. M.P.	25 Ethyl Propionate 99.17° −73.0°	26 Water 100.00° 0°	27 Formic Acid 100.6° +8.35°	28 Propyl Acetate 102.00° ..
800 mm.	100.80°	101.44°	102.60°	103.66°
700	96.52	97.70	97.75	99.31
600	91.83	93.49	92.80	94.55
500	86.33	88.70	87.20	88.96
400	79.71	82.95	80.60	82.23
300	71.71	75.87	72.15	74.11
200	61.31	66.44	61.50	63.55
100	45.26	51.57	44.55	47.25
90	43.02	49.44	42.25	44.97
80	40.42	47.10	39.75	42.33
70	37.72	44.48	36.90	39.59
60	34.67	41.53	33.60	36.48
50	31.15	38.11	29.40	32.92
40	26.72	34.05	24.80	28.41
30	21.22	28.97	19.50	22.82
20	14.39	22.14	11.24	15.89
15	9.60	17.50	..	11.02
10	3.64	11.23	..	4.97
8	+0.38	7.92	..	+1.66
6	−3.69	3.76	..	−2.47
5	−6.21	+1.21	..	−5.03
4	−9.27	−1.84	.	−8.14

B.P. 760 mm. M.P.	29 Methyl *n*-Butyrate 102.86° .	30 Dimethyl Ethyl Carbinol 102.9° −8.4°	31 Acetal 103.54°	32 Isobutyl Alcohol 107.34° −108°
800 mm.	104.52°	104.25°	105.23°	108.73°
700	100.16	100.75	100.70	105.12
600	95.38	..	..	101.07
500	89.77	..	..	96.45
400	83.01	..	..	90.91
300	74.85	..	..	84.09
200	64.25	..	..	74.89
100	47.89	..	47.56	60.67
90	45.60	..	..	58.62
80	42.95	..	..	56.36
70	40.20	..	..	53.83
60	37.08	..	36.56	50.99
50	33.49	45.09	32.91	47.70
40	28.98	41.30	28.30	43.78
30	23.36	36.56	22.59	38.63
20	16.40	30.18	15.50	32.31
15	11.51	25.84	10.52	27.83
10	5.43	19.99	+4.33	21.79
8	+2.11	16.90	..	18.60
6	−2.04	13.01	..	14.59
5	−4.60	10.63	−5.89	12.13
4	−7.73	7.78	..	9.20

B.P. 760 mm. M.P.	33 Ethyl Isobutyrate 109.88°	34 Toluene 110.56° −94.5°	35 Pyridine 115.50° .	36 Isobutyl Acetate 116.09° .
800 mm.	111.55°	112.34°	117.24°	117.79°
700	107.15	107.65	112.68	113.33
600	102.38	102.56	107.69	109.43
500	96.76	96.00	101.82	102.69
400	89.99	88.32	94.77	95.78
300	81.82	80.61	86.25	87.43
200	71.20	69.28	75.18	76.59
100	54.79	51.78	58.08	59.84
90	52.50	49.33	55.69	57.50
80	49.85	46.50	52.93	54.79
70	47.09	43.55	50.05	51.97
60	43.97	40.22	46.79	48.78
50	40.37	36.39	43.05	45.11
40	35.84	31.51	38.33	40.49
30	30.22	25.55	32.47	34.65
20	23.24	18.11	25.19	27.62
15	18.35	12.89	20.09	22.63
10	12.26	6.38	13.74	16.41
8	8.93	+2.83	10.23	13.00
6	4.77	−1.60	5.90	8.76
5	+2.20	−3.35	3.26	6.13
4	−0.94	−7.69	0.00	2.83

B.P. 760 mm. M.P.	37 *n*-Butyl Alcohol 117.01° −79.9°	38 Methyl Valerate 118.46°	39 Acetic Acid 118.7° 16.6°	40 Ethyl Butyrate 119.61°
800 mm.	118.44°	120.24°	120.85°	121.34°
700	114.82	115.55	116.30	116.80
600	110.75	110.45	114.20	111.83
500	106.11	104.44	106.03	105.99
400	100.55	97.21	99.30	98.96
300	93.70	88.45	91.05	90.47
200	84.58	77.13	80.70	79.45
100	70.20	59.60	63.45	62.42
90	68.14	57.16	61.05	60.04
80	65.87	54.32	58.45	57.28
70	63.34	51.37	55.75	54.42
60	60.49	48.03	52.45	51.18
50	57.18	44.19	48.40	47.45
40	53.25	39.35	43.25	42.74
30	48.34	33.35	37.80	36.90
20	41.73	25.89	29.90	29.66
15	37.24	20.66	..	24.58
10	31.17	14.15	16.40	18.26
8	27.98	10.59	..	14.80
6	23.95	6.15	..	10.48
5	21.49	3.41	..	7.82
4	18.53	0.06	..	4.56

B.P. 760 mm. M.P.	41 Isopropyl Isobutyrate 120.78°	42 Propyl Propionate 121.36°	43 Isoamyl Formate 123.97°	44 *n*-Octane 125.44° −56.5°
800 mm.	122.55°	124.14°	125.77°	127.24°
700	117.93	119.59	121.05	122.52
600	112.85	114.60	115.88	117.35
500	106.[illegible]0	108.75	109.82	111.28
400	99.74	101.51	102.52	103.98
300	91.10	93.19	93.70	94.16
200	79.86	82.14	82.25	83.69
100	62.51	65.06	64.56	66.00
90	60.09	62.68	62.09	63.52
80	57.28	59.91	59.23	60.66
70	54.36	57.04	56.25	57.68
60	51.05	53.79	52.88	54.31
50	47.25	50.05	49.01	50.43
40	42.46	45.33	44.12	45.54
30	36.51	39.48	38.06	39.48
20	29.13	30.21	30.54	31.95
15	23.95	27.12	25.26	26.67
10	17.51	20.78	18.69	20.09
8	13.98	17.31	..	16.50
6	9.58	12.98	..	12.01
5	6.87	10.31	.	9.24
4	3.55	7.04	.	5.86

B.P. 760 mm. M.P.	45 Isoamyl Alcohol 130.58° Cong. P. −117°	46 Propyl Isobutyrate 133.9°	47 Ethyl Isovalerate 134.35° −99.3°	48 Acetic Anhydride 136.4° ..
800 mm.	132.20°	135.69°	136.16°	138.08°
700	128.40	130.99	131.41	133.67
600	124.10	125.70	126.30	128.90
500	119.20	119.50	120.10	123.20
400	113.30	112.40	112.80	116.30
300	106.20	102.40	104.20	108.15
200	96.65	92.10	92.75	97.65
100	81.60	74.50	74.70	80.70
90	79.45	72.00	72.20	78.70
80	77.10	69.30	69.30	75.85
70	74.50	66.30	66.35	73.00
60	71.40	63.00	63.00	70.00
50	68.05	59.15	59.20	66.55
40	64.00	54.30	54.25	61.70
30	58.80	48.25	48.20	56.05
20	51.95	40.90	40.70	49.10
15	47.30	35.50	35.15	44.05
10	41.00	28.82	28.40	37.91
8	37.55	25.24	24.88	34.54
6	33.35	20.77	20.38	30.55
5	30.77	18.00	17.59	27.76
4	27.69	14.62	14.20	24.60

B.P. 760 mm. M.P.	49 Isobutyl Propionate 136.8° ..	50 *p*-Xylene 138.40°	51 Ethyl Sulfide 138.6° ..	52 *m*-Xylene 139 00° −53.6°
800 mm.	138.60°	140.27°	140.49°	140.84°
700	133.88	135.36	135.53	136.00
600	128.70	129.99	130.00	130.70
500	122.60	123.60	123.70	124.40
400	115.35	116.10	116.20	116.90
300	106.60	106.85	105.75	107.80
200	95.25	95.00	94.90	96.20
100	77.45	76.50	76.55	77.80
90	75.10	74.00	74.00	75.35
80	72.30	71.20	71.20	72.45
70	69.16	67.90	68.00	69.20
60	65.80	64.40	64.60	65.80
50	61.93	60.50	60.50	62.00
40	57.45	55.35	55.35	56.90
30	51.39	49.00	49.00	50.75
20	43.47	41.30	41.50	43.00
15	38.21	35.75	34.80	37.50
10	31.65	28.85	27.89	30.70
8	28.05	25.05	24.11	27.00
6	23.58	20.52	19.40	22.70
5	21.93	17.64	16.48	19.80
4	17.43	14.12	12.93	16.30

B.P. 760 mm. M.P.	53 Propionic Acid 140.35° −24°	54 Propyl Butyrate 142.47°	55 *o*-Xylene 142.64°	56 Styrene 144.0° ..
800 mm.	141.98°	144.26°	144.52°	145.83°
700	137.74	139.56	139.59	141.02
600	132.90	134.40	133.90	135.80
500	127.50	128.40	126.50	129.50
400	121.00	121.10	120.15	122.15
300	113.10	112.20	109.50	113.20
200	102.40	100.75	98.90	101.40
100	85.60	83.00	80.20	83.10
90	83.10	80.55	77.75	80.70
80	80.40	77.70	74.80	77.90
70	77.50	74.65	71.90	74.80
60	74.70	71.50	68.30	71.30
50	70.05	67.45	64.25	67.55
40	65.45	62.60	59.25	62.45
30	59.90	56.70	52.95	56.20
20	52.05	49.20	45.20	48.60
15	46.90	43.93	39.55	43.10
10	39.55	37.38	32.69	36.30
8	35.70	33.78	29.94	32.55
6	31.10	29.31	24.26	28.10
5	28.30	26.54	21.36	25.25
4	25.00	23.16	17.73	21.70

B.P. 760 mm. M.P.	57 Isobutyl Isobutyrate 146.51° ..	58 n-Nonane 149.48° ..	59 Allyl Isothiocyanate 150.70° .	60 Anisole 153.80° −37.2°
800 mm.	148.38°	151.38°	152.56°	155.70°
700	143.48	146.40	147.68	150.72
600	138.30	141.00	142.40	..
500	131.95	134.60	136.10	..
400	124.30	126.90	128.50	..
300	115.25	117.80	119.55	..
200	103.25	105.60	107.60	..
100	84.60	86.80	89.00	..
90	81.95	84.20	86.60	..
80	79.20	81.20	83.60	..
70	74.75	77.95	80.45	..
60	72.50	74.45	77.00	..
50	68.75	70.50	73.10	74.90
40	63.50	66.20	68.00	69.60
30	57.15	59.00	61.70	63.20
20	49.50	51.10	54.10	55.45
15	44.00	45.37	48.51	46.75
10	37.00	38.40	41.60	42.80
8	33.35	34.60	37.90	38.90
6	28.80	30.00	33.50	34.40
5	25.91	27.20	30.65	31.55
4	22.41	23.45	27.00	27.75

B.P. 760 mm. M.P.	61 Isobutyric Acid 154.35° −47°	62 α-Pinene 154.75° .	63 Heptaldehyde 155° .	64 Propyl Isovalerate 155.87° .
800 mm.	155.97°	156.72°	156.92°	157.73°
700	151.76	151.55	151.88	152.85
600	147.01	145.88	..	147.60
500	141.60	139.23	..	141.30
400	135.15	131.23	..	133.80
300	127.30	121.55	..	124.80
200	116.40	108.98	..	112.85
100	99.80	89.48	..	94.35
90	97.30	86.87	..	91.80
80	94.50	83.72	..	88.95
70	91.70	80.46	..	85.80
60	88.30	76.76	..	82.30
50	85.40	72.51	74.90	78.50
40	79.75	67.15	69.68	73.35
30	74.30	60.50	63.10	67.05
20	66.50	52.25	55.16	59.40
15	61.38	46.46	49.52	53.80
10	54.40	39.25	42.50	47.00
8	50.60	35.31	38.66	43.20
6	46.00	30.39	33.87	38.70
5	43.00	27.35	30.91	35.90
4	39.59	23.64	27.30	32.30

B.P. 760 mm. M.P.	65 Isobutyl Butyrate 156.9°	66 Camphene 159.5° 51–52°	67 Amyl Propionate 160.36° .	68 n-Butyric Acid 162.20° −6.7°
800 mm.	158.76°	161.44°	162.36°	163.86°
700	153.87	156.35	157.27	159.54
600	148.50	..	151.83	154.60
500	142.21	..	145.50	149.10
400	134.60	..	137.70	142.55
300	125.45	..	128.40	134.35
200	113.60	..	116.35	123.40
100	95.10	95.45	97.67	106.30
90	92.55	92.80	95.05	103.80
80	89.60	89.80	92.10	101.10
70	86.50	86.50	88.90	98.05
60	83.00	82.80	85.35	94.65
50	79.00	78.70	81.33	90.65
40	74.00	73.40	76.07	85.95
30	67.70	66.85	69.70	80.20
20	59.90	58.80	61.80	72.20
15	54.40	53.10	56.25	66.85
10	47.60	46.00	49.30	59.60
8	43.90	..	45.50	55.90
6	39.35	..	40.80	51.10
5	36.47	..	37.95	48.10
4	32.95	.	34.40	44.60

B.P. 760 mm. M.P.	69 Dimethyl Oxalate 164.24°	70 Propionic Anhydride 167.0°	71 Ethyl Trichloracetate 168.19°	72 Furfuryl Alcohol 169.35°
800 mm.	166.01°	168.87°	169.87°	171.24°
700	161.36	163.96	165.16	166.28
600	156.20	158.58	..	..
500	150.20	152.35	..	..
400	142.95	144.70	..	..
300	132.05	135.55	..	..
200	123.05	123.65	..	..
100	105.55	105.25	..	..
90	103.20	102.65	..	..
80	100.30	99.60	..	..
70	97.30	96.55	..	..
60	94.00	93.00	..	..
50	90.20	89.00	90.34	96.20
40	85.40	83.90	85.27	91.40
30	79.45	77.50	78.97	85.45
20	72.00	69.70	71.15	77.35
15	66.80	64.18	65.67	71.90
10	60.35	57.40	58.85	64.50
8	56.80	53.65	..	60.55
6	52.47	49.00	..	56.00
5	49.70	46.20	47.58	52.70
4	46.40	42.70	44.09	..

B.P. 760 mm M.P.	73 Isobutyl Valerate 169.39°	74 Amyl Isobutyrate 169.78°	75 Phenetol 170.72°	76 Methyl Hexyl Ketone 171°
800 mm.	171.34°	171.72°	172.64°	172.82°
700	166.22	166.62	167.60	168.04
600	160.60	161.04	162.09	.
500	154.00	147.20	155.70	..
400	146.07	146.55	147.90	..
300	136.45	137.00	138.55	..
200	124.05	124.70	131.40	..
100	104.80	105.50	107.60	..
90	102.20	102.90	105.00	..
80	99.05	99.80	102.00	..
70	95.80	96.50	98.85	..
60	92.10	92.90	95.25	98.90
50	87.90	88.70	91.10	94.85
40	82.60	83.40	86.00	90.10
30	76.00	76.80	79.60	83.95
20	67.90	68.80	71.60	76.30
15	62.13	63.00	60.60	71.00
10	55.10	56.00	59.20	64.40
8	51.20	..	55.40	60.75
6	46.35	..	50.80	56.20
5	43.35		47.90	53.40
4	39.80	..	42.25	50.05

B.P. 760 mm M.P.	77 Phthalic Acid 171.5° ..	78 Myrcene 171.5°	79 β-Crotonic Acid 171.9° 15.45°–15.5°	80 Amyl Ether 172.75° ..
800 mm.		173.51°	173.61°	174.72°
700	170.00°	168.24	169.17	169.55
600	168.20	..	..	..
500	166.00	..	..	..
400	162.90	..	.	..
300	158.40	..	.	..
200	152.30	..	..	..
100	143.20	..	.	..
90	141.50	..	.	..
80	139.50	..	..	..
70	136.00	..	.	..
60	130.20	92.40	.	..
50	112.80	88.00	98.40	90.50
40	96.00	82.60	93.60	85.15
30	78.30	75.85	87.60	78.70
20	60.00	67.50	79.40	70.30
15	50.00	61.70	74.00	64.45
10	40.00	54.35	66.60	57.20
8	30.00	50.40	62.60	53.25
6	..	45.45	57.77	48.40
5	..	42.40	..	45.40
4	..	38.60	..	41.70

B.P. 760 mm. M.P.	81 *n*-Decane 173° .	82 Methyl Heptenone 173.11° ..	83 Methyl Cyclohexanol 174.5°	84 *n*-Heptyl Alcohol 175 17°
800 mm.	174.99°	175.15°	176.32°	176.82°
700	169.77	170.00	171.55	172.54
600	164.10	164.35	..	..
500	157.40	157.70	..	..
400	149.40	149.80	..	..
300	139.60	140.15	..	..
200	126.80	127.60	..	..
100	107.20	108.30	..	..
90	104.40	105.70	..	..
80	101.20	102.60	..	..
70	97.95	99.40	..	..
60	94.20	95.70	..	..
50	89.80	91.45	98.69	104.25
40	84.50	86.15	93.75	99.60
30	77.70	79.60	87.62	93.75
20	69.45	71.60	80.01	86.00
15	63.64	65.86	74.67	80.70
10	56.35	58.73	68.03	73.50
8	52.35	54.82	64.39	69.70
6	47.40	49.95	59.86	65.05
5	44.40	46.94	57.06	62.07
4	40.65	43.27	.	..

B.P. 760 mm. M.P.	85 α-Phellandrene 175.79° .	86 Cineole 176.4° Cong. P. +1°	87 *p*-Cymene 176.8° ..	88 Dipentene 177 6° −96 6°
800 mm.	177.75°	178.47°	178.85°	179.65°
700	172.60	173.04	173.48	174.27
600	..	167.10	..	168.40
500	..	160.10	..	161.50
400	..	151.70	..	153.15
300	..	141.50	..	143.05
200	..	128.35	..	129.95
100	..	107.90	..	109.75
90	..	105.10	..	106.90
80	..	101.80	..	103.60
70	..	98.40	..	100.2[illegible]
60	98.10	94.60	95.80	96.40
50	93.80	90.10	91.40	92.00
40	88.50	84.50	85.80	86.40
30	81.90	77.40	78.95	79.50
20	73.70	68.95	70.30	71.00
15	68.00	62.75	64.30	64.90
10	60.80	55.15	56.90	57.50
8	56.90	51.00	52.80	53.35
6	52.05	45.95	47.65	48.25
5	49.05	42.70	44.60	45.10
4	46.60	38.90	40.80	41.30

B.P. 760 mm. M.P.	89 Benzaldehyde 178.07° .	90 *n*-Amyl Butyrate 178.6° ..	91 Ethyl Acetoacetate 180.42° ..	92 Phenol 182.24° 43°
800 mm.	180.08°	180.56°	182.35°	184.09°
700	174.81	175.41	177.29	179.23
600	169.00	169.70	171.70	174.00
500	162.20	163.05	165.25	167.75
400	154.10	155.00	157.40	161.10
300	144.25	145.45	147.90	151.05
200	131.60	133.10	135.70	139.20
100	111.80	113.60	116.70	120.90
90	109.00	110.90	114.05	118.35
80	105.70	107.80	110.95	115.40
70	102.45	104.50	107.70	112.30
60	98.70	100.80	104.50	108.80
50	94.40	96.60	100.00	104.75
40	89.00	91.30	94.70	99.80
30	82.20	84.70	88.20	93.55
20	73.80	76.55	80.10	85.80
15	67.89	70.83	74.45	80.36
10	60.60	63.70	67.45	73.57
8	56.55	59.70	63.55	68.60
6	51.55	54.80	58.80	65.20
5	48.45	51.80	55.80	62.35
4	44.75	..	52.20	58.90

B.P. 760 mm. M.P.	93 Aniline 184.10° −6.2°	94 *n*-Valeric Acid 184.8° −34.5°	95 α-Crotonic Acid 185°	96 Diethyl Oxalate 185° −40.6°
800 mm.	186.06°	186.99°	186.71°	186.71°
700	180.91	182.10	182.27	182.22
600	175.30	177.30	..	177.35
500	168.55	171.60	..	171.20
400	160.60	165.00	..	164.60
300	151.10	156.65	..	156.20
200	138.55	145.50	..	145.25
100	119.25	129.20	..	128.40
90	116.60	125.50	..	126.00
80	113.40	122.75	..	123.30
70	110.20	119.70	..	120.40
60	106.50	116.30	..	117.20
50	102.25	112.30	111.40	113.50
40	97.00	107.50	106.70	108.90
30	88.10	101.70	100.70	103.10
20	82.20	93.65	92.65	95.90
15	76.40	88.10	87.10	90.90
10	69.20	80.80	79.60	84.60
8	65.30	76.90	75.75	81.20
6	60.40	72.05	70.80	76.95
5	57.45	69.05	67.80	74.30
4	50.05	65.50	64.20	72.15

B.P. 760 mm. M.P.	97 Benzyl Ethyl Ether 185° ..	98 Crotyl Sulfide 186.5° ..	99 Monochloracetic Acid 186.95° 61.9°	100 o-Cresol 190.67° 31°
800 mm.	187.03°	188.38°	188.61°	192.60°
700	181.70	183.45	184.30	187.54
600	..	..	179.40	182.00
500	..	..	173.90	175.40
400	..	..	167.30	167.60
300	..	..	159.00	158.20
200	..	.	140.20	145.90
100	..	..	131.00	127.90
90	..	..	128.60	124.30
80	..	..	125.90	121.20
70	..	..	122.80	118.05
60	104.55	112.22	119.40	114.40
50	100.20	108.17	115.50	110.30
40	94.65	103.07	110.80	105.20
30	87.80	96.73	104.90	98.60
20	79.30	88.87	97.00	90.60
15	73.30	83.36	91.66	85.00
10	65.90	76.61	84.40	77.90
8	61.85	..	80.60	74.95
6	56.80	..	75.80	69.30
5	53.75	65.16	72.30	66.35
4	49.90	..	69.30	62.70

B.P. 760 mm. M.P.	101 Trimethyl Phosphate 190.68° .	102 Dimethyl Aniline 193.19° .	103 Fenchone 193.53° +5° to +6°	104 Methyl Aniline 194.36° ..
800 mm.	192.67°	195.22°	195.80°	196.32°
700	187.45	189.90	190.20	191.18
600	181.74	184.07	184.10	185.56
500	175.00	177.25	176.80	179.00
400	166.90	169.00	168.10	171.00
300	157.10	159.05	157.60	161.40
200	144.40	146.15	143.90	148.90
100	125.00	126.25	122.80	129.65
90	122.15	123.45	119.90	127.00
80	118.95	120.20	116.50	123.90
70	115.70	116.90	113.00	120.70
60	112.00	113.10	109.00	117.00
50	107.75	108.70	104.35	112.80
40	102.30	103.15	98.60	107.45
30	95.60	96.30	91.40	100.80
20	87.40	87.90	82.40	92.70
15	81.60	81.86	76.11	86.94
10	74.40	74.50	68.30	79.75
8	70.40	70.50	64.10	75.85
6	65.36	65.40	58.80	71.00
5	62.30	62.30	55.50	68.00
4	59.56	58.55	51.45	64.30

B.P. 760 mm. M.P.	105 Maleic Anhydride 196° 60°	106 *trans*-Trimethyl-cyclohexanol 196.04° 34.5°	107 Salicyl Aldehyde 196.70° .	108 Ethylene Glycol 197.10° −11.5°
800 mm.	198.05°	197.88°	198.78°	198.86°
700	192.67	193.04	193.31	194.30
600	..	..	187.32	189.10
500	..	..	180.05	183.25
400	..	..	171.50	176.30
300	.	..	161.25	167.70
200		..	148.00	155.95
100	.	.	127.35	138.05
90	.	..	124.50	135.40
80	..	..	121.20	132.50
70	..	..	117.70	129.40
60	114.46	..	113.80	125.80
50	110.58	.	109.25	121.60
40	104.99	114.31	103.60	116.65
30	98.07	108.10	96.50	108.10
20	89.50	100.40	87.80	102.20
15	83.48	95.00	81.61	96.50
10	76.00	88.27	74.00	88.80
8	71.88	84.60	69.80	84.80
6	66.78	80.01	64.60	79.90
5	63.63	77.17	61.40	76.75
4			57.50	73.00

B.P. 760 mm M.P	109 *o*-Toluidine 198.12° −24.4°	110 Methyl Benzoate 198.13° −12.5°	111 Linalool 198.3°	112 Isocaproic Acid 199.7°
800 mm.	200.11°	200.18°	200.15°	201.42°
700	194.88	194.80	195.05	196.96
600	191.00	188.90	189.50	191.95
500	184.15	182.00	182.90	186.20
400	175.80	173.60	175.00	179.40
300	165.90	163.55	165.40	170.95
200	153.00	150.50	163.10	159.65
100	133.00	130.30	134.00	142.00
90	130.30	127.50	131.30	139.45
80	127.10	124.30	128.35	136.60
70	123.75	120.90	125.00	133.55
60	119.90	117.00	121.40	130.00
50	115.60	112.60	117.20	125.90
40	110.00	107.10	111.90	121.10
30	103.40	100.10	105.40	115.10
20	94.90	91.60	97.20	106.90
15	89.00	85.50	91.49	101.38
10	81.40	78.00	84.40	93.90
8	77.45	73.90	80.55	89.90
6	72.40	68.80	75.80	85.00
5	69.30	65.70	72.80	81.95
4	65.60	61.85	69.10	78.35

B.P. 760 mm. M.P.	113 *n*-Nonyl Aldehyde 200.3°	114 *m*-Cresol 200.5° +4°(?)	115 *p*-Toluidine 200.54° ..	116 *β*-Thujone ca. 201° decomp.
800 mm.	..	..	202.56°	..
700	..	198.00°	197.26	..
600	..	192.50	191.47	193.00°
500	..	186.30	184.70	185.80
400	..	179.00	176.50	177.05
300	..	169.45	166.70	166.40
200	..	157.00	153.80	152.80
100	128.00°	137.80	133.90	131.60
90	125.00	135.10	131.10	128.60
80	121.45	132.00	127.90	125.15
70	117.80	128.70	124.60	121.65
60	113.70	125.05	120.80	117.60
50	109.00	120.80	116.40	113.00
40	103.10	115.50	111.00	107.10
30	95.65	108.90	104.15	99.90
20	86.45	100.70	95.60	90.91
15	80.00	94.90	89.70	84.79
10	72.05	87.80	82.35	76.73
8	67.65	83.90	78.35	72.40
6	62.10	79.00	73.30	67.10
5	58.80	76.00	70.20	63.80
4	.	72.40	66.50	59.75

B.P. 760 mm. M.P.	117 *p*-Cresol 201.1° 36°	118 Fenchyl Alcohol 201.5° 45°	119 Acetophenone 202.38° 20°	120 *cis*-Trimethyl-cyclohexanol 203°
800 mm.	..	203.40°	204.47°	204.94°
700	198.50°	198.41	198.98	199.84
600	193.00	..	192.95	..
500	186.80	..	185.88	..
400	179.40	..	177.40	..
300	170.00	..	167.10	..
200	157.60	..	153.75	..
100	138.40	..	133.10	..
90	135.75	..	130.20	..
80	132.60	..	126.85	..
70	129.40	..	123.40	..
60	125.80	..	119.50	..
50	121.45	122.20	114.90	121.90
40	116.25	117.00	109.30	116.60
30	109.70	113.00	102.20	112.50
20	101.55	102.65	93.45	101.90
15	95.80	97.00	87.30	96.30
10	88.60	90.05	79.60	89.20
8	84.70	86.30	75.50	85.25
6	79.90	81.50	70.25	80.45
5	76.90	78.65	67.10	77.40
4	73.30	..	63.10	..

B.P. 760 mm. M.P.	121 *m*-Toluidine 203.3°	122 Camphor 204° 175°	123 Ethyl Aniline 204.0° ..	124 Benzyl Alcohol 204.50° ..
800 mm.	205 31°	205.99°	206.03°	206.40°
700	200.03	200.76 liquid	200.69	201.35
600	194.24	..	194.84	195.85
500	187.50	..	187.95	189.40
400	179.30	.	179.60	181.50
300	169.35	173.00 solid	169.60	172.20
200	156.50	159.65	156.60	160.10
100	136.60	139.05	136.60	141.30
90	133.95	136.10	133.75	138.70
80	130.75	132.90	131.80	135.65
70	127.40	129.40	127.20	132.50
60	123.60	125.50	123.30	128.90
50	119.20	120.85	119.10	124.85
40	113.75	115.30	113.55	119.70
30	106.95	108.20	106.70	113.20
20	98.50	99.50	99.20	105.20
15	92.66	93.30	92.23	99.59
10	85.30	85.75	83.90	92.60
8	81.25	81.60	80.60	88.80
6	76.25	76.40	75.50	84.00
5	73.20	73.20	72.50	81.00
4	69.20	69.25	68.60	77.40

B.P. 760 mm. M.P.	125 Bornyl Ethyl Ether 204.77°	126 Guaiacol 205.1° ..	127 *n*-Caproic Acid 205 7° ..	128 Citronellal 206.93° ..
800 mm.	206.84°	207.02°	207.44°	208.90°
700	201.41	201.93	202.91	203.40
600	..	..	197.80	197.30
500	.	..	192.00	190.20
400	..	..	185.15	181.60
300	.	..	176.70	171.35
200	.	..	165.10	158.00
100	..	..	147.10	137.30
90	..	..	144.45	134.40
80	..	..	141.70	131.00
70	..	..	138.60	127.60
60	122.60	128.90	134.90	123.65
50	118.20	124.80	130.80	119.10
40	112.60	119.60	125.85	113.40
30	105.60	113.00	119.55	106.30
20	97.00	105.00	111.45	97.60
15	90.00	99.30	105.80	91.42
10	83.30	92.25	98.20	83.65
8	79.20	88.45	94.20	79.50
6	72.80	83.60	89.20	74.30
5	..	80.70	..	71.30
4	..	..	..	67.10

B.P. 760 mm. M.P.	129 n-Decyl Aldehyde 208.25° ..	130 Menthone ca. 209° decomp. .	131 Nitrobenzene 209.79° +5.72°	132 Borneol 212° 203°
800 mm.	210.26°	..	211.91°	213.93°
700	204.99	206.20°	206.34	208.86 liquid
600	..	200.00	200.24	..
500	..	192.60	193.10	201.00 solid
400	..	183.80	184.45	193.10
300	..	173.00	174.00	183.55
200	..	159.05	160.50	171.20
100	..	137.48	139.60	152.20
90	..	134.40	136.65	149.55
80	..	130.90	133.25	146.45
70	..	127.30	129.70	143.30
60	128.50	123.20	125.75	139.65
50	124.20	118.45	121.20	135.50
40	118.70	112.50	115.40	130.20
30	112.00	105.10	108.25	123.70
20	103.50	96.05	99.30	115.60
15	97.55	89.61	93.10	110.00
10	90.25	81.60	85.40	102.90
8	86.20	77.25	81.10	99.05
6	81.10	71.80	75.90	94.25
5	78.10	68.40	72.70	91.30
4	..	64.30	68.75	87.65

B.P. 760 mm. M.P.	133 Ethyl Menthyl Ether 212°	134 Ethyl Benzoate 212.08° −34.2°	135 Methyl Chavicol 215.5°	136 Benzyl Acetate 215.54° ..
800 mm.	214.15°	214.20°	217.58°	217.40°
700	208.50	208.70	212.12	212.00
600	..	202.60	..	206.00
500	..	195.40	..	199.00
400	..	186.80	..	190.50
300	..	176.30	..	180.40
200	..	162.80	..	167.25
100	140.70	141.90	..	146.70
90	137.70	138.90	..	143.80
80	134.40	135.50	..	140.55
70	130.85	132.00	..	137.10
60	126.80	128.10	..	133.20
50	122.20	123.50	128.65	128.75
40	116.30	117.80	123.00	123.10
30	109.10	110.60	115.95	116.10
20	100.10	101.80	107.20	107.50
15	93.80	95.57	101.10	101.40
10	85.90	87.80	93.45	93.80
8	81.65	83.50	89.30	89.60
6	76.30	78.25	84.10	84.40
5	..	75.05	80.90	81.20
4	..	71.05	..	77.40

B.P. 760 mm. M.P.	137 *l*-Menthol 216° 41°	138 α-Terpineol 217.5° 35°	139 Naphthalene 217.96° 80.4°	140 Menthyl Formate 219° 9°
800 mm.	218.00°	219.44°	220.19°	221.12°
700	212.70	214.35	214.33	215.55
600	..	..	207.93	..
500	..	.	200.40	..
400	..	.	191.35	..
300	..	.	180.41	..
200	..	.	166.20	..
100	..		144.25	..
90	.		141.19	..
80	..	..	137.64	..
70	..	.	133.94	..
60	..	..	129.76	..
50	..	136.70	124.96	130.15
40	..	131.50	118.90	124.35
30	..	124.90	111.37	117.16
20	111.00	116.80	102.04	108.24
15	105.30	111.10	95.49	101.99
10	98.20	104.00	87.44	94.20
8	94.25	100.20	82.88	89.94
6	89.45	95.30	77.32	84.62
5	86.50	92.40	73.88	81.34
4	..	..	69.68	..

B.P. 760 mm M.P.	141 1.2-Bromstyrol (ω) 219° ..	142 Phenyl Isothiocyanate 220.41° ..	143 Tetrahydrocarveol 220.5° ..	144 Tetrahydrocarvone 220.5°
800 mm.	221.12°	222.60°	222.49°	222.69°
700	215.55	216.84	217.26	216.94
600	..	210.54	..	..
500	..	203.10	..	..
400	..	194.15	.	..
300	..	183.40	..	..
200	..	169.35	..	..
100	..	147.70	..	..
90		144.65	..	..
80	..	141.15	..	..
70	..	137.50	..	..
60	..	133.40	..	..
50	130.40	128.70	137.20	129.05
40	124.60	122.70	131.80	123.10
30	117.45	115.40	125.05	115.70
20	108.55	106.20	116.80	106.50
15	102.30	99.86	110.90	100.00
10	94.30	91.80	103.60	92.05
8	90.30	87.40	99.65	87.70
6	85.10	82.00	94.60	82.25
5	81.80	78.55	91.55	78.90
4	..	74.50	..	..

	145 Phenylethyl Alcohol	146 Dihydrocarvone	147 Bornyl Acetate	148 Methyl Salicylate
B.P. 760 mm.	222.02°	222.40°	223°	223.03°
M.P.	..	..	29°	−8.6°
800 mm.	224.07°	224.63°	225.12°	225.21°
700	218.70	218.78	219.55	219.49
600	..	..	..	213.34
500	..	..	..	206.00
400	..	..	..	197.10
300	..	..	..	186.40
200	..	..	.	172.50
100	..	.	..	151.00
90	..	..	..	148.00
80	..	..	..	144.50
70	..	..	..	140.90
60	141.10	..	139.10	136.80
50	136.70	129.30	134.50	132.10
40	131.10	123.20	128.80	126.25
30	124.30	115.70	121.20	118.90
20	115.75	106.40	112.70	109.90
15	109.70	99.90	106.50	103.57
10	102.30	91.75	98.80	95.55
8	98.20	87.30	94.55	91.15
6	93.00	81.70	89.25	85.70
5	90.00	78.30	86.00	82.40
4	..	..	82.00	78.30

	149 Pulegone	150 Citronellol	151 Dihydrocarveol	152 Menthen-1-one-6
B.P. 760 mm.	ca. 224° decomp.	224.42°	224.5°	227.50°
M.P.	..	..	..	.
800 mm.	..	226.40°	226.52°	229.70°
700	221.05°	221.20	221.22	223.92
600	214.80	215.50	..	..
500	207.20	208.90	..	..
400	198.20	200.90	..	..
300	187.30	191.20	..	..
200	173.15	178.85	..	..
100	151.30	159.30	..	..
90	148.20	156.60	..	..
80	144.60	153.40	..	..
70	140.90	150.15	..	..
60	136.80	146.50	..	..
50	132.00	142.20	140.20	135.60
40	126.00	136.80	134.70	129.60
30	118.50	130.10	127.90	122.20
20	109.20	122.00	119.40	113.00
15	102.70	116.16	113.45	106.60
10	94.60	108.90	106.10	98.60
8	90.10	104.85	102.00	94.20
6	84.70	100.00	96.95	88.65
5	81.20	96.90	93.90	85.35
4	77.00	93.20	..	81.20

B.P. 760 mm. M.P.	153 Citral ca 228° decomp.	154 Acetyl Methyl Hexyl Ketone 228.5° −6°	155 *n*-Propyl Benzoate 229.5°	156 Geraniol 229.65° ..
800 mm.	..	230.44°	231.55°	231.30°
700	225.10°	225.35	226.16	226.00
600	219.10	..	..	..
500	211.90	..	..	..
400	203.20	..	..	..
300	192.85	..	..	..
200	179.40	..	..	..
100	158.60	..	..	162.70
90	155.60	..	..	159.90
80	152.20	..	..	156.75
70	148.70	..	..	153.40
60	144.70	..	140.60	149.60
50	140.10	147.74	135.70	145.25
40	134.40	142.48	131.00	139.80
30	127.30	135.94	122.05	133.00
20	118.40	127.83	112.70	124.50
15	112.15	122.16	106.05	118.65
10	104.40	115.07	97.90	111.20
8	100.10	..	93.40	107.30
6	94.90	..	87.75	102.30
5	91.65		84.30	99.15
4	87.60		80.10	95.35

B.P. 760 mm. M.P.	157 Methyl Nonyl Ketone 230.65° .	158 Carvone 230.84°	159 *n*-Decyl Alcohol 231°	160 Thymol 231.32° 51°
800 mm.	232.81°	232.08°	232.95°	233.42°
700	227.15	227.20	227.88	227.90
600	..	220.80	..	..
500	..	213.30	..	..
400	..	204.25	..	..
300		193.20	..	..
200		179.00	..	..
100	..	156.90	..	161.70
90	..	153.85	..	158.80
80	..	150.20	..	155.50
70	..	146.50	..	152.00
60	145.40	142.30	..	148.00
50	140.75	137.40	146.85	143.50
40	134.89	131.40	141.40	137.80
30	124.93	123.80	134.60	130.65
20	118.59	114.60	125.35	121.90
15	112.27	107.80	119.00	115.75
10	104.39	99.60	110.50	108.05
8	..	95.20	106.50	103.85
6	..	89.40	100.50	98.60
5	..	86.20	99.25	95.40
4	..	82.00	..	91.50

B.P. 760 mm. M.P.	161 Carvenone 233.28° ..	162 Ethyl Salicylate 233.75° 	163 Safrole 234.5° 11°	164 Phenylpropyl Alcoho 235° ..
800 mm.	235.53°	235.96°	236.69°	237.15°
700	229.63	230.16	230.95	231.50
600	223.16	223.71	224.66	..
500	215.50	216.25	217.28	..
400	206.30	207.25	208.40	..
300	195.30	196.40	197.65	..
200	181.00	182.30	183.73	..
100	158.70	160.55	162.21	..
90	155.60	157.50	159.20	..
80	152.00	154.00	155.72	..
70	148.30	150.40	152.09	..
60	142.70	146.20	148.00	..
50	139.30	141.45	143.28	145.00
40	133.25	135.45	137.33	139.20
30	125.70	128.00	129.96	131.95
20	116.40	118.80	120.80	123.00
15	109.70	112.35	114.38	116.65
10	101.60	104.30	106.38	108.50
8	97.00	99.95	102.01	104.50
6	91.45	94.45	96.56	99.10
5	88.00	91.00	93.18	95.85
4	83.80	87.00	89.07	91.20

B.P. 760 mm. M.P.	165 Cuminic Aldehyde 235.5°	166 Ethyleneglycol Monophenyl Ether 237°	167 *p*-Isobutyl Phenol 237° .	168 Isobutyl Benzoate 237.0° .
800 mm.	237.74°	239.00°	239.06°	239.17°
700	231.86	234.75	233.64	233.47
600	..	228.01	..	..
500	..	221.20	..	..
400	..	213.00	..	..
300	..	203.25	..	..
200	..	190.50	..	..
100	..	170.90	..	..
90	..	168.10	..	..
80	..	165.00	..	..
70	..	161.70	..	..
60	..	157.95	..	..
50	..	153.65	150.80	146.45
40	..	148.20	145.20	140.60
30	128.20	141.50	138.20	134.60
20	118.90	133.20	129.40	124.20
15	112.30	127.30	123.50	117.90
10	104.00	120.05	116.05	109.95
8	99.80	..	112.00	104.20
6	94.10	..	106.80	100.20
5	90.70	..	103.70	96.90
4	..	..	..	..

B.P. 760 mm. M.P.	169 Carvacrol 237.7° 1°	170 Anethole 239.5° 22.5 to 23°	171 p-Cuminic Alcohol 246.6° ..	172 Bornyl n-Butyrate 247° ..
800 mm.	239.83°	..	248.64°	249.17°
700	234.23	..	243.39	243.48
600	228.10	..	..	..
500	220.80	..	..	..
400	212.00	..	..	..
300	201.60	..	..	..
200	188.00	..	..	..
100	167.10	165.30°	..	..
90	164.10	162.10	..	..
80	160.80	158.55	..	..
70	157.20	154.80	..	..
60	153.20	150.50	..	..
50	148.60	145.70	161.60	156.60
40	142.80	139.65	156.10	150.65
30	135.60	134.80	149.20	143.70
20	126.80	122.65	140.75	134.35
15	120.50	116.05	134.80	128.00
10	112.70	108.00	127.30	120.10
8	108.55	103.30	123.30	115.80
6	105.85	97.80	118.20	110.40
5	99.95	94.30	115.10	107.10
4	95.95	90.10	111.30	103.00

B.P. 760 mm. M.P.	173 Methyl Undecylenate 248°	174 Anisic Aldehyde 248.35° 0°(?)	175 Benzoic Acid 249° 120°	176 Cinnamic Aldehyde 251.00° −7.5°
800 mm.	250.11°	250.25°	250.90°	253.29°
700	244.57	244.73	244.97	247.27
600	.	..	240.43	240.68
500	.	..	234.12	232.95
400	..	..	226.56	223.60
300	.	..	217.30	212.60
200	.	..	204.80	197.70
100	.	..	185.30	175.10
90	..	..	182.40	172.00
80	..	170.55	179.30	168.30
70	..	167.00	175.90	164.50
60	..	162.90	172.00	160.20
50	159.84	158.20	167.50	155.25
40	154.09	152.30	162.20	149.05
30	146.96	145.00	155.50	141.30
20	138.17	136.00	146.40	131.70
15	131.90	129.60	140.40	125.00
10	124.15	121.70	132.10	116.60
8	..	..	..	112.00
6	..	..	..	106.30
5	111.41	..	..	102.80
4	..	..	..	98.50

	177 Eugenol	178 Diphenyl	179 Betelphenol	180 Isosafrole
B.P. 760 mm.	252.66°	254.48°	254.5°	254.95°
M.P.	..	69.0°	8.5°	..
800 mm.	254.86°	256.74°	256.68°	257.24°
700	249.09	250.81	250.96	251.22
600	242.78	244.31	..	244.62
500	235.30	236.58	..	236.67
400	226.30	227.49	..	227.60
300	215.60	216.40	..	216.25
200	201.60	201.99	.	201.80
100	180.10	179.74	..	179.10
90	177.00	176.63	..	175.90
80	173.40	173.03	.	172.20
70	169.80	169.28	..	168.40
60	165.60	165.04	..	164.20
50	160.90	160.17	163.70	159.20
40	154.95	154.02	157.80	152.95
30	147.60	146.39	150.40	145.20
20	138.48	136.93	141.35	135.60
15	132.03	130.29	135.00	129.00
10	124.00	122.02	127.00	120.50
8	119.60	117.50	122.70	116.00
6	114.15	111.86	117.25	110.15
5	110.80	108.37	114.00	106.61
4	106.70	104.12	109.90	102.29

	181 Methyl Anthranilate	182 Cinnamic Alcohol	183 Ethyl Undecylenate	184 Ethyl Anthranilate
B.P. 760 mm.	ca. 255°	257.5°	259°	ca. 260°
M.P.	25.5°		.	
800 mm.	..	..	261.16°	..
700	..	..	255.48	..
600	..	248.50°	..	..
500	..	241.30	..	..
400	..	232.60	..	..
300	..	222.20	..	.
200	..	208.60	..	..
100	..	187.60	..	..
90	..	182.70	..	..
80	..	181.25	..	..
70	..	177.80	..	..
60	..	173.80	..	..
50	163.40°	169.20	168.71	171.20°
40	157.45	163.40	162.83	162.35
30	150.10	156.10	155.52	154.90
20	140.90	147.10	146.46	145.80
15	134.45	140.93	140.10	139.35
10	126.45	133.10	132.19	131.30
8	122.05	129.00	..	126.95
6	116.60	123.70	..	121.50
5	113.30	120.45	119.13	118.10
4	..	116.50	..	..

B.P. 760 mm. M.P.	185 Caryophyllene 260.5°	186 Isoamyl Benzoate 260.93°	187 Methyl Cinnamate 261.58°	188 Allylphenylacetic Acid ca. 262° 34°
800 mm.	..	263.23°	263.90°	..
700	.	257.20	257.80	..
600	..	250.60	251.15	..
500	..	242.80	243.10	..
400	..	233.55	233.80	..
300	.	222.25	222.30	..
200	.	207.65	207.65	..
100	.	185.00	184.90	195.30°
90	..	181.80	181.80	192.30
80	..	178.30	178.00	189.00
70	..	174.40	174.20	185.35
60	.	170.15	170.00	181.30
50	160.85°	165.15	164.85	176.50
40	154.40	158.95	158.60	170.90
30	146.40	151.20	150.80	164.00
20	136.50	141.60	141.20	157.25
15	129.50	134.76	134.43	148.00
10	120.80	126.50	126.00	139.30
8	116.10	122.00	121.30	138.00
6	110.00	116.30	115.60	..
5	106.60	112.80	112.00	..
4	.	108.50	107.70	..

B.P. 760 mm M.P.	189 Heliotropin ca. 264.5° 37°	190 Phenylacetic Acid 265.5° 76°	191 Isoeugenol 266.52°	192 *n*-Capric Acid ca. 268° 31.3°
800 mm.	..	267.50°	268.96°	..
700	..	262.30	262.80	..
600	..	256.44	256.20	..
500	..	249.80	248.40	..
400	..	241.60	239.50	..
300	..	231.80	227.70	..
200	..	218.70	213.05	..
100	190.80°	198.10	190.50	201.25°
90	189.20	195.10	187.30	198.60
80	185.70	191.90	183.60	195.75
70	182.25	188.20	179.80	191.60
60	177.90	184.05	175.50	188.90
50	173.20	179.30	170.80	184.80
40	167.50	173.45	164.30	179.80
30	159.90	166.70	156.60	173.55
20	150.70	157.15	147.00	165.20
15	144.30	150.80	140.23	159.50
10	136.30	142.00	131.80	151.85
8	131.95	137.40	127.20	147.70
6	126.50	131.65	120.10	142.60
5	123.10	128.20	118.00	139.55
4	..	123.90	113.70	135.80

B.P. 760 mm. M.P.	193 Zingiberene ca. 269.5° ..	194 Ethyl Cinnamate 271° .	195 *d*-Cadinene ca. 274° .	196 Amyl Salicylate 277.47°
800 mm.	..	273.32°	..	279.76°
700	..	267.23	..	273.74
600	..	..	..	..
500	..	..	..	..
400	..	..	..	..
300	..	..	..	..
200	..	..	..	..
100	..	..	..	..
90	..	..	..	..
80	..	..	..	..
70	..	..	..	..
60	..	..	177.35°	186.75
50	167.95°	174.25	172.05	181.80
40	161.30	168.00	165.40	174.70
30	152.60	160.20	157.20	167.82
20	142.90	150.50	147.05	158.08
15	135.70	143.65	139.90	151.49
10	126.80	135.20	131.00	143.10
8	122.00	130.60	126.15	..
6	115.95	124.80	120.10	..
5	..	121.25	..	129.25
4	..	..	..	.

B.P. 760 mm. M.P.	197 α-Naphthol 279° 94°	198 Dibenzyl 284° .	199 Dillapiole ca. 285°	200 Vanillin 285° 82 to 84°
800 mm.	281.32°	286.51°	..	287.09°
700	275.23	279.92	..	281.60
600	..	..	..	..
500	..	..	..	..
400	..	..	..	..
300	..	..	..	..
200	..	..	..	..
100	..	..	..	215.70
90	..	..	..	212.80
80	..	..	..	209.50
70	..	..	..	206.00
60	..	..	..	202.00
50	182.05	179.35	195.95°	197.60
40	175.70	172.40	190.10	191.90
30	168.00	163.95	182.90	184.90
20	158.30	153.40	174.10	175.80
15	151.45	146.00	167.85	170.00
10	142.95	136.80	160.00	162.15
8	138.40	131.80	155.80	158.35
6	132.60	126.90	150.45	153.00
5	129.00	121.60	147.20	149.80
4	..	116.00	..	145.90

B.P. 760 mm. M.P.	201 β-Naphthol 286° .	202 Phenyl Benzyl Ether 286.5° ..	203 Glycerin 290° 17°	204 Phthalide 290° 83°(?)
800 mm.	288.44°	288.92°	292.01°	292.48°
700	282.04	282.57	286.79	285.97
600	..	..	280.91	..
500	..	.	274.23	..
400	..	.	266.20	..
300	..	..	256.32	..
200	..	..	243.16	..
100	..	..	222.41	..
90	..	..	219.44	..
80	.	..	216.17	..
70	..	..	212.52	..
60	..	..	208.40	..
50	185.70	185.80	203.62	186.40
40	179.10	179.20	197.96	179.70
30	171.00	171.00	190.87	171.40
20	161.05	160.80	181.34	161.00
15	154.00	153.60	174.86	153.80
10	145.25	144.40	166.11	144.70
8	140.50	139.85	161.49	139.80
6	134.50	133.80	155.69	133.60
5	..	130.00	152.03	129.80
4	..	..	147.87	125.50

B.P. 760 mm M.P.	205 Phenyl Sulfide 296°	206 p-Dinitrobenzene 297.96° 171 to 172°	207 Coumarin 301.72° 70°	208 m-Dinitrobenzene 301.88° 89.72°
800 mm.	298.54°	300.41°	304.24°	304.34
700	291.88	293.98	297.62	297.89
600	..	..	290.37	290.82
500	..	..	281.85	..
400	..	..	271.61	..
300	..	..	259.23	..
200		..	243.15	..
100	..	216.99	218.31	220.65
90	..	..	214.84	..
80	..	..	210.82	..
70	.	..	206.64	..
60	.	201.07	201.91	204.06
50	190.20	195.71	196.48	199.53
40	183.30	189.19	189.62	192.70
30	174.70	181.84	181.10	184.40
20	164.10	170.67	170.54	174.13
15	156.60	163.41	163.13	166.90
10	148.40	..	153.90	157.92
8	..	..	148.86	..
6	..	..	142.57	..
5	..	..	138.67	143.09
4	..	..	133.93	..

B.P. 760 mm. M.P.	209 α-Santalol 301.99° 	210 Isoapiole (Parsley) 304° 55 to 56°	211 Benzophenone 305.89° 47.2°	212 Stilbene 306.5° ..
800 mm.	304.46°	306.47°	308.36°	308.95°
700	297.97	299.99	301.87	302.52
600	..	..	294.77	..
500	..	..	286.42	..
400	..	..	276.38	..
300	..	..	264.26	..
200	..	..	248.50	..
100	..	..	224.17	..
90	..	..	220.77	..
80	..	..	216.83	..
70	..	..	212.74	..
60	..	..	208.10	..
50	198.95	201.00	202.77	199.10
40	192.20	194.30	196.05	193.30
30	183.30	186.00	187.71	186.20
20	173.35	175.80	177.36	177.30
15	166.10	168.50	170.10	171.10
10	159.90	159.60	161.07	163.35
8	152.10	154.60	156.12	159.10
6	145.90	147.00	149.96	153.80
5	142.10	144.60	146.14	150.60
4	..	..	141.49	..

B.P. 760 mm. M.P.	213 β-Santalol 309° .	214 Phenyl Benzoate 314° .	215 o-Dinitrobenzene 318.14° 116.5°	216 Benzyl Benzoate 323.5° 21°
800 mm.	311.48°	316.65°	320.69°	326.04°
700	304.97	309.69	314.00	319.37
600	..	..	..	..
500	..	..	..	..
400	..	..	..	..
300	..	..	..	..
200	..	..	.	..
100	..	..	233.85	..
90	..	..	..	..
80	..	..	..	..
70	..	..	..	..
60	..	.	217.28	..
50	205.70	203.00	211.78	217.41
40	198.95	196.00	204.85	210.50
30	190.50	187.00	196.25	201.93
20	180.15	176.00	185.57	191.29
15	172.80	168.30	178.08	183.82
10	163.80	158.60	168.76	174.54
8	158.80	153.20	..	169.45
6	152.60	146.60	..	163.02
5	148.75	124.50	153.37	158.58
4	..	..	..	153.02

B.P. 760 mm. M.P.	217 Dibenzyl Ketone 330.5° 40°(?)	218 Triphenylmethane 355.73° 92°	219 Azelaic Acid 356.77° 106°	220 Sebacic Acid 364.4°
800 mm.	333.04°	358.45°	..	..
700	326.37	351.31	..	.
600	319.07	..	.	..
500	310.49	..	..	..
400	300.17	..	..	..
300	287.55	..		..
200	271.40	.	.	..
100	246.60	.	..	..
90	243.00	..	..	..
80	239.00	..	..	..
70	234.80	..	..	..
60	230.05	..	..	..
50	224.50	242.00	267.30°	274.40°
40	217.59	234.60	261.40	268.50
30	209.02	225.50	254.20	261.20
20	198.38	214.10	244.30	251.20
15	191.91	206.20	237.50	244.50
10	181.62	196.20	228.50	235.40
8	176.48	190.90	223.70	230.60
6	170.21	184.00	217.70	224.60
5	..	179.90	..	..
4	.	..	..	..

IV. CONVERSION TABLES

Degrees Centigrade to Fahrenheit

(Formula: °C × 1.8 + 32 = °F)

°C.	°F.	°C.	°F.	°C.	°F.	°C.	°F.	°C.	°F.	°C.	°F.	°C.	°F.
−50	−58.0	8	46.4	66	150.8	124	255.2	182	359.6	240	464.0	298	568.4
49	56.2	9	48.2	67	152.6	125	257.0	183	361.4	241	465.8	299	570.2
48	54.4	10	50.0	68	154.4	126	258.8	184	363.2	242	467.6	300	572.0
47	52.6	11	51.8	69	156.2	127	260.6	185	365.0	243	469.4	301	573.8
46	50.8	12	53.6	70	158.0	128	262.4	186	366.8	244	471.2	302	575.6
45	49.0	13	55.4	71	159.8	129	264.2	187	368.6	245	473.0	303	577.4
44	47.2	14	57.2	72	161.6	130	266.0	188	370.4	246	474.8	304	579.2
43	45.4	15	59.0	73	163.4	131	267.8	189	372.2	247	476.6	305	581.0
42	43.6	16	60.8	74	165.2	132	269.6	190	374.0	248	478.4	306	582.8
41	41.8	17	62.6	75	167.0	133	271.4	191	375.8	249	480.2	307	584.6
40	40.0	18	64.4	76	168.8	134	273.2	192	377.6	250	482.0	308	586.4
39	38.2	19	66.2	77	170.6	135	275.0	193	379.4	251	483.8	309	588.2
38	36.4	20	68.0	78	172.4	136	276.8	194	381.2	252	485.6	310	590.0
37	34.6	21	69.8	79	174.2	137	278.6	195	383.0	253	487.4	311	591.8
36	32.8	22	71.6	80	176.0	138	280.4	196	384.8	254	489.2	312	593.6
35	31.0	23	73.4	81	177.8	139	282.2	197	386.6	255	491.0	313	595.4
34	29.2	24	75.2	82	179.6	140	284.0	198	388.4	256	492.8	314	597.2
33	27.4	25	77.0	83	181.4	141	285.8	199	390.2	257	494.6	315	599.0
32	25.6	26	78.8	84	183.2	142	287.6	200	392.0	258	496.4	316	600.8
31	23.8	27	80.6	85	185.0	143	289.4	201	393.8	259	498.2	317	602.6
30	22.0	28	82.4	86	186.8	144	291.2	202	395.6	260	500.0	318	604.4
29	20.2	29	84.2	87	188.6	145	293.0	203	397.4	261	501.8	319	606.2
28	18.4	30	86.0	88	190.4	146	294.8	204	399.2	262	503.6	320	608.0
27	16.6	31	87.8	89	192.2	147	296.6	205	401.0	263	505.4	321	609.8
26	14.8	32	89.6	90	194.0	148	298.4	206	402.8	264	507.2	322	611.6
25	13.0	33	91.4	91	195.8	149	300.2	207	404.6	265	509.0	323	613.4
24	11.2	34	93.2	92	197.6	150	302.0	208	406.4	266	510.8	324	615.2
23	9.4	35	95.0	93	199.4	151	303.8	209	408.2	267	512.6	325	617.0
22	7.6	36	96.8	94	201.2	152	305.6	210	410.0	268	514.4	326	618.8
21	5.8	37	98.6	95	203.0	153	307.4	211	411.8	269	516.2	327	620.6
20	4.0	38	100.4	96	204.8	154	309.2	212	413.6	270	518.0	328	622.4
19	2.2	39	102.2	97	206.6	155	311.0	213	415.4	271	519.8	329	624.2
18	−0.4	40	104.0	98	208.4	156	312.8	214	417.2	272	521.6	330	626.0
17	+1.4	41	105.8	99	210.2	157	314.6	215	419.0	273	523.4	331	627.8
16	3.2	42	107.6	100	212.0	158	316.4	216	420.8	274	525.2	332	629.6
15	5.0	43	109.4	101	213.8	159	318.2	217	422.6	275	527.0	333	631.4
14	6.8	44	111.2	102	215.6	160	320.0	218	424.4	276	528.8	334	633.2
13	8.6	45	113.0	103	217.4	161	321.8	219	426.2	277	530.6	335	635.0
12	10.4	46	114.8	104	219.2	162	323.6	220	428.0	278	532.4	336	636.8
11	12.2	47	116.6	105	221.0	163	325.4	221	429.8	279	534.2	337	638.6
10	14.0	48	118.4	106	222.8	164	327.2	222	431.6	280	536.0	338	640.4
9	15.8	49	120.2	107	224.6	165	329.0	223	433.4	281	537.8	339	642.2
8	17.6	50	122.0	108	226.4	166	330.8	224	435.2	282	539.6	340	644.0
7	19.4	51	123.8	109	228.2	167	332.6	225	437.0	283	541.4	341	645.8
6	21.2	52	125.6	110	230.0	168	334.4	226	438.8	284	543.2	342	647.6
5	23.0	53	127.4	111	231.8	169	336.2	227	440.6	285	545.0	343	649.4
4	24.8	54	129.2	112	233.6	170	338.0	228	442.4	286	546.8	344	651.2
3	26.6	55	131.0	113	235.4	171	339.8	229	444.2	287	548.6	345	653.0
2	28.4	56	132.8	114	237.2	172	341.6	230	446.0	288	550.4	346	654.8
−1	30.2	57	134.6	115	239.0	173	343.4	231	447.8	289	552.2	347	656.6
0	+32.0	58	136.4	116	240.8	174	345.2	232	449.6	290	554.0	348	658.4
+1	33.8	59	138.2	117	242.6	175	347.0	233	451.4	291	555.8	349	660.2
2	35.6	60	140.0	118	244.4	176	348.8	234	453.2	292	557.6	350	662.0
3	37.4	61	141.8	119	246.2	177	350.6	235	455.0	293	559.4		
4	39.2	62	143.6	120	248.0	178	352.4	236	456.8	294	561.2		
5	41.0	63	145.4	121	249.8	179	354.2	237	458.6	295	563.0		
6	42.8	64	147.2	122	251.6	180	356.0	238	460.4	296	564.8		
7	44.6	65	149.0	123	253.4	181	357.8	239	462.2	297	566.6		

Degrees Fahrenheit to Centigrade

$$\left(\text{Formula: } \frac{°F - 32}{1.8} = °C\right)$$

°F.	°C.	°F.	°C.	°F.	°C.	°F.	°C.	°F.	°C.	°F.	°C.	°F.	°C.
-50	-45.56	7	21.67	36	2.22	79	26.11	122	50.00	165	73.89	208	97.78
49	45.00	6	21.11	37	2.78	80	26 67	123	50.56	166	74.44	209	98.33
48	44.44	5	20.56	38	3.33	81	27.22	124	51.11	167	75.00	210	98.89
47	43.89	4	20.00	39	3 89	82	27.78	125	51 67	168	75.56	211	99.44
46	43.33	3	19.44	40	4 44	83	28 33	126	52 22	169	76.11	212	100.00
45	42.78	2	18 89	41	5 00	84	28 89	127	52.78	170	76.67	213	100 56
44	42.22	1	18.33	42	5 56	85	29.44	128	53 33	171	77.22	214	101.11
43	41 67	0	17.78	43	6.11	86	30 00	129	53 89	172	77.78	215	101.67
42	41 11	+1	17 22	44	6 67	87	30 56	130	54.44	173	78.33	216	102.22
41	40.56	2	16 67	45	7.22	88	31.11	131	55.00	174	78 89	217	102.78
40	40.00	3	16.11	46	7 78	89	31.67	132	55 56	175	79.44	218	103 33
39	89 44	4	15.56	47	8 33	90	32 22	133	56.11	176	80.00	219	103.89
38	38.89	5	15.00	48	8 89	91	32.78	134	56.67	177	80 56	220	104.44
37	38.33	6	14.44	49	9.44	92	33.33	135	57.22	178	81 11	221	105 00
36	37.78	7	13.89	50	10 00	93	33.89	136	57.78	179	81 67	222	105.56
35	37.22	8	13.33	51	10 56	94	34 44	137	58 33	180	82 22	223	106 11
34	36 67	9	12.78	52	11.11	95	35 00	138	58 89	181	82 78	224	106 67
33	36.11	10	12 22	53	11 67	96	35 56	139	59.44	182	83 33	225	107 22
32	35.56	11	11.67	54	12 22	97	36 11	140	60 00	183	83 89	226	107.78
31	35.00	12	11 11	55	12.78	98	36 67	141	60.56	184	84 44	227	108 33
30	34.44	13	10.56	56	13 33	99	37 22	142	61 11	185	85 00	228	108.89
29	33 89	14	10 00	57	13 89	100	37.78	143	61 67	186	85 56	229	109.44
28	33 33	15	9 44	58	14.44	101	38 33	144	62 22	187	86.11	230	110 00
27	32.78	16	8 89	59	15 00	102	38 89	145	62 78	188	86.67	231	110.56
26	32.22	17	8 33	60	15.56	103	39.44	146	63 33	189	87.22	232	111.11
25	31.67	18	7.78	61	16.11	104	40 00	147	63 89	190	87.78	233	111.67
24	31 11	19	7 22	62	16 67	105	40 56	148	64 44	191	88 33	234	112 22
23	30.56	20	6 67	63	17 22	106	41.11	149	65 00	192	88 89	235	112.78
22	30.00	21	6 11	64	17.78	107	41 67	150	65 56	193	89 44	236	113 33
21	29.44	22	5.56	65	18 33	108	42 22	151	66 11	194	90 00	237	113 89
20	28 89	23	5 00	66	18 89	109	42 78	152	66 67	195	90.56	238	114 44
19	28 33	24	4 44	67	19.44	110	43 33	153	67 22	196	91.11	239	115.00
18	27.78	25	3.89	68	20 00	111	43 89	154	67 78	197	91.67	240	115.56
17	27 22	26	3 33	69	20 56	112	44 44	155	68 33	198	92 22	241	116.11
16	26.67	27	2.78	70	21.11	113	45 00	156	68 89	199	92.78	242	116.67
15	26.11	28	2 22	71	21.67	114	45.56	157	69 44	200	93 33	243	117.22
14	25.56	29	1 67	72	22 22	115	46.11	158	70 00	201	93.89	244	117.78
13	25.00	30	1.11	73	22 78	116	46.67	159	70 56	202	94.44	245	118.33
12	24.44	31	−0.56	74	23.33	117	47.22	160	71 11	203	95.00	246	118 89
11	23.89	32	0.00	75	23.89	118	47.78	161	71 67	204	95 56	247	119.44
10	23.33	33	+0.56	76	24.44	119	48.33	162	72 22	205	96 11	248	120.00
9	22.78	34	1.11	77	25.00	120	48 89	163	72.78	206	96.67	249	120.56
8	22.22	35	1.67	78	25.56	121	49.44	164	73.33	207	97.22	250	121.11

Reduction Table

Dilution of Alcohol to Lower Strengths

Figures in each column indicate the units of water (by volume) to be added to each 100 units of alcohol of the strengths shown to produce the strengths listed in left-hand column.

Desired Strength % x Vol.	Strengths to be Reduced (% x Volume)											
	95%	94%	92%	90%	85%	80%	75%	70%	65%	60%	55%	50%
90%	6.4	5.1	2.5									
85%	13.3	11.9	9.2	6.58								
80%	20.9	19.5	16.6	13.8	6.83							
75%	29.5	27.9	24.9	21.9	14.5	7.2						
70%	39.1	37.5	35.9	34.2	23.1	15.3	7.64					
65%	50.2	48.4	45.0	41.5	33.0	24.6	16.4	8.15				
60%	63.0	61.2	57.3	53.6	44.2	35.4	26.5	17.6	8.76			
55%	78.0	76.0	71.9	67.8	57.9	48.0	38.3	28.6	19.0	9.5		
50%	95.9	93.6	89.2	84.8	73.9	63.1	52.4	41.7	31.3	20.5	10.4	
45%	118.0	115.0	110.0	105.0	93.3	81.3	64.5	57.8	48.0	34.5	22.9	11.4
40%	144.0	142.0	136.0	131.0	117.0	104.0	90.8	77.6	64.5	51.4	38.5	25.6
35%	179.0	176.0	169.0	163.0	148.0	133.0	118.0	103.0	88.0	72.0	56.3	43.6
30%	224.0	221.0	213.0	206.0	189.0	171.0	154.0	136.0	119.0	102.0	85.0	67.5
25%	287.0	283.0	275.0	266.0	245.0	224.0	204.0	183.0	162.0	142.0	121.0	101.0
20%	382.0	377.0	366.0	356.0	330.0	304.0	277.0	253.0	227.0	200.0	176.0	150.0
15%	540.0	533.0	519.0	505.0	471.0	437.0	403.0	369.0	335.0	300.0	267.0	234.0

Note: This table takes into account the shrinkage that results from mixing alcohol and water.

Method of Determining Pounds per Gallon @ 20°C.

Determine the specific gravity of the item at 20° C./20° C. Multiply this figure by 8.330. This gives the weight in pounds of one gallon at 20° C.

Example:

Determine the number of pounds in one gallon of Ethyl Acetate Lot "B."

$$\text{Specific gravity of ethyl acetate @ } \frac{20^\circ \text{ C.}}{20^\circ \text{ C.}} = 0.900$$

$$8.330 \times 0.900 = 7.497$$

1 gallon Ethyl Acetate Lot "B" weighs 7.5 lb. @ 20° C.

Equivalents for U. S. System and Metric System

Length

U.S.	Metric	Metric	U.S.
1 mile	1.6094 km.	1 km.	0.6214 mile
1 yd.	0.9144 m.	1 m.	1.0936 yd.
1 ft.	30.4801 cm.	1 m.	3.2808 ft.
1 in.	2.5400 cm.	1 cm.	0.3937 in.

Area

U.S.	Metric	Metric	U.S.
1 sq. mile	2.5900 sq. km.	1 sq. km.	0.3861 sq. mile
1 sq. mile	258.9998 hectares	1 sq. km.	247.1044 acres
1 acre	0.4047 hectare	1 hectare	2.4710 acres
1 acre	4046.8730 sq. m.	1 sq. m.	10.7639 sq. ft.

Weight

U.S.	Metric	Metric	U.S.
1 ton (short)	0.9072 ton	1 ton	1.1023 ton (short)
1 lb. (avoir.)	0.4536 kg.	1 kg.	2.2046 lb. (avoir.)
1 lb. (avoir.)	453.5924 g.	1 g.	0.0353 oz. (avoir.)
1 oz. (avoir.)	28.3495 g.		

Volume

U.S.	Metric	Metric	U.S.
Liquid		*Liquid*	
1 gal.	3.7853 liters	1 liter	0.2642 gal.
1 qt.	0.9463 liter	1 liter	1.0567 qt.
1 pt.	0.4732 liter	1 liter	2.1134 pt.
1 oz. (fl.)	29.5729 ml.	1 liter	33.8147 oz. (fl.)
		1 ml.	0.0338 oz. (fl.)
Dry		*Dry*	
1 bu. (dry)	0.0352 cu. m.	1 cu. m.	28.3776 bu.
Capacity		*Capacity*	
1 cu. yd.	0.7646 cu. m.	1 cu. m.	1.3079 cu. yd.
1 cu. ft.	0.0283 cu. m.	1 cu. m.	35.3144 cu. ft.
1 cu. in.	16.3872 cu. cc.	1 cu. cm.	0.0610 cu. in.

INDEX